Thomas Scott, George Brook

List of Marine Fauna Collected at the Tarbert Laboratory during

1885

Part 1

.

Thomas Scott, George Brook

List of Marine Fauna Collected at the Tarbert Laboratory during 1885
Part 1

ISBN/EAN: 9783337269814

Printed in Europe, USA, Canada, Australia, Japan

Cover: Foto ©berggeist007 / pixelio.de

More available books at **www.hansebooks.com**

1. Appendix F.--No. XVII.
   List of the Marine Fauna collected at the Tarbert
   Laboratory during 1885.
   Appendices to 4th Ann. Rpt. Fish. Bd. Scotland, Pt. I,
   pp. 231-242.

2. Description of a New Copepod.
   Appendices to 6th Ann. Rpt. Fish. Bd. Scotland, Pt. 1a
   pp. 233-234, pl. VIII, figs. 7-12.

3. Notes on the Contents of the Stomachs of Herring and
   Haddocks.
   Appendices to 6th Ann. Rept. Fish. Bd. Scotland, pp. 225-
   231.

4. A Revised List of the Crustacea of the Firth of Forth.
   Appendices to 6th Ann. Rpt. Fish. Bd. Scotland, pp. 235-
   262.

5. Notes on Interesting Fishes, &c., Sent to the University
   of Edinburgh.
   Ann. Rpt. Fish. Bd. Scotland, pp. 264, 265.

6. VII.--Additions to the Fauna of the Firth of Forth.
   8th Ann. Rpt. Fish. Bd. Scotland, Pt. 3, pp. 312-333,
   pls. XII, XIII.

7. No. VIII.--The Invertebrate Fauna of Inland Waters.
   1.--Report on Loch Coulter and the Coulter Burn, Stirling-
   shrie.
   8th Ann. Rpt. Fish. Bd. Scotland, Pt. 3, pp. 334-347.

8. III.--The Invertebrate Fauna of the Inland Waters of
   Scotland.
   9th Ann. Rpt. Fish. Bd. Scotland, Pt. 3, pp. 269-296,
   pls. V, VI.

9. V.--Additions to the Fauna of the Firth of Forth.
   9th Ann. Rpt. Fish. Bd. Scotland, Pt. 3, pp. 300-310.

10. III.--Additions to the Fauna of the Firth of Forth. Pt. IV.
    10th Ann. Rpt. Fish. Bd. Scotland, pp. 244-272, pls. VII-
    XIII.

11. II.--Additions to the Fauna of the Firth of Forth. Pt. V.
    11th Ann. Rpt. Fish. Bd. Scotland, pp. 197-219, pls. II-V.

12. III.--The Invertebrate Fauna of the Inland Waters of
    Scotland.
    11th Ann. Rpt. Fish. Bd. Scotland, Pt. 3, pp. 220-238,
    pls. VI, VII.

(over)

ANNUAL REPORT

OF THE

FISHERY BOARD.

SCOTLAND.

1886 - 1894.

————————

CRUSTACEA, ETC.

VOL. 1.

T.Scott.

LIST of the MARINE FAUNA collected at the Tarbert Laboratory during 1885. Part I. By George Brook and Thomas Scott.

The present list does not pretend to include a complete list of the forms known to inhabit Lochfyne and the adjacent waters, but simply includes those species which have been collected and identified in connection with the enquiries at present in progress in Lochfyne under the direction of the Fishery Board.

The large group of Amphipods have not been included in the present list of Crustacea, as many species have not yet been identified. Other groups, such as the Cœlenterates, Sponges, Annelids, Tunicates, &c., have not received special attention, and will be worked up later.

Our thanks are due to the Rev. A. M. Norman for the very valuable assistance he has rendered us in the identification of the Crustacea, who contributes a paper on some of the more interesting forms.

In order to add to the value of the present list, an asterisk has been placed in front of each species which is known to form a part of the food of fishes.

PISCES.

TELEOSTEI.

Cottus scorpius (Bloch) (father lasher), common.
     ,,   ,,   var. *grœnlandicus*, frequent.
     ,,   *bubalis* (Euph.) (long-spined cottus), common.
Trigla lineata (Gmel.) (streaked gurnard), not common.
     ,,   *gurnardus*, L. (gray gurnard), common in East Loch Tarbert.
Agonus cataphractus, L. (pogge), occasionally.
Scomber scomber, L. (mackerel), abundant in the summer, particularly
                 along the west shore of Lochfyne ; captured in large
                 quantities during the herring fishing in the seine nets.
Zeus faber, L. (John Dory), usually enter Tarbert Harbour in September
                 to feed on the herring offal. Are present in the Sound
                 of Kilbrannan considerably earlier, and are frequently
                 taken off Carradale Pier.
Gobius ruthensparri (Euphr.) (two spotted goby), common amongst the
                 Zostera beds in East Loch Tarbert.
     ,,   *minutus* (Gmel.) (one spotted goby), frequent in sandy spots in
                 East Loch Tarbert but not so numerous as the preceding
                 species. In confinement this species frequently buries
                 itself in the sand.
Callionymus lyra, L. (dragonet), both males and females (= *Callionymus*
                 dracunculus) are frequent but small.
Cyclopterus lumpus, L. (lump-sucker), we have frequently met with
                 immature forms, some not over one inch long, but have not
                 as yet dredged any large specimens.
Liparis vulgaris, L. (sea-snail), between tide marks.
     ,,   *montagui* (Donov.) (Montague's sucker), between tide marks.
Lepadogaster sp., Sunidale Bay, 10 fathoms.
Carelophus ascanii (Walb.), one taken amongst boulders at low water,
                 East Loch Tarbert, also occurs in Rothesay Bay.
Centronotus gunnellus, L. (butter-fish), plentiful between tide marks.
Atherina presbyter (Jenyns) (atherine), frequent in a Zostera bed in
                 East Loch Tarbert in the spring, but not met with later.
Gasterosteus spinachia, L. (fifteen-spined stickleback), abundant in
                 Zostera beds in East Loch Tarbert.
Labrus maculatus (Bl.) (ballan wrasse), frequent in the autumn.
     ,,   *mixtus*, L. (striped wrasse), occasionally at the mouth of East
                 Loch Tarbert.
Ctenolabrus rupestris, L., common, particularly near Skate Island, in 10
                 to 12 fathoms. Does not appear to frequent shallower
                 water.
Centrolabrus exoletus, occasionally in Lochfyne.
Gadus morhua, L. (cod), not very abundant in Lochfyne, nor are the
                 fish in good condition. The rock variety is found nearer
                 shore. The principal cod fishing industry in this district
                 is in the Sound of Gigha during the winter months.
     ,,   *œglifinus*, L., haddock, not abundant.
     ,,   *luscus* (Will.) (bib), occasionally in Tarbert Harbour.
     ,,   *minutus*, L. (power-cod), frequent in Tarbert Harbour.
     ,,   *merlangus*, L. (whiting), not abundant. Neither the whiting nor
                 the haddock afford any systematic fishery in Lochfyne.
                 A few are caught on hand lines in or near the harbours.

Gadus virens, L. (coal fish or saith), very abundant.

,, *pollachius*, L. (pollack or lythe), frequents rocky parts and is fairly abundant but usually small.

Molva vulgaris (Flem.) (ling), frequently brought in by the fishermen.

Motella mustela, L. (five-bearded rockling), rare, between tide marks, East Loch Tarbert.

Ammodytes lanceolatus (Lesaw.) (larger sand eel), occasionally. There is not, however, much suitable ground for this species in the neighbourhood.

Hippoglossus vulgaris (Flem.) (holibut), occasionally in Lochfyne. Larger specimens are captured in the Sound of Bute and Sound of Kilbrannan.

Hippoglossoides limandoides (Bloch) (long rough dab), taken during the winter fishing and from the Sound of Gigha.

Rhombus maximus, L. (turbot), occasionally.

Zeugopterus unimaculatus (Risso), we have secured a few specimens of this rare species from near Barmore.

Pleuronectes platessa, L. (plaice), fairly plentiful in suitable parts of Lochfyne.

,, *microcephalus*, frequent in suitable localities. There is a regular fishery for this species around the Island of Bute.

,, *limanda*, L. (common dab), Skipness and Barmore.

,, *flesus*, L. (flounder), common in Tarbert Harbour.

Solea vulgaris (Quen.) (sole), Barmore Bay, small.

Salmo salar, L. (salmon), occasionally taken in herring nets between Tarbert and Barmore.

,, *trutta*, L. (sea-trout), regularly taken in small quantities just outside East Loch Tarbert.

Clupea harengus, L. (herring), enters Lochfyne in May, and the fishing usually lasts until November.

,, *sprattus*, L.? (sprat), the sprat is not usually regarded as being present in the Firth of Clyde, but specimens sent from Girvan have been identified by Mr Matthews as sprats.

Anguilla vulgaris (Turt.) (common eel), affords a small but regular fishery in Tarbert Harbour.

Conger vulgaris (Cuv.) (conger-eel), frequent but small.

Siphonostoma typhle, L. (broad-nosed pipe fish), Zostera bed, East Loch Tarbert.

Syngnathus acus, L. (greater pipe fish), common.

Nerophis æquoreus, L. (ocean pipe fish), Zostera bed, East Loch Tarbert.

,, *lumbriciformis*, L. (worm pipe fish), Zostera bed, East Loch Tarbert.

GANOIDEI.

Acipenser sturio, L. (sturgeon), fine specimens are often noticed in Lochfyne during the herring fishery, but these are seldom captured.

ELASMOBRANCHII.

Pristiurus melanostomus (Bonah.) (black mouthed dog fish), occasionally brought in amongst the Acanthii in the winter fishing. Taken also in Rothesay Bay.

Acanthias vulgaris (Risso) (picked dog fish), very abundant and a perfect nuisance to the fishermen while prosecuting the herring fishery.

Raia batis, L. (common skate), fished principally in the winter and from the West Loch and Sound of Gigha.

„   *clavata*, L. (thornback ray), frequent during the winter fishing.

The above are all the species of fishes that have come under our notice at Tarbert during the past year ; there are doubtless many more still to be recorded. The list of Elasmobranchs is very imperfect, no special attention having been given to this group.

BRACHIOPODA.

Terebratula caput-serpentis, L., fairly common off Battle Island in 40 fathoms, and in other localities.

Crania anomala (Müll.), abundant and large, near Maol-Dubh Point, 12 to 15 fathoms. Frequent also in other parts of the loch, usually at a depth of less than 20 fathoms.

MOLLUSCA.

LAMELLIBRANCHIATA.

ASIPHONIDA, Flem. (1828).

*Anomia ephippium, L., common.

Ostrea edulis, L., generally distributed but small and scarce.

Pecten pusio (L.), a few off Skate Island, 14 fathoms, and East Loch Tarbert.

*   „   *varius*, L., East Loch Tarbert.

„   *opercularis* (L.), common. The lower (convex) valve is frequently covered with a soft pink sponge, amongst which are specimens of *Tubularia*, &c.

„   *septemradiatus*, Müll., common and of fair size, but local. Exceedingly plentiful in the deep water (104 fathoms) off Skate Island.

*   „   *tigrinus*, Müll., not uncommon off Battle Island (40 fathoms) and in other localities.

„   „   var. *costata*, Jeff. in 40 fathoms, off Battle Island.

„   *striatus*, Müll., Furlong Bay and Maol-Dubh Point. Not common.

„   *maximus* (L.), off Skate Island in 14 fathoms and in other localities, but not very plentiful.

Lima subauriculata (Mont.), dead shells off Battle Island in 40 fathoms.

„   *hians* (Gmel.), off Battle Island in 40 fathoms.

*Mytilus edulis, L., Fairly plentiful in East and West Loch Tarbert generally.   In the latter it is larger and more abundant.

„   *modiolus*, L., common.

Modiolaria marmorata (Forb.), imbedded in the mantle of *Ascidia mentula*, which is abundant on the long bank (10 to 16 fathoms) off the mouth of East Loch Tarbert.

„   *discors*, L., attached to roots of *Laminaria*, &c., East Loch Tarbert and other localities.

*Nucula nitida (G. B. Sow), Battle Island, not common.

*   „   *tenuis* (Mont.), occasional, off Battle Island.

*Leda minuta (Müll.), off Battle Island.

Pectunculus glycymeris, L., single valve of a dead shell in coralline bed, 19 fathoms, Sunidale Bay.

SIPHONIDA, Flem. (1828).

Leptom nitidum, Turt., Sunidale Bay.
Montacuta bidentata (Mont.), East Loch Tarbert.
Lasæa rubra (Mont.), var. *pallida* (Jeff.), East Loch Tarbert.
Lucina borealis, L., dead shells frequent on the shore.
Axinus flexuosus (Mont.), East Loch Tarbert.
  ,,   *ferruginosus* (Forb.), off Tarbert, 16 to 18 fathoms.
Cyamium minutum (Fabr.), White Shore, East Loch Tarbert.
Cardium echinatum, L., generally distributed.
  ,,   *exiguum*, Gmel., East Loch Tarbert.
  * ,,   *fasciatum*, Mont.
  ,,   *edule*, L., more common in West than in East Loch Tarbert.
  ,,   *norvegicum*, Spenz., single valves in East Loch Tarbert and in Lochfyne.
Cyprina islandica, L., dead shells, but fresh near Laggan; 30 fathoms.
Astarte sulcata (Da C.), Laggan Bay and East Loch Tarbert.
  ,,   ,,   var. *elliptica* (Bro.), Laggan Bay.
Circe minima (Mont.), single valves and fresh dead shells.
Venus exoleta, L., dead shells, Sunidale Bay.
  ,,   *fasciata* (Da C.), common.
  ,,   *casina*, L., single valve, Sunidale Bay.
  ,,   *ovata*, Penn., off Battle Island, muddy bottom, and in East Loch Tarbert.
  * ,,   *gallina*, L., White Shore, East Loch Tarbert.
Tapes virgineus (L.), dead shell, Sunidale Bay
  ,,   *pullastra* (Mont.), common, White Shore, East Loch Tarbert.
Tellina crassa (Penn.), single valve, Sunidale Bay.
  ,,   *balthica*, head of West Loch Tarbert.
  ,,   *tenuis*, Da C., common.
Psammobia tellinella, Lamk., Sunidale Bay.
Mactra subtruncata, Da C., frequent, East Loch Tarbert.
Scorbicularia alba (Wood), off Battle Island, in muddy sand.
Solen siliqua, L., var. *arcuata*, Jeff., East Loch Tarbert.
Lyonsia norvegica (Chem.), Lochfyne, not common.
Thracia papyracea (Poli.), var. *villosiuscula* (Macg.), Sunidale Bay, recent, dead.
Neæra abbreviata, Forb., in the deeper portions of Lochfyne, rare.
  ,,   *costellata* (Desh.), off Battle Island, 40 fathoms, rare.
  ,,   *cuspidata* (Olivi), off Battle Island, rare.
Corbula gibba (Olivi), Lochfyne and East Loch Tarbert, not common.
Mya arenaria, L., West Loch Tarbert.
  ,,   *truncata*, L., dead shells, East Loch Tarbert.
Saxicava rugosa, L., shores of East Loch Tarbert.
Xylophaga dorsalis (Turt.), about two hundred living specimens taken out of a small piece of wood dredged in East Loch Tarbert, also in similar situation off Skate Island.

SCAPHOPODA, Bronn (1862).

Dentalium entalis, L., common off Barmore and in other localities. Some of the dead shells are inhabited by Sipunculids.

GASTEROPODA, Cuvier (1798).

CYCLOBRANCHIATA (Cuvier).

Chiton fascicularis, L., rare, East Loch Tarbert.
„   *marginatus*, Penn., White Shore, East Loch Tarbert.
„   *ruber*, Lowe, under stones between tide marks.
„   *marmoreus*, Fabr., East Loch Tarbert, Sunidale Bay.
„   *sp.*, Sunidale Bay.

PECTINIBRANCHIATA, (Cuvier).

Patella vulgata, L., common and of large size.
Helcion pellucidum (L.), var. *lævis* (Penn.), White Shore, East Loch
    Tarbert.
Tectura testudinalis (Müll.), White Shore, East Loch Tarbert.
„   *virginea* (Müll.), Sunidale Bay.
Puncturella Noachina (L.), Laggan Bay.
Emarginula fissura (L.), off Battle Island and in other localities, dead
    shells.
„   *crassa*, J. Sow., off Battle Island, dead.
Capulus hungaricus (L.), Furlong Bay, 15 fathoms.
Trochus helicinus, Fabr., East Loch Tarbert.
„   *magus*, L., White Shore and Lochfyne.
* „   *tumidus*, Mont., East Loch Tarbert.
„   *cinerarius*, L., common.
„   *umbilicatus* (Mont.), common near low-water mark.
„   *millegranus*, Phil., off Battle Island.
„   *zizyphinus*, L., Laggan Bay, and in other localities.
Lacuna dicaricata (Fabr.), off Battle Island, and in East Loch Tarbert.
„   „   var. *canalis* (Mont.), East Loch Tarbert. Both
    variety and type occur amongst Zostera.
„   *pallidula* (Dac.), on Laminaria.
„   „   var. *albescens*, Jeff., E. Loch Tarbert.
Littorina obtusata (L.), common.
„   „   var. *ornata*, Jeff., White Shore, East Loch Tarbert.
„   *rudis*, Maton., common.
„   *littorea*, L., common.
Rissoa reticulata (Mont.), off Battle Island.
„   *abyssicola* (Mont.), deep water, 50 fathoms.
„   *parva* (Da C.).
„   *membranacea* (Ad.), var. *elata*, Phil., East Loch Tarbert.
„   *violacea*, Desm., East Loch Tarbert.
„   *striata* (Ad.), common.
„   *cingillus* (Mont.), common.
Hydrobia ulvæ (Penn.)? West Loch Tarbert, in quantity.
Skenea planorbis (Fabr.), common.
Homalogyra atomus (Phil.), scarce, East Loch Tarbert.
Cæcum glabrum (Mont.), Lochfyne and East Loch Tarbert, not common.
Turritella terebra (L.), dead shells, East Loch Tarbert.
Odostomia conspicua, Ald.
„   *unidentata* (Mont.).
„   *spiralis* (Mont.).
Eulima distorta (Desh.), Sunidale Bay and East Loch Tarbert.

Natica Alderi, Forb., East Loch Tarbert and off Battle Island.

    „   *Montacuti*, Forb., off Battle Island.

Lamellaria perspicua (L.), on stones between tide marks, East Loch Tarbert, frequent.

* *Velutina lævigata* (Penn.), East Loch Tarbert.

* *Trichotropis borealis*, Brod. and S., Furlong Bay.

* *Aporrhais pes-pelecani* (L.), Furlong Bay, off Battle Island, &c.

SIPHOBRANCHIATA.

Cerithiopsis tubercularis (Mont.), East Loch Tarbert and Sunidale Bay.

Purpura lapillus (L.), common.

* *Buccinum undatum*, L., East Loch Tarbert.

* *Fusus gracilis* (Da C.), Lochfyne.

Nassa incrassata (Str.), East Loch Tarbert.

    „   *reticulata*, L., East Loch Tarbert.

Defrancia linearis (Mont.), dead shells, off Battle Island and in East Loch Tarbert.

* *Pleurotoma turricula* (Mont.), East Loch Tarbert (living), off Battle Island (dead shells).

Cypraea europea (Mont.), on rocks at extreme low water, Lochfyne.

PLEUROBRANCHIATA.

Utriculus truncatulus (Brug.), East Loch Tarbert.

    „   *hyalinus* (Turt.), East Loch Tarbert.

Actæon tornatilis (L.), Laggan Bay,

Scaphander lignarius (L.), Furlong Bay and off Maol-Dubh Point in 15 fathoms.

* *Aplysia punctata*, Cuv., White Shore, East Loch Tarbert.

Pleurobranchus plumula (Mont.), between tide marks, and dredged, East Loch Tarbert.

NUDIBRANCHIATA.

PELLIBRANCHIATA.

Elysia viridis, Mont., common in Zostera bed, East Loch Tarbert. Varies from pale green or brown to a dark sage green.

POLYBRANCHIATA.

Hermaea bifida (Mont.), East Loch Tarbert.

Eolis papillosa (L.), between tide marks, frequent.

Eolis Drummondi, Thomp., more common than preceding species. Immature forms are abundant amongst Zostera.

    „   *alba*, A. and H., frequent in Zostera bed, East Loch Tarbert.

    „   *Farrani*, A. and H., one specimen under stones, East Loch Tarbert.

Dendronotus arborescens (Müll.), frequent in Lochfyne, 10 to 20 fathoms.

ACANTHOBRANCHIATA.

Ægirus punctilucens (d'Orb.), under stones, East Loch Tarbert.

Triopa claviger, Müll., between tide marks, East Loch Tarbert.

Polycera quadrilineata (Müll.), East Loch Tarbert.

    „   „   *black* var. East Loch Tarbert.

Goniodoris nodosa (Mont.), East Loch Tarbert.
Doris tuberculata, L., East Loch Tarbert, frequent.
  ,,   *Johnstoni*, A. and H., East Loch Tarbert, frequent.
  ,,   *repanda*, A. and H., rare, East Loch Tarbert.
  .,   *pilosa*, Müll., occasionally, East Loch Tarbert.

PULMONIBRANCHIATA.

Melampus bidentatus (Mont.), common, East Loch Tarbert.

CEPHALOPODA.

*_Sepiola Rondeletii_, Leach, Laggan Bay, eggs frequent, attached to
    Ascidians, &c.
Rossia macrosoma (Del. Ch.), Laggan Bay.
Octopus vulgaris, Lmk., in herring nets in the summer, Laggan Bay.
*_Eledone cirrosa_ (Lmk.), young, taken in herring nets in Laggan Bay in
    December.

CRUSTACEA.

PODOPTHALMATA.

BRACHYURA.

Stenorhynchus tenuirostris, not common in Lochfyne.
*_Inachus dorsettensis_, frequent in 10 to 20 fathoms, Lochfyne.
*_Hyas araneus_, common between tide marks.
* ,,   *coarctatus*, common in the off-shore waters of Lochfyne, but not so
    abundant as on some parts of the east coast of Scotland.
*_Cancer pagurus_, frequent on the rocky shores of Lochfyne, but not large
    so far as our observation goes. There is no systematic
    crab fishery.
*_Carcinus mænas_, common.
Portunus puber, frequent in 10 to 15 fathoms or more in Lochfyne.
  ,,   *depurator*, frequent.
* ,,   *marmoreus*, frequent.
  ,,   *pusillus*, frequent.
Ebalia bryerii, Sunidale Bay, 20 fathoms.

ANOMOURA.

Lithodes maia, common in Lochfyne, mostly medium size.
*_Pagurus bernhardus_, common.
  ,,   *prideauxii*, common. Often has the beautiful anemone Peachia
    attached to it.
* ,,   *ferrugineus* (Norm.), East Loch Tarbert.
  ,,   *lævis*, frequent.
*_Porcellana longicornis_, not very plentiful, between tide marks.
*_Galathea squamifera_, common.
  ,,   *intermedia* (Lill.), East Loch Tarbert and Buck Bay.

MACRURA.

Palinurus vulgaris, a single specimen taken in herring nets by Ardrishaig
   fishermen in the spring.
*Calocaris M'Andrew, occurs sparingly in the deeper portions of Lochfyne,
   60 to 90 fathoms.
Homarus vulgaris, generally distributed. There is a small summer
   fishery in Buck Bay, which is carried on by Ardrossan
   fishermen.
*Crangon vulgaris, by no means plentiful.
  „ *spinosus,* rare.
* „ *Almani,* frequent in the off-shore water.
Crangon neglectus (Sars), East Loch Tarbert and at Barmore, new to
   Britain.
*Virbius (Hippolyte) varians, Zostera bed, East Loch Tarbert, abundant.
  „ *fasciger* (Gosse.), East Loch Tarbert.
Hippolyte pandaliformis, East Loch Tarbert.
  „ *securifrons* (Norm.), common in 20 to 40 fathoms, Lochfyne.
Pandalus annulicornis, common.
  „ *brevirostris* (Rathke), East Loch Tarbert.

SCHIZOPODA.

*Boreophausia Raschii (G. O. Sars), Lochfyne.
*Nyctiphanes norvegica (M. Sars), Lochfyne ; also found in stomachs of
   herring and *Acanthias.*
Erythrops pygmaea (Sars), Barmore, and East Loch Tarbert, new to
   Britain.
Mysidopsis gibbosa (G. O. Sars), East Loch Tarbert, new to Britain.
  „ *angusta* (G. O. Sars), in 4 fathoms, Barmore.
Leptomysis linguara (G .O. Sars), East Loch Tarbert.
Mysis chamaeleon (J. V. Thomps.), Zostera bed, East Loch Tarbert.
  „ *inermis* (Rathke), Zostera bed, East Loch Tarbert.
  „ *arenosa* (G. O. Sars), Zostera bed, East Loch Tarbert.
  „ *lamornae* (Couch.), East Loch Tarbert.
*Siriella Clausii (G. O. Sars), East Loch Tarbert.
  „ *Brooki* (Norm.), East Loch Tarbert.
  „ *armata* (M. Edw.), East Loch Tarbert.

CUMACEA.

Vaunthompsonia cristata (Ball.), East Loch Tarbert.
Lamprops uniplicata, (?) East Loch Tarbert.
Hemilamprops fasciata (G. O. Sars), at low water, East Loch Tarbert.
Pseudocuma cercaria (Van Ben.), East Loch Tarbert.
Diastylis rugosa (Sars), East Loch Tarbert.

COPEPODA.

See separate list given by Mr Calderwood.

OSTRACODA.

Pontocypris mytiloides (Norm.), East Loch Tarbert.
  „ *trigonella* (Sars), East Loch Tarbert.
Cythere lutea (Müll.), East Loch Tarbert.

Cythere viridis (Müll.), East Loch Tarbert.
,,   *pellucida* (Baird), East Loch Tarbert.
,,   *crispata* (Brady), East Loch Tarbert.
,,   *albo-maculata* (Baird), East Loch Tarbert.
,,   *convexa* (Baird), East Loch Tarbert.
,,   *tuberculata* (Sars), East Loch Tarbert.
,,   *concinna* (Jones), East Loch Tarbert.
,,   *angulata* (Sars), East Loch Tarbert.
,,   *villosa* (Sars), East Loch Tarbert.
,,   *antiquata* (Baird), Lochfyne.
,,   *Jonesii* (Baird), Lochfyne.
Cytheridea papillosa (Bosq.), Lochfyne.
,,   *punctillata* (Brady), Lochfyne.
Eucythere argus (Sars), Lochfyne.
Krithe Bartonensis (Jones), East Loch Tarbert.
Loxoconcha impressa (Baird), East Loch Tarbert.
,,   *granulata* (Sars), East Loch Tarbert.
,,   *tamarindus* (Jones), East Loch Tarbert.
,,   *multiflora* (Norm.), East Loch Tarbert.
Xestoleberis aurantia (Baird), East Loch Tarbert.   X.
Cytherura nigrescens (Baird), East Loch Tarbert.
,,   *angulata* (Brady) East Loch Tarbert.
,,   *striata* (Sars), East Loch Tarbert.
,,   *undata* (Sars), East Loch Tarbert.
,,   *gibba* (Müll.), East Loch Tarbert.
,,   *cornuta* (Brady), East Loch Tarbert.
,,   *cellulosa* (Norm.), East Loch Tarbert.
,,   *flavescens* (Brady), East Loch Tarbert.
Cytheropteron latissimum (Norm.), East Loch Tarbert.
,,   *arcuatum,* (B. C. and R.), East Loch Tarbert.
,,   *nodosum* (Brady), East Loch Tarbert.
,,   *angulatum* (B. and R.), East Loch Tarbert.
Bythocythere simplex (Norm.), East Loch Tarbert.
,,   *turgida* (Sars), Furlong Bay.
Pseudocythere caudata (Sars), Lochfyne.
Sclerochilus contortus (Norm.), Lochfyne.
Xiphichilus tenuissima (Norm.), Lochfyne.
Paradoxostoma variabile (Baird), East Loch Tarbert.
,,   *pulchellum* (Sars), Furlong Bay.
Philomedes interpuncta (Baird), East Loch Tarbert.
Asterope Maria (Baird), East Loch Tarbert).
,,   *teres* (Norm.), East Loch Tarbert.
Polycope orbicularis (Sars), East Loch Tarbert.

CIRRIPEDIA.

Balanus balanoides, L., on stones and shells, East Loch Tarbert.
,,   *crenatus* (Brug.), on stones and shells, East Loch Tarbert.
Verruca Stroemia (Müll.), on stones and shells, East Loch Tarbert.

ECHINODERMATA.

CRINOIDEA.

Antedon rosaceus, common near the east shore of Lochfyne.

OPHIUROIDEA.

*Amphiura filiformis (Müll.), off Buck Island.
    ,,   elegans (Leach), frequent between tide marks.
Ophiactis Ballii (Thomp.) ? Zostera bed, East Loch Tarbert.
*Ophiocoma nigra, (O.-F. M.), very abundant up to 15 to 20 fathoms.
*Ophioglypha albida (Forb.), common.
    ,,   lacertosa (Linck.), frequent, East Loch Tarbert.
Ophiopholis aculeata (O. F. M.), common near east shore of Lochfyne.
*Ophiothrix rosula (Linck.), common, particularly near the east shore of
    Lochfyne

ASTEROIDEA.

Asterias rubens (L.), abundant.
    ,,   violacea (O. F. M.), with the preceding.
    ,,   glacialis (L.), frequent in 20 to 30 fathoms in Lochfyne, some
    of large size.
Solaster papposus (L.), frequent in Lochfyne, smaller specimens between
    tide marks.
    ,,   endeca, L., frequent some distance from shore, and also occurs
    at low water.
*Cribrella oculata (Linck.), East Loch Tarbert.
*   ,,   rosea (Müll), East Loch Tarbert.
Hippasteria plana (Linck.), frequent in shallow water.
Luidia savignyi (Aud.), Sunidale Bay, 20 fathoms.

ECHINOIDEA.

Echinus esculentus (Penn.), very abundant in some parts of Lochfyne,
    particularly in the Channel between Skate Island and the
    Mainland.
*   ,,   miliaris (L.), between tide marks, East Loch Tarbert.
*Echinocyamus pusillus (Müll.), Sunidale Bay.
*Spatangus purpureus (Leske), Sunidale Bay.
*Echinocardium cordatum (Penn.), very abundant in the sand at White
    Shore Bay, East Loch Tarbert.

HOLOTHUROIDEA.

Several species which are not yet identified.

FORAMINIFERA.

Cornuspira foliacea (Phil.), East Loch Tarbert.
Biloculina ringens (Lam.), Lochfyne, near Tarbert.
    ,,   depressa (d'Orb.), East Loch Tarbert.
Triloculina trigonula (Lam.), Lochfyne, near Tarbert.
    ,,   tricarinata (d'Orb.), Lochfyne, near Tarbert.
Miliolina seminulum (L.) East Loch Tarbert.
    ,,   subrotunda (Mont.), East Loch Tarbert.
    ,,   secans (d'Orb.), East Loch Tarbert.
    ,,   Ferussacii (d'Orb.), East Loch Tarbert.
    ,,   agglutinans (d'Orb.), East Loch Tarbert.
Spiroloculina limbata (d'Orb.), Lochfyne.
    ,,   canaliculata (d'Orb.), Lochfyne.

Trochammina, sp., East Loch Tarbert.
Lagena sulcata (W. and J.), East Loch Tarbart.
,, *lævis* (Mont.), Lochfyne.
,, *gracillina* (Sagz.), Lochfyne.
,, *globosa* (Mont.), Lochfyne.
,, *striata* (d'Orb.), Lochfyne.
,, *marginata* (W. and J.), East Loch Tarbert.
,, *melo* (d'Orb.), Lochfyne.
,, *squamosa* (Mont.), East Loch Tarbert.
,, *hexagona* (Will.), Lochfyne.
,, *Jeffreysii* (Brady), Lochfyne and East Loch Tarbert.
Nodosaria scalaris (Batsch.) Lochfyne, frequent.
Polymorphina lactea (W. and J.), East Loch Tarbert.
,, *tubulosa* (d'Orb.), Lochfyne.
Dentalina communis (d'Orb.), Lochfyne.
Orbulina universa (d'Orb.), East Loch Tarbert.
Textularia sagittula (Defrance), East Loch Tarbert.
Bulimina marginata (d'Orb.), East Loch Tarbert.
Discorbina rosacea (d'Orb.), East Loch Tarbert.
Truncatulina lobatula (Walker), East Loch Tarbert.
Rotalia Beccarii, L., East Loch Tarbert, common.
,, *nitida* (Will.), East Loch Tarbert, rare.
Patellina corrugata (Will.), Lochfyne.
Polystomella crispa (L.), East Loch Tarbert, common.
,, *striato-punctata* (F. and M.), East Loch Tarbert, common.
Nonionina asterizans (F. and M.), East Loch Tarbert, common.
,, *depressula* (W. and J.), East Loch Tarbert, common.
Planorbulina Mediterranensis (d'Orb.), East Loch Tarbert, common.

DESCRIPTION OF A NEW COPEPOD. By Thomas Scott.
(Pl. VIII. figs. 7–12.)

Artotrogus papillatus, n. sp.—In general appearance this species is not very unlike *A. Normani*, Brady and Robertson.

The first segment is equal in length to rather more than half of the whole body, or as 21 is to 19.

The anterior antennæ are eight-jointed, stout, bearing comparatively few setæ, and each is terminated by a long curved olfactory appendage. The relative lengths of the joints are as follows :—

$$\frac{1 \quad 2 \quad 3 \quad 4 \quad 5 \quad 6 \quad 7 \quad 8}{14 \cdot 44 \cdot 14 \cdot 9 \cdot 12 \cdot 11 \cdot 14 \cdot 29}.$$

The terminal joint has a distinct curved band stretching across it near the middle, which in certain positions gives it the appearance of being two-jointed.

The last joint of the posterior antennæ is about as long as the two preceding ones, and terminates in a rather long slender spine.

The maxillæ resemble very closely those of *Dyspontius striatus*, Thorell.

The second pair of foot-jaws are rather powerfully clawed, and bear each a single spine at the distal end and on the inner aspect of the second-last joint.

The outer margins of the joints of the outer branches of the swimming feet are fringed with minute close-set spines, and the outer terminal angles of the first and second joints are armed with one stout lancet-shaped, and one small spine; the third joint has three such pairs of spines on its outer edge, and terminates in a sabre-like spine which has its outer edge finely serrate. The inner margins of these joints are clothed somewhat similar to those of *A. magniceps*, Brady.

The integument of the cephalothorax is thickly covered with small conical papillæ.

This species has, like *A. Normani*, an elongated abdomen; the last joint is dilated distally, and is rather shorter than the third, while the second-last is shorter and narrower than either; the caudal appendages are stout and about as long as the last abdominal segment; each bears three setæ, the inner one is short and slender, and the other two long, the middle one being considerably the longest; on the outer edge of each appendage, and about a fourth of its length from the free end, there is inserted in a kind of notch a short plumose seta. Length, one-twentieth of an inch (1·3 mm.).

One specimen only of this form was obtained among material dredged in the vicinity of Inchkeith. This solitary specimen was a female, and does not agree with any species known to me. Dr G. S. Brady, F.R.S., and Mr J. C. Thomson, F.L.S., to whom I submitted the specimen, have also been unable to assign it to any described species. Mr Thomson, who has recently been working very largely at Copepoda, both of British and foreign seas, was at first of opinion that it is not impossible that it may be found to belong to the genus *Dyspontius*; but the differential characters cannot be completely made out without dissection, and I do not wish to destroy the specimen for that purpose.

Professor Ewart has suggested that the species should be named *A. papillatus*, which has been adopted.

I am indebted to Mr W. L. Calderwood for the drawings which accompany this description.

EXPLANATION OF THE PLATE.

Figs. 1–6. *Cyclops Ewarti*, n. sp.

Fig. 1. Male ; magnified 40 times.
Fig. 2. Anterior antenna of male ; magnified 90 times.
Fig. 3. Anterior antenna of female ; magnified 90 times.
Fig. 4. Foot of first pair ; magnified 90 times.
Fig. 5. Foot of fifth pair ; magnified 200 times.
Fig. 6. Abdomen of male ; magnified 90 times.

Figs. 7–12. *Artotrogus papillatus*, n. sp.

Fig. 7. Adult female ; ventral view.
Fig. 8. Anterior antenna.
Fig. 9. Posterior antenna.
Fig. 10. Maxilla.
Fig. 11. Second foot-jaw.
Fig. 12. Foot of third pair of swimming feet.

PLATE VIII.

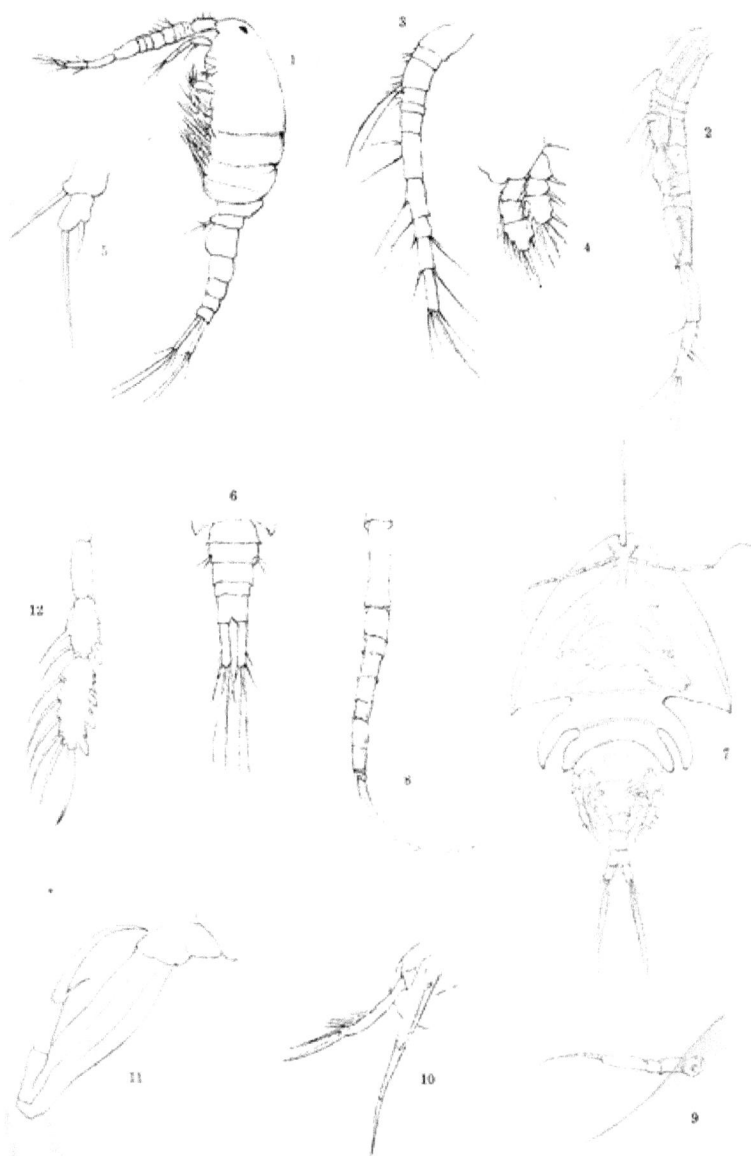

Figs. 1—6. Cyclops Ewarti, n. sp.

„ 7—12. Artotrogus papillatus, n. sp.

NOTES ON THE CONTENTS OF THE STOMACHS OF HERRING AND HADDOCKS. By Thomas Scott.

In the following tables are given the results of the examination of a number of herring's and haddock's stomachs which had been sent to the Central Laboratory, from various localities, by officers of the Fishery Board.

A good deal of the material was almost useless, partly from overcrowding of the specimens and partly from bad preservation. Unless the spirit be frequently renewed the more delicate forms become more or less decomposed, and much more difficult to identify. In many instances, indeed, the whole of the contents, excepting such hard structures as the shells of Molluscs, the plates and tests of Echinoderms, and the limbs and other hard parts of Crustaceans, are destroyed and entirely useless for purposes of identification.

Herring's Stomachs.

In the following table the results of the examination of the stomachs of 164 herrings are given :—

TABLE I.

Locality.	Date.	Number of Stomachs.	Number of stomachs empty.	Schizopods.	Amphipods.	Copepods.	Fish.	Remarks.
BROADFORD— Loch Ainort, .	4.9.86	5	...			A few		All contained the remains of Copepods ; species not distinguishable.
,,	29.7.86	1	1	...				
,,	29.9.86	3		Numerous				The stomach was filled with young *Nyctiphanes*.
Crenlin Sound,	14.7.86	1	1	...				
,,	4.10.87	6	...	Fairly numerous				All contained *Nyctiphanes*.
Loch Eshort, .	16.8.87	5	5	...				
Sound of Scalpa,	26.10.86	3	3	...				...
Loch Slapan, .	7.10.86	1	1	...				
,,	9.10.86	5	4	...				1 contained a little food; could not be identified.
	26.8.87	6	6					
CARRADALE .	5.11.86	6	1	Fairly numerous				*Nyctiphanes* (nearly full grown).
,,	12.11.86	5	5					...
Off King's Cove, Arran,	19.11.86	5		Numerous		Numerous		All distended with food; 1 contained *Nyctiphanes* only, the others had the lower half filled with Copepods (*Calanus*) and the upper with *Nyctiphanes*.
BALLANTRAE.	13.1.87	1		Numerous	Numerous			*Nyctiphanes* and *Parathemisto* in about equal numbers.
,,	22.1.87	4		Numerous	Numerous			1 contained *Nyctiphanes* only, 1 *Parathemisto* only, 2 about equal numbers of both.
,,	9.2.87	1		Numerous	A few			Organisms similar to last.
LOCH NELL.	12.7.87	3	4				A few	Young fish partly digested.
MACHRIE, . .	16.12.86	2		Fairly numerous				*Nyctiphanes*.
OFF MAY ISLAND,	23.2.84	4		...				All contained some very much decomposed stuff.
STORNOWAY— Off Butt of Lewis,	12.5.87	3					Many	Very young fish—'20 per cent. gorged with food.'
,,	21.6.87	2		Numerous			A few	About 10 per cent. of the Catch contained food ; 1 of the stomachs contained Schizopods only, the other very young fish.

TABLE I.—*continued.*

Locality.	Date.	Number of Stomachs.	Number of Stomachs empty.	Schizopods.	Amphipods.	Copepods.	Fish.	Remarks.
Off Tiumpan Head,	21.5.87	2	1	Many			..	About 5 per cent. of the Catch contained food—*Nyctiphanes*(very young).
,,	25.5.87	4	2	Numerous		Many	...	About 3 per cent. of the Catch only contained food; 1 contained *Nyctiphanes* (very young) the other, *Nyctiphanes* and Copepods(*Calanus* mostly).
,,	18.6.87	2		Numerous			A few	1 contained very young fishes, 1 Schizopods; about 11 per cent. of the Catch contained food.
,,	28.6.87	4		A few			A few	1 contained a young fish, the others a few very young fish and Schizopods.
Off Keboch Head,	7.7.87	1					2	40 per cent. of the Catch contained more or less food; small fish very much digested.
Off Tolsta Head,	31.5.87	3		Numerous		Many	..	2 contained Schizopods and Copepods (*Calanus*), the other Schizopods only; about 15 per cent. of the Catch contained food.
,,	17.6.87	2		...			Remains	Both contained remains of young fish; about 10 per cent. of the Catch contained food.
STONEHAVEN,	26.1.87	6		Numerous	Numerous			3 contained Schizopods (*Nyctiphanes*) only; 3 Schizopods and Amphipods (*Parathemisto*).
TARBERTNESS,	14.1.87	20	18	Numerous				Schizopods only—*Nyctiphanes* (young).
,,	19.1.87	23	22	Numerous	Numerous			The contents comprised a great many *Nyctiphanes* and *Parathemisto oblivia.*
,,	20.1.87	20	17	Numerous	Numerous			
WICK	2.2.84	1	1	...				

This table shows that 70 stomachs (42·7 per cent.) contained food, of which Schizopods, particularly *Nyctiphanes*, formed by far the greater bulk ; and the stomachs were those of fish taken not only during the winter months, but also in the middle of summer.

Amphipods were not of so frequent occurrence ; those present all belonged to the Hyperiidæ, and so far as could be made out, to one species, *Parathemisto oblivia*, Kröyer. Of *Hyperia galba*, Mont. (*H. medusarum*, Mull.) which has been so frequently recorded as being found in the stomachs of herring I did not observe a single specimen.

Copepods only occurred in four samples of stomachs, which were all sent from the West Coast, and, so far as could be made out from the decomposed state of their contents, were almost all specimens of *Calanus finmarchicus.*

Young fish (sprats or herring) were observed in a number of stomachs, principally from Stornoway.

The results here brought out agree generally with those of previous statistics. It is shown by this table, as by those given in former papers, that Schizopods and pelagic Amphipods form the chief food of the herring on the East Coast. No Copepods were observed in the above cases in any of the stomachs sent from East Coast districts ; possibly the season of the year when the stomachs were collected, viz., January and February, may partly account for this total absence. On the other hand, Copepods were observed among the contents of the stomachs of only four of the samples from the West Coast, while Schizopods were very plentiful. The absence of Copepods from so many of the stomachs from the West Coast cannot be ascribed to the season of the year, for it will be seen from the table that the herring were caught during May, June, July, August, October, November and December ; and, besides, the stomachs of one of the four samples in which Copepods were observed, and in which they were abundant, were those of herring caught off King's Cove, Arran, about the middle of November. Thus the results brought out here do not quite correspond with the general statement that, 'during summer 'and autumn the Copepods supply the all-important food (of the herring) 'on the West Coast.'[1] Copepods certainly formed a large part of the contents of the herrings' stomachs examined at Tarbert, and the results brought out there would seem to apply to Loch Fyne generally. In the statistics given in the *Fourth Annual Report,*[2] it is shown that in the stomachs of herrings caught in the Loch Fyne district Copepods were observed much more frequently than Schizopods during the summer months, while in those examined in September, specimens of both these groups occurred in nearly an equal number of stomachs. It appears so far to be satisfactorily determined that, as regards the Loch Fyne district, Copepods do form the greater part of the food of the herring during the summer, and probably also during the later spring and early autumn months ; the examination of the stomachs of herring from other parts of the West Coast yield, however, somewhat different results. In the above table, the list of the contents of the stomachs sent from the Broadford district, which were collected during July, August, September and October, show that the herring there had been feeding mostly upon Schizopods ; and the stomachs sent from the Stornoway district, and collected during May, June and July, give a similar result ; so that, according to this evidence, Copepods hold a subordinate place as the food of the herring in certain districts of the West Coast even during summer. In the previously published statistics above referred to, it will be observed that

[1] *Fourth Ann. Rep. Fish. Board for Scot.*, Appendix F, p. 128 (1886).
[2] *Loc. cit.*

of the herrings' stomachs from the Stornoway district collected in April, May and June, Schizopods were more or less abundant in twenty-eight, and Copepods in sixteen, of the samples examined; and in those from Loch Broom district, collected in March, October, November and December, there were no Copepods at all. This, again, shows a preponderance of Schizopods over Copepods as constituting the food of the herring; so that, with the exception of the Loch Fyne district, it would appear from these statistics that on the West as well as on the East Coast of Scotland Schizopods are at least of no less importance than the Copepods, as forming part of the food of the herring, even during the summer months.

Another point relating to the herring and its food may be noticed here. The experiments with the tow-net round a considerable part of both the East and the West Coasts of Scotland, carried out by instructions from the Fishery Board during recent years, have shown that the young Euphausidæ and the pelagic Amphipods and Copepods are not nearly as plentiful in deep water as they are near the surface, or at least within a few fathoms of the surface; and these being the organisms that the herring almost exclusively feed on, it follows that when herring are feeding they do not go into deep water, but keep near the surface where their food is most abundant.

An interesting fact brought out by the published statistics already referred to, is the total absence of the Hyperiidæ in the stomachs from the West Coast—not a single specimen of these Amphipods being observed among the contents of any of those examined; and in the table given here these Amphipods are also equally unrepresented. Why they should be so frequent on the East Coast, and so very rare on the West, is a rather puzzling question.

Another fact perhaps worth noticing, is that in all but one of the stomachs which came from King's Cove, Arran, and which were distended with food, the lower part (nearly half) was filled with Copepods only, while the upper part was exclusively filled with Schizopods. This abrupt and complete change in the nature of the food is interesting, for from the well-preserved condition of the organisms, little time can have elapsed between the relinquishing of the one, and the taking to the other kind of food. A possible explanation is that the Copepods were being attacked by a shoal of *Nyctiphanes* when the herrings fell in with them, and that the latter after passing through among the Copepods met with the Schizopods, feeding freely on each in turn.

Dr Möbius has suggested 'that the herring does not discriminate ' between one form of food and another when feeding on the more minute ' species, but that the swarms of microscopic animals which are diffused ' through the sea are drawn into the mouth along with the water of ' respiration, and are retained while the water passes out through the ' opercula.' While it is quite possible that some of the organisms observed in the stomachs of herrings may be captured in that way, yet there can be no doubt that herrings intentionally pursue and feed on such minute creatures as the Copepoda. I have frequently seen over 200 herrings feeding freely on them, pursuing and capturing them. I have also on many occasions, when at Rothesay, gone out with the tow-net and collected such forms for the purpose of feeding the herrings kept alive in the tanks; and it was interesting to observe how quickly they noticed the presence of the minute Crustaceans, even though transferred to their tank with as little disturbance as possible, and how after all the organisms seemed to have been captured, a few herring, possibly more hungry than the others, would continue to search about the tank, and eagerly pounce on any Copepod that had previously escaped notice.

Haddock's Stomachs.

The stomachs of 80 haddocks have been examined, and the results are summarised in the following table, in which the particulars and the approximate number of the organisms are given :—

TABLE II.

Locality.	Date.	Number of Stomachs.	Zoophytes.	Echinodermata.	Annelida.	Crustacea.	Mollusca.	Fishes.	Remarks.
Buckie,	11.10.86	5	...	10	...	6 and Remains	2 and Remains	...	...
Fraserburgh,	19.2.87	4	...	Remains	...	2	...	...	2 empty; others contained very little food
Lybster,	25.10.86	12	...	23	2 and Remains	11	20	...	1 stomach empty.
,,	23.11.86	22	1	Numerous	...	3 and Remains	34	...	Contents more or less fragmentary.
,,	27.11.86	15	...	Numerous	1	13	12	...	...
	24.5.84	2	...	...	5 and Remains	Remains	4	...	Fragmentary.
,,	1.6.87	3	...	3 and Remains.	1	2	6	...	Contents very fragmentary.
Montrose,	12.10.86	8	...	31	1	1	...	...	...
Wick, .	12.10.86	9	...	3 and Remains.	1	Remains	5	...	...

In the next table the names of the different species identified are given, and the number of stomachs in which each was observed :—

TABLE III.

Name of Organisms.	Number of Stomachs in which observed.	Remarks.
Hydrozoa —		
Diphasia,	1	A fragment, possibly accidentally present.
Echinodermata—		
Ophioglypha albida, Forbes, .	26	Numerous but fragmentary.
Ophiopholis aculeata, Lin., .	27	... ...
Ophiothrix pentaphyllum, Penn.,	2	... ...
Antedon rosacea, Linck., .	10	Numerous fragments.
Echinocyamus pusillus, Müll,	5	... ...
Echinus miliaris, Lin., .	1	Fragments of test.
Annelida—		
Aphrodite sp.,	3	Very much broken up.
Nereis sp., .	1	
Crustacea—		
Hyas sp.,	1	Fragmentary.
Pinnotheres pisum, Penn.,	1	Four fairly whole specimens.
Pagurus bernhardus, Lin.,	1	...
Crangon sp., .	1	Too much digested to be identified.
Hippolyte sp.,	1	
Anonyx holbolli, Kröyer, .	4	Several stomachs contained fragments of Amphipods but not perfect enough to be identified.
Ampelisca lævigata, B. & W.,	1	
Corophium crassicorne, Bruz.,	1	

TABLE III.—*continued.*

Name of Organisms.		Number of Stomachs in which observed.	Remarks.
Mollusca —			
Pecten similis, Laskey,	.	6	Mostly in stomachs from Lybster. A good many specimens.
„ *tigrinum*, Müll.,	. .	1	...
Cardium sp., .	. .	4	Young.
Parmacobia tellinella, Lam.,	. .	3	Young.
Mactra stultorum, Lin.,	. .	4	Young.
Saxicava rugosa, Lin.,	. .	1	... ...
Chiton sp.,	. .	5	... ...
Trochus tumidus, Mont.,	. .	1	... ...
Velutina lævigata, Pess.,	. .	3	... ...
Natica sp.,	. .	2	Young.
Nassa incrassata, Strom.,	. .	1	A few specimens in which hermit crabs had been living.

There is nothing brought out by Tables II. and III. that seems to call for special remark, unless it be the occurrence of a considerable number of the rather uncommon Bivalve Mollusc *Pecten similis* in some of the stomachs sent from Lybster. This species of *Pecten*, which is a very small one, though not very commonly met with, is gregarious in its habits and therefore likely to occur in considerable numbers in places frequented by it. Echinoderms, especially the *Ophiuroids*, are shown here, as in previous statistics, to form the largest part of the haddock's food. Crustacea and Mollusca rank next in importance. Among the Annelids, the 'Sea Mouse' seems to be specially relished by the haddock. It will also be observed that all the *Molluscs* were either young or, if adult, were small species.

In previous Reports there are several extensive accounts given of the nature of the food of fishes, which was a subject that early called for attention. In the *Fourth Annual Report*[1] there is a special paper on the food of the haddock, to which this may be regarded in some respects as supplementary. The tables given in the paper referred to are extensive but the results of these later investigations are in general accord with the conclusions given.

[1] *Report of Fishery Board*, p. 128, 1885.

A REVISED LIST OF THE CRUSTACEA OF THE FIRTH OF FORTH. By THOMAS SCOTT.

In 1880-81 a list of the Invertebrate Fauna of the Firth of Forth was published by Leslie and Herdman in the *Proceedings of the Royal Physical Society of Edinburgh*,[*] in which were enumerated all the species of Crustacea belonging to the Cirripedia, Amphipoda, Isopoda, Cumacea, Schizopoda, and Decapoda, then known to occur in the estuary, but leaving out the Copepoda and Ostracoda.

In December 1884, Professor J. R. Henderson, in a paper read before the same Society,[†] recorded several additions to Leslie and Herdman's list, the Entomostraca being again left out. Professor Henderson, however, expressed the hope that he would yet be able 'to form lists of these 'more minute, though not the less interesting organisms;' but this hope does not appear to have been realised.

Seeing that the Entomostraca, and especially the Copepoda, form such an important part of the food of certain fishes, it is rather surprising that they should have been passed over from time to time by local naturalists.

During the latter half of 1887 I was at various times engaged on board the 'Garland,' assisting in collecting data required in connection with the trawling experiments carried on under the direction of Professor Ewart and Sir James Maitland, Bart. Excellent facilities were thus afforded for gaining a knowledge of the variety and distribution of the fauna of the Forth, and, acting on the instructions of Professor Ewart, a record was made of all the organisms observed, including the Copepoda and Ostracoda; and as further additions to the lists of Crustacea already published have also been made, I propose in the following paper to give a record of the species of Crustacea now known to occur in the Forth estuary—the parasitic Copepoda (fish-lice) excepted, which will be dealt with in a separate paper later on.

It will be understood that this record is not to be considered exhaustive; the Forth will have to be more thoroughly investigated before the preparation of an exhaustive list can be attempted. Meantime, in order to make this one as complete as possible, several species recorded in the papers previously referred to are included, though they have not yet been observed during the Board's investigations, and some additional stations for a few of the rarer forms have also been added on the same authority. In preparing the list of Copepoda the classification and nomenclature adopted by Dr G. S. Brady in his monograph [‡] have been followed. For the Ostracoda, the *British Entomostraca*, by Dr Baird, and various lists by David Robertson, F.L.S., and others, have been consulted. The valuable monograph of the *British Sessile-eyed Crustacea*, by Bate and Westwood, is still the standard work on the British Amphipoda and Isopoda; but as within recent years considerable changes have been made in the nomenclature of these groups, it has been thought advisable to follow that adopted in Part III. of the *Museum Normanianum*,—a series of catalogues which are being published for private distribution by the Rev. Canon Norman, D.C.L., F.L.S. Advantage has also been taken of these catalogues to correct the nomenclature of the other groups. The *British Stalk-eyed Crustacea*, by Professor Bell, the *Popular History of British Crustacea*, by White, and other works, have also been consulted. I wish also to acknowledge

* *Proc. Roy. Phys. Soc. Edin.*, vol. vi. pp. 215, 313 (1880–81).
† *Proc. Roy. Phys. Soc. Edin.*, vol. viii. p. 307 (1883–85).
‡ British Copepoda, 1878–80.

the continued kindness of Dr G. S. Brady, F.R.S., and the Rev Canon Norman, in examining doubtful species. One of the staff of the Board has at various times visited the coast about Joppa, Cramond, and other places, and has collected material for me which has proved very useful by adding to our knowledge of the distribution of various forms, two species of Schizopoda having been added to the Forth fauna in this way. Finally, Professor Cossar Ewart has afforded me encouragement and help in many ways in prosecuting the study of marine organisms, which I feel it to be my duty to acknowledge here.

CIRRIPEDIA.*

PELTOGASTRIDÆ.

Peltogaster paguri, Rathke.
> *Peltogaster paguri*, Anderson, Proc. Roy. Phys. Soc. Edin., vol. ii. Occasionally on *Pagurus Bernhardus*, in the littoral zone, and also in deep water. Joppa (Dr J. Anderson); east of May Island (S.F.B.).

Peltogaster carcini, Anderson.
> *Peltogaster carcini*, Anderson, Proc. Roy. Phys. Soc. Edin., vol. ii. Joppa (Dr J. Anderson).

Sacculina carcini, Thompson.
> *Sacculina carcini*, Leslie and Herdman, Proc. Roy. Phys. Soc. Edin., vol. vi. p. 215. Occasionally attached to the abdomen of *Carcinus mœnas*. Firth of Forth (Dr Anderson); Gulland Bay (S.F.B.).

Sacculina triangularis, Anderson.
> *Sacculina triangularis*, Anderson, Proc. Roy. Phys. Soc. Edin., vol. ii. Firth of Forth (Dr J. Anderson).

LEPADIDÆ.

Lepas anatifera, Linné.
> *Lepas anatifera*, Darwin, Mon. Lepadidæ, p. 73, Tab. i. fig. 1, 1851. Attached to floating timber, Firth of Forth (Ed. Mus., L. & H.).

Conchoderma virgata, Spengler.
> *Conchoderma virgata*, Darwin, Mon. Lepadidæ, p. 146, Tab. iii. fig. 2, 1851 On floating timber, Firth of Forth (Ed. Mus., L. & H.).

Conchoderma auritum, Linné.
> On floating timber, Firth of Forth (Ed. Mus., L. & H.).

BALANIDÆ.

Balanus balanoides, Linné.
> *Balanus balanoides*, Darwin, Mon. Balanidæ, p. 267, pl. vii. figs. 2a–2d, 1854. Common on rocks and stones between tide-marks and also in deep water.

Balanus porcatus, Da Costa.
> *Balanus porcatus*, Darwin, Mon. Balanidæ, p. 256, pl. vi. figs. 4a–4c, 1854. "Not uncommon, attached to stones," &c. (L. & H.).

---

* In the following list the letters S.F.B. indicate that the species and locality are discovered during the investigations of the Scottish Fishery Board; the letters L. & H. indicate that the authority is that of Leslie and Herdman, in the above-mentioned work.

Balanus crenatus, Bruguière.
> *Balanus crenatus,* Darwin, Mon. Balanidæ, p. 261, pl. vi. figs.
> 60–69, 1853. Portobello (Ed. Mus., L. & H.) frequent;
> east of Inchkeith (S.F.B.).

Balanus hameri, Ascanius.
> *Balanus hameri,* Darwin, Mon. Balanidæ, p. 277, pl. vii. figs.
> 5a–5c. Largo Bay and west of Inchkeith, single specimens,
> and occasionally in colonies, attached to pieces of wood, but
> not very common (S.F.B.).

Verruca stromia, Müller.
> Frequent on stones brought up in the dredge east of Inchkeith
> (S.F.B.).

Remarks.—Balanus tintinnabulum, Lin., included in the *Invertebrate Fauna of the Firth of Forth,* though frequently brought to our
shores attached to ships' bottoms, has no claim to be considered a British
species.

COPEPODA.

CALANIDÆ, Dana.

Calanus finmarchicus, Gunner.
> *Calanus finmarchicus,* Brady, Mon. Brit. Cop., vol. i. p. 38, 1872.
> Frequent in surface tow-net gatherings in all parts of the
> estuary, but does not seem to be so common as it is in the
> Firth of Clyde.

Pseudocalanus elongatus, Boeck.
> *Pseudocalanus elongatus,* Brady, Mon. Brit. Cop., vol. i. p. 45,
> 1872.
> *Clausia elongata,* Norman, Mus. Nor., pt. iii., 1886. Frequent
> in surface tow-net gatherings; distribution similar to the last.

Dias longiremis, Lilljeborg.
> *Dias longiremis,* Brady, Mon. Brit. Cop., vol. i. p. 51, pl. v. figs.
> 1–14, 1872. Equally frequent with the last two.

Temora longicornis, Müller.
> *Temora longicornis,* Brady, Mon. Brit. Cop., vol. i. p. 54, pl. iii.
> figs. 10–19.
> *Temora longicaudata,* Norman (Lubbock), Mus. Nor., pt. iii., 1886.
> Also of frequent occurrence.

Temora velox, Lilljeborg.
> *Temora velox,* Brady, Mon. Brit. Cop., vol. i. p. 56, pl. vi. figs.
> 1–5, 1878.
> *Temorella velox,* Norman, Mus. Nor., pt. iii., 1886. Very frequent
> in surface gatherings taken off Bo'ness, and inshore about
> South Queensferry; this is rather a brackish water than a
> marine species, and its occurrence in the upper parts of the
> estuary may be due to the large admixture of fresh water there.

Centropages typicus, Kröyer.
> *Centropages typicus,* Brady, Mon. Brit. Cop., vol. i. p. 65, pl. viii.
> figs. 1–10, 1878. Occasionally in surface gatherings, but
> not very common (S.F.B.).

Centropages hamatus, Lilljeborg.
> *Centropages hamatus,* Brady, Mon. Brit. Cop., vol. i. p. 67, pl. viii.
> figs. 11–13, 1878. More frequent than the last, though not
> very common; it has been observed in surface material from
> Gulland Bay, east of Inchkeith, and vicinity of May Island
> (S.F.B.).

Parapontella brevicornis, Lilljeborg.

> *Parapontella brevicornis*, Brady, Mon. Brit. Cop., vol. i. p. 69, pl. xi. figs. 11–13, 1878. In surface nettings taken off Bo'ness; rather scarce (S.F.B.).

Anomalocera patersonii, Templeton.

> *Anomalocera patersonii*, Templeton, Trans. Entom. Soc.; p. 35, t. v. figs. 1–3, 1837.
>
> *Ireneus patersonii*, Goodsir, Edin. New Phil. Journ., vol. xxxv. p. 339, t. iv. figs. 12–17 ; t. iv. figs. 1–9, 1843.
>
> This species was taken in considerable numbers with the surface net east of May Island. Its colours were light and dark green, no red being present as in many of the Clyde specimens (S.F.B.).

CYCLOPIDÆ, Baird (in part)

Oithona spinifrons, Boeck.

> *Oithona spinifrons*, Brady, Mon. Brit. Cop., vol. i. p. 90, pl. xiv. figs. 1–9, and pl. xxiv. figs. 1, 2, 1878.
>
> *Oithina spinirostris*, Norman (Claus), Mus. Nor., pt. iii. p. 23. Frequent in surface gatherings east of Inchkeith (S.F.B.).

Cyclopina littoralis, Brady.

> *Cyclopina littoralis*, Brady, Nat. Hist. Trans., Northumberland and Durham, vol. iv. p. 427, pl. xvii. figs. 9–14, 1872. Mon. Brit. Cop., vol. i. p. 92, pl. xv. figs. 1–9, 1878. A few specimens in rock pools near low water at Cramond Island (S.F.B.).

Cyclops æquoreus, Fischer.

> *Cyclops æquoreus*, Fischer, Abhandl. der Akad. der Wissenschaft, Bd. viii. p. 654, Taf. xx. figs. 26–29, 1860.
>
> *Cyclops æquoreus*, Brady, Mon. Brit. Cop., vol. i. p. 119, pl. xix. figs. 8–10, and pl. xxi. figs. 10–17, 1880. One specimen, obtained in a pool above high-water mark, Cramond Island. This is the only British species of Cyclops having six-jointed antennæ, with the exception of one recently described by J. C. Thomson, F.L.S., which was taken off Puffin Island, near Anglesea. In some material from the mussel beds at the mouth of the River Eden in Fife, sent to me by Mr Simpson, Leuchars, there were several specimens of this species.

Cyclops Ewarti, n. sp.

> This species, which was taken some miles west of Queensferry, is described by Dr Brady at p.   .

NOTODELPHYIDÆ, Thorell.

Notodelphys Allmanni, Thorell.

> *Notodelphys Allmanni*, Brady, Mon. Brit. Cop., vol. i. p. 126, pl. xxv. figs. 1–10, 1878. In branchial cavities of Ascidians dredged in various places, as about Inchkeith and Granton Harbour ; not unfrequent (S.F.B.).

Notodelphys agilis, Thorell.

> *Notodelphys agilis*, Brady, Mon. Brit. Cop., vol. i. p. 130, pl. xxvi figs. 1–10, 1876. In branchial cavities of Ascidians, attached to the pier at Granton (S.F.B.).

Ascidicola rosea, Thorell.
 Ascidicola rosea, Brady, Mon. Brit. Cop., vol. i. p. 145, pl. xxx.
  figs. 1–10, 1887. Occasionally in branchial cavities of
  Ascidians, but not common (S.F.B.).
Doropygus sp. (possibly *porcicauda*, Brady).
 One specimen among dredged material from vicinity of Inchkeith
  (S.F.B.).

HARPACTICIDÆ, Claus (in part).

Longipedia coronata, Claus.
 Longipedia coronata, Brady, Mon. Brit. Cop., vol. ii. p. 6, pls.
  xxxiv., xxxv., 1880. This is rather a common species in the
  Firth of Forth, occurring at times in considerable numbers, as
  among dredged material from a little west off Inchkeith,
  vicinity of Bass Rock, &c. Also frequent in Cromarty Firth,
  near Invergordon (S.F.B.).
Ectinosoma spinipes, Brady.
 Ectinosoma spinipes, Brady, Mon. Brit. Cop., vol. ii. p. 9, pl.
  xxxvi. figs. 1–10, 1880. In dredged material from off North
  Berwick ; a few specimens (S.F.B.).
Robertsonia tenuis, Brady and Robertson.
 Robertsonia tenuis, Brady, Mon. Brit. Cop., vol. ii. p. 25, pl. xli
  figs. 1–14, 1880. In the same material as last ; a few speci-
  mens (S.F.B.).
Amymone sphærica, Claus.
 Amymone sphærica, Brady, Mon. Brit. Cop., vol. ii. p. 28, pl. xlix.
  figs. 1–11, 1880. A few specimens in material dredged in
  Gulland Bay and at west side of Inchkeith. Also not
  uncommon in material dredged near Invergordon, Cromarty
  Firth (S.F.B.).
Stenhelia ima, Brady.
 Stenhelia ima, Brady, Mon. Brit. Cop., vol. ii. p. 35, pl. xliii. figs.
  1–14, 1880. A few specimens in material dredged off North
  Berwick, and from rock pools between tide-marks, Cramond
  Island (S.F.B.).
Mesochra lilljeborgii, Boeck.
 Mesochra lilljeborgii, Boeck, Oversigt Norges Copepoder, p. 51,
  1864.
 Paratachidius gracilis, B. and R., Ann. and Mag. Nat. Hist.,
  ser. 4, vol. xii. p. 131, pl. viii. figs. 8–16, 1873.
 Mesochra lilljeborgii, Brady, Mon. Brit. Cop., vol. ii. p. 62,
  pl. xli. figs. 15–21, and pl. xlvii. figs. 16–21, 1880. Fre-
  quent in rock pools near high-water mark, Cramond Island
  (S.F.B.).
Laophonte similis (Claus).
 Cleta similis, Claus, Die Copepoden-Fauna von Nizza, t. v. p. 23,
  figs. 13, 14, 1866.
 (?) ——— *forcipata*, Norman, Last Shetland Dredging Report, p. 29,
  1868.
 Laophonte similis, Brady, Mon. Brit. Cop., vol. ii. p. 78, pl. lxxv.
  figs. 1–14, 1880. Several specimens in rock pools between
  tide-marks, Cramond Island. Also in Cromarty Firth
  (S.F.B.).

Laophonte thoracica, Boeck.

>*Laophonte thoracica*, Brady, Mon. Brit. Cop., vol. ii. p. 76, pl. lxxvii. figs. 1–8, 1880. Not unfrequent in material dredged off the west side of Inchkeith. This species is easily distinguished by the very long and slender second foot-jaws and first pair of fee. Also observed in Cromarty Firth, but not common (S.F.B.).

Laophonte curticauda, Boeck.

>*Laphonte curticauda*, Brady, Mon. Brit. Cop., vol. ii. p. 80, pl. lxxiii. figs. 15–18, and pl. lxxvi. figs. 1–9, 1880. Two or three specimens among material dredged off North Berwick ; rock pools, about high-water mark, Cramond Island (S.F.B.).

Laophonte lamellifera, Claus.

>*Cleta lamellifera*, Claus, Die freilebenden Copepoden, p. 123, Taf. xv. figs. 21–25, 1863.
>*Laophonte lamellifera*, Brady, Mon. Brit. Cop., vol. ii. p. 83, pl. lxxv. figs. 15–23, 1880. A few specimens in material collected between tide-marks, near Joppa, and in rock pools Cramond Island (S.F.B.).

Cletodes limicola, Brady.

>*Cletodes limicola*, Brady, Mon. Brit. Cop., vol. ii. p. 90, pl. lxxix. figs. 1–12, 1880. Among material dredged off North Berwick in 10 to 14 fathoms, and Gulland Bay ; rather scarce (S.F.B.).

Cletodes propinqua, Brady and Robertson.

>*Cletodes propinqua*, Brady, Mon. Brit. Cop., vol. ii. p. 94, pl. lxxvii. figs. 9–17, 1880. In pools among heaped-up stones between tide-marks, Newhaven ; between tide-marks, Cramond Island (S.F.B.). Also observed near Invergordon, in Cromarty Firth.

Dactylopus tisboides, Claus.

>*Dactylopus tisboides*, Brady, Mon. Brit. Cop., vol. ii. p. 106, pl. liv. figs. 1–16, 1880. Of frequent occurrence in material collected between tide-marks about Joppa ; rock pools, high-water mark, Cramond Island (S.F.B.).

Thalestris rufocincta, Norman.

>*Thalestris rufocincta*, Brady, Mon. Brit. Cop., vol. ii. p. 125, pl. lvii. figs. 1–9, 1880. Frequent in material collected between tide-marks about Joppa and Granton (S.F.B.). Also observed in Cromarty Firth (S.F.B.).

Thalestris longimana, Claus.

>*Thalestris longimana*, Brady, Mon. Brit. Cop., vol. ii. p. 136, pl. lx. figs. 1–13, 1880. One or two specimens, in material dredged off west side of Inchkeith , rock pools above high-water mark, Cramond Island (S.F B.).

Westwoodia nobilis (Baird).

>*Harpacticus obilis*, Baird, Trans. Berw. Nat. Club, vol. ii. p. 155, 1845 ; Nat. Hist. Brit. Entom., p. 214, tab. xxviii. figs. 2–2a–e, 1850.
>*Westwoodia nobilis*, Brady, Mon. Brit. Cop., vol. ii. p. 141, pl. lxiii. figs. 1–13, 1880. Not infrequent in rock pools, between tide-marks, about Cramond Island. Also near Invergordon, Cromarty Firth (S.F.B.).

Harpacticus chelifer (Müller).

>*Harpacticus chelifer*, Brady, Mon. Brit. Cop., vol. ii. p. 146,

pl. lxv. figs. 1-15, and pl. lxiv. figs. 19, 20, 1880. One or
two specimens in material collected between tide-marks west
of Granton (S.F.B.). Also observed in Cromarty Firth
(S.F.B.).

Harpacticus fulvus, Fischer.

> *Harpacticus fulvus,* Fischer, Abhandl. der König. Bayer. Akad.
> Bd. viii. p. 656, Taf. i. figs. 30–33 ; Taf. ii. figs. 34–39,
> 1860.
>
> *Harpacticus fulvus,* Brady, Mon. Brit. Cop., vol. ii. p 149,
> pl. lxiv. figs. 1–11. Very common in rock pools about high-
> water mark at Cramond Island (S.F.B.). This species occurs
> in rather unexpected places. Dr G. S. Brady has recorded
> it from 35 fathoms off the Yorkshire coast, and from Kinny
> Lough (a fresh-water loch, near the sea, in County Donegal).
> He explains its occurrence here by storms or excessively high
> tides carrying sea water into the loch, rendering it more or
> less brackish. Besides being recorded from several European
> localities it occurs at Kerguelen Island.

Zaus spinatus, H. Goodsir.

> *Zaus spinatus,* Brady, Mon. Brit. Cop., vol. ii. p. 153, pl. lxvi.
> figs. 1–9, 1880. Firth of Forth (H. Goodsir), in material
> collected between tide-marks about Joppa ; frequent
> (S.F.B.).

Alteutha depressa, Baird.

> *Alteutha depressa,* Baird, Trans. Berw. Nat. Club, vol. ii. p. 155,
> 1845.
>
> *Peltidium depressum,* Brady, Mon. Brit. Cop., vol. ii. p. 160,
> pl. lxxii. figs. 1–5, 1880.
>
> *Alteutha purpurocincta,* Norman, Brit. Assoc. Rep., p. 298, 1868.
> Several specimens among material dredged off west side of
> Inchkeith, and among material collected between tide-marks
> about Joppa (S.F.B.). Also of frequent occurrence in
> Cromarty Firth (S.F.B.).

Alteutha interrupta (Goodsir).

> *Alteutha interrupta, see* Brady, Rep. Scot. Fish. Board, p. 328,
> 1886.
>
> *Peltidium interruptum,* Brady, Mon. Brit. Cop., vol. ii. p. 162,
> pl. lxxi. figs. 4–15. 1880.
>
> *Peltidium depressum,* Norman, Mus. Norm., part iii. Frequent in
> the Forth in shore and dredged material ; about Joppa,
> Inchkeith, &c. (S.F.B.). Also observed in Cromarty Firth
> (S.F.B.).

Idya furcata (Baird).

> *Idya furcata,* Brady, Mon. Brit. Cop., vol. ii. p. 172, pl. lxvii.
> figs. 1–11, 1880. Frequent in shore material collected about
> Joppa, west of Granton ; rock pools, about high-water mark,
> Cramond Island (S.F.B.).

SAPPHIRINIDÆ, Thorell.

Lichomolgus fucicolus, Brady.

> *Lichomolgus fucicolus,* Brady, Mon. Brit. Cop., vol. iii. p. 41,
> pl. lxxxv. figs 1–11, 1880. One specimen, in material
> dredged in Gulland Bay (S.F.B.).

ARTOTROGIDÆ, Brady.

Artotrogus Boeckii, Brady.

 Artotrogus Boeckii, Brady, Mon. Brit. Cop., vol. iii. p. 60, pl. xci.
 figs. 1–9, 1880. A number of specimens of this species were
 taken in Granton Harbour; they were obtained by washing
 a specimen of *Chalinula oculata* found attached to the pier at
 extreme low water; they had probably been seeking shelter
 about the Sponge (S.F.B.).

Artotrogus magniceps, Brady.

 Artotrogus magniceps, Brady, Mon. Brit. Cop., iii.
 pl. xciii. figs. 1–9, 1880. A few specimens, in material
 dredged west of Queensferry (S.F.B.).

Artotrogus papillatus, n. sp.

 This species was dredged in deep water a little west of Inchkeith, and is described on p.   .

Acontiophorus scutatus, Brady and Robertson.

 Acontiophorus scutatus, Brady, Mon. Brit. Cop., vol. iii. p. 69,
 pl. xc. figs. 1–10, 1880. Occasionally among dredged
 material from the vicinity of Inchkeith (S.F.B.).

Remarks.—The preceding are all the species of Copepoda that have
been identified, but there are several others, which, from their general
distribution around our shores, may be expected to be yet found in the
Forth. The following additional species have been observed in Cromarty
Firth, near Invergordon—*Laophonte curticanda*, Boeck; *Porcellidium
fimbriatum*, Claus; *Cyclopicera nigripes*, B. and R., and *Cyclopicera
Gracilicanda*, Brady—which do not appear to have been previously
recorded from that district.

The Calanidæ, which include most of the free swimming or pelagic
species, are the most important of the Copepoda in relation to the food
supply of fishes and especially of young fishes. The majority of the
others keep near the bottom, and seek shelter and possibly food among
the forests of Algæ and Zoophytes. It is probable that Shrimps, Prawns,
Schizopods, and other small Crustaceans feed on these, and they thus
become, though indirectly, of nearly equal importance with the others.
The greater number of the Calanidæ, and some of the Harpacticidæ, have
already been recognised among the contents of the stomachs of herring
and of the young of several species of Gadidæ; but a thorough and
practical acquaintance with the group in their living state is required to
enable one to identify them with anything like certainty in the stomach
of fishes.

OSTRACODA.

CYTHERIDÆ.

Cythere lutea, Müller.

 Cythere lutea, Brady, Trans. Lin. Soc., vol. xxvi. p. 395, pls.
 xxviii., xxxix., 1870. Gulland Bay, vicinity of Inchkeith;
 frequent (S.F.B.).

Cythere pellucida, Baird.

 Cythere pellucida, Brady, *op. cit.*, p. 397, pls. xxviii., xxxviii.
 Largo Bay, Gulland Bay; not infrequent (S.F.B.).

Cythere tenera, Brady.

 Cythere tenera, Brady, *op. cit.*, p. 399, pl. xxviii. Gulland Bay,
 deep water, west of May Island; not very common (S.F.B.).

Cythere crispata, Brady.
> *Cythere crispata*, Brady,   .   Vicinity of Inchkeith, west side ; scarce (S.F.B.).

Cythere Robertsoni, Brady.
> *Cythere robertsoni*, Brady   .   Gulland Bay, vicinity of Inchkeith ; not very common (S.F.B.).

Cythere villosa (G. O. Sars).
> *Cythere villosa*, Brady, *op. cit.*, p. 411, pl. xxix.   Gulland Bay, off Inchkeith, and other places ; a common species (S.F.B.).

Cythere convexa, Baird.
> *Cythere convexa*, Brady, *op. cit.*, p. 401, pls. xxix., xxxix.   One or two specimens dredged near Inchkeith (S.F.B.).

Cythere albomaculata, Baird.
> *Cythere albomaculata*, Brady, *op. cit.*, p. 402, pls. xxviii., xxxix. In material dredged off Bo'ness ; a few (S.F.B.).

Cythere limicola (Norman).
> *Cythere limicola*, Brady, *op. cit.*, p. 405, pl. xxxi.   A few specimens dredged in Gulland Bay, and in deep water west of May Island (S.F.B.).

Cythere tuberculata (G. O. Sars).
> *Cythere tuberculata*, Brady, *op. cit.*, p. 406, pl. xxx.   In the same localities as last ; frequent (S.F.B.).

Cythere concinna, Jones.
> *Cythere concinna*, Brady, *op. cit.*, p. 808, pls. xxvi., xxxviii. With the two previous ; frequent (S.F.B.).

Cythere angulata (G. O. Sars).
> *Cythere angulata*, Brady, *op. cit.*, p. 409, pl. xxvi.   A few specimens from deep water, west of May Island (S.F.B.).

Cythere quadridentata, Baird.
> *Cythere quadridentata*, Brady, *op. cit.*, p. 313, pl. xxxi.   One specimen, among material dredged off the west side of Inchkeith (S.F.B.).

Cythere Dunelmensis (Norman).
> *Cythere Dunelmensis*, Brady, *op. cit.*, p. 416, pl. xxx.   A few specimens from deep water, west of May Island (S.F.B.).

Cythere antiquata (Baird).
> *Cythere antiquata*, Brady, *op. cit.*, p. 417, pl. xxx.   Gulland Bay and vicinity of Inchkeith ; not common (S.F.B.).

Cythere Jonesii (Baird).
> *Cythere Jonesii*, Brady, *op. cit.*, p. 418, pl. xxx.   Occasionally in dredged material from Gulland Bay and deep water west of May Island (S.F.B.).

Cythere cyamos, Norman.
> *Cythere viridis*, Brady, (*non* Müller), *op. cit.*, p. 397, pls. xxviii. xxxviii.
> *Cythere cyamos*, Norman, Mus. Nor., part iii. p. 21, 1886.   Gulland Bay ; rather frequent (S.F.B.).

Cytheridea papillosa, Bosquet.
> *Cytheridea papillosa*, Brady, *op. cit.*, p. 423, pls. xxviii., xl. Gulland Bay, 10 to 12 fathoms ; off North Berwick, frequent. This species sometimes occurs in considerable abundance (S.F.B.).

Cytheridea elongata, Brady.
> *Cytheridea elongata*, Brady, *op. cit.*, p. 421, pls. xxviii., xl.   A few from deep water, west of Inchkeith (S.F.B.).

Eucythere declivis (Norman).
  Eucythere declivis, Brady, *op. cit.*, p. 430, pl. xxvii.
  Eucythere argus, G. O. Sars (var.), Mus. Nor., part iii.
  Eucythere anglica, Brady (var). A few from same locality as last (S.F.B.).

Loxoconcha guttata (Norman).
  Loxoconcha guttata, Brady, *op. cit.*, p. 436, pl. xxvii. Gulland Bay, 10 to 12 fathoms; off North Berwick, frequent (S.F.B.).

Loxoconcha tamarindus (Jones).
  Loxoconcha tamarindus, Brady, *op. cit.*, p. 435, pl. xxv. Vicinity of Inchkeith, Gulland Bay, off North Berwick; frequent (S.F.B.).

Xestoleberis depressa, G. O. Sars.
  Xestoleberis depressa, Brady, *op. cit.*, p. 438, pl. xxvii. One or two among material dredged a little west of Inchkeith (S.F.B.)

Cytherura striata, G. O. Sars.
  Cytherura striata, Brady, *op. cit.*, p. 441, pl. xxxii.
  Cytherura quadrata (fem.), Norman, Mus. Nor., p. iii. Gulland Bay, both forms; not unfrequent (S.F.B.).

Cytherura nigrescens (Baird).
  Cytherura nigrescens, Brady, *op. cit.*, p. 440, pls. xxxii., xxxix. Gulland Bay; common (S.F.B.).

Cytherura acuticostata, G. O. Sars.
  Cytherura acuticostata, Brady, *op. cit.*, p. 445, pl. xxxii. Among material dredged a little west of Inchkeith (S.F.B.).

Cytherura angulata, Brady.
  Cytherura angulata, Brady, *op. cit.*, p. 440, pl. xxxii.
  Cytherura insolita, Brady, Norman, Mus. Nor., pt. iii. Among material dredged west of Inchkeith (S.F.B.).

Cytherura cuneata, Brady.
  Cytherura cuneata, Brady, *op. cit.*, p. 442, pl. xxxii.
  Cytherura flavescens, Brady (fem.), Norman, Mus. Nor., pt. iii. Gulland Bay; frequent (S.F.B.)

Cytherura undata, G. O. Sars.
  Cytherura undata, Brady, *op. cit.*, p. 443, pl. xxxii. Among material dredged west of Inchkeith (S.F.B.).

Cytherura clathrata, G. O. Sars.
  Cytherura clathrata, Brady, *op. cit.*, p. 446, pl. xxix. Among material dredged off Bo'ness and a little west of Inchkeith; not very common (S.F.B.).

Cytherura cellulosa (Norman).
  Cytherura cellulosa, Brady, *op. cit.*, p. 446, pl. xxix. Gulland Bay; not very common (S.F.B.).

Cytheropteron nodosum, Brady.
  Cytheropteron nodosum, Brady, *op. cit.*, p. 448, pl. xxxiv. Gulland Bay; not unfrequent (S.F.B.).

Cytheropteron latissimum (Norman).
  Cytheropteron latissimum, Brady, *op. cit.*, p. 448, pl. xxxiv. Gulland Bay, and in deep water west of May Island; frequent (S.F.B.).

Bythocythere simplex (Norman).
  Bythocythere simplex, Brady, *op. cit.*, p. 450, pls. xxxiii., xl. Occasionally in the same localities as last (S.F.B.).

Cytherideis subulata, Brady.

> *Cytherideis subulata,* Brady, *op. cit.,* p. 454, pl. xxxv. A few
> specimens, among material dredged a little west of Inchkeith
> (S.F.B.).

Sclerochilus contortus, Norman.

> *Sclerochilus contortus,* Brady, *op. cit.,* p. 455, pls. xxxiv., xli.
> Gulland Bay, and in deep water (26 to 28 fathoms) west of
> May Island ; frequent (S.F.B.).

Xiphichilus tenuissimus, Norman.

> *Xiphichilus tenuissimus,* Norman, Mus. Nor., pt. iii., p. 21.
> In deep water west of May Island ; not common (S.F.B.).

Paradoxostoma abbreviatum, G. O. Sars.

> *Paradoxostoma abbreviatum,* Brady, *op. cit.,* p. 458, pl. xxxv.
> Gulland Bay ; a few specimens (S.F.B.).

Paradoxostoma ensiforme, Brady.

> *Paradoxostoma ensiforme.* Brady, *op. cit.,* p. 460, pl. xxxv. Among
> material dredged a little west of Inchkeith (S.F.B.).

Paradoxostoma flexuosum, Brady.

> *Paradoxostoma flexuosum,* Brady, *op. cit.,* p. 461, pl. xxxv. Gul-
> land Bay ; among material dredged off Bo'ness ; frequent
> (S.F.B.).

CYPRIDINIDÆ.

Philomedes interpuncta (Baird).

> *Philomedes interpuncta,* Brady, *op. cit.,,* p. 463, pl. xxxiii. Occa-
> sionally among dredged material, Gulland Bay, west of Inch-
> keith, &c., but seldom in material taken with surface-net
> (S.F.B.).

Remarks.—The Ostracoda are not so important as the Copepoda as a
source of food for fishes. Only in a very few instances have I noticed
them among the contents of fishes' stomachs, and these were the
stomachs of ground feeders, such as haddock and cod, and the only species
of Ostracod observed was the one last recorded in the preceding list, viz.,
Philomedes interpuncta. The young of *Balanus* in an early stage (the
Ostracod stage) are frequently found in the stomachs of herring and other
fishes, and are liable to be mistaken for a species of Ostracod, which may
account for such Entomostracans being recorded as occurring in the contents of fishes' stomachs more often than is really the case. Most of the
species live on or in the mud at the bottom, or among the Algæ and
Zoophytes which grow there within certain limits.

A small Entomostracan, *Evadne Nordmannii,* Loven, is frequently
noticed among the material collected by the surface-net in the seaward
part of the Forth, but not in so great abundance as it occurs in the Firth
of Clyde, the stomachs of herring taken there being sometimes found to
contain considerable numbers of these organisms. Whether the herring
purposely seek for and capture them, or whether they are swallowed in
a sort of indiscriminate way, as suggested by Dr Mobius, has not been
satisfactorily ascertained.

AMPHIPODA.*

ORCHESTIIDÆ.

Talitrus locusta, Linn. (Pallas?).

> *Talitrus locusta,* Bate and Westwood, Brit. Sess.-eyed Crust.,
> vol. i. p. 16, 1863. 'Very abundant about high-tide mark,
> 'frequent among stones, sea-weed, &c.' (L. & H.).

* The arrangement and nomenclature of Part iii. of Museum Normaniauum are
followed here.

Orchestia gammarellus, Pallas.
> *Orchestia gammarellus*, Norman, Mus. Nor., p. iii. p. 13, 1886.
> *Orchestia littorea*, B. & W., Brit. Sess.-eyed Crust., vol. i. p. 27,
> 1863. Common among decaying sea-weed about high-water
> mark, west of Granton.

Hyale nilssoni (Rathke).
> *Hyale nilssoni*, Norman, Mus. Nor., pt. iii. p. 14, 1886.
> *Allorchestes Nilssoni*, B. & W., Brit. Sess.-eyed Crust., vol. i.
> p. 40, 1863. Granton Quarry, common ; May Island, near
> high water ' (Henderson). Surface-net, Society Bank, west
> of Queensferry, one specimen (S.F.B.).

GAMMARIDÆ.

Orchomene serrata, Boeck.
> *Orchomene serrata*, Norman, Mus. Nor., pt. iii. p. 14, 1886.
> *Anonyx Edwardsii*, B. & W. (*non* Kröyer), Brit. Sess.-eyed Crust.,
> vol. i. p. 94.
> Between tide-marks, near Joppa, one or two specimens (S.F.B.).

Bathyporeia pilosa, Lindstrom.
> *Bathyporeia pilosa*, Norman, Mus. Nor., pt. iii. p. 14, 1886.
> *Bathyporeia pilosa*, *pelagica*, and *Robertsoni*, B. & W., Brit. Sess.-
> eyed Crust., vol. i. pp. 304-7-9. Near low-water mark,
> west of Granton ; one or two specimens (S.F.B.).

Harpina plumosa (Kröyer).
> *Harpina plumosa*, Norman, Mus. Nor., pt. iii. p. 14, 1886.
> *Phoxus plumosus*, B. & W., Brit. Sess.-eyed Crust., vol. i. p. 146.
> Gulland Bay, and in deep water west of May Island (S.F.B.).

Amphilochus manudens, Bate.
> *Amphilochus manudens*, B. & W., Brit. Sess.-eyed Crust., vol. i.
> p. 179. Gulland Bay, one or two specimens (S.F.B.).

Stenothoe monoculoides (Mont.).
> *Stenothoe monoculoides*, Norman, Mus. Nor., pt. iii. p. 14, 1886.
> *Montagui monoculoides*, Mont., B. & W., Brit. Sess.-eyed Crust.,
> vol. i. p. 54. Granton Harbour, taken with surface-net ; not
> common (S.F.B.).

Metopa pollexiana (Bate).
> *Metopa pollexiana*, Norman, Mus. Nor., pt. iii. p. 14, 1886.
> *Montagui pollexiana*, B. & W., Brit. Sess.-eyed Crust., vol. i.
> p. 64.
> *Stenothoe pollexiana*, Henderson, Proc. Roy. Phys. Soc. Edin.,
> vol. viii. p. 310, 1884. Newhaven, from fishermen's lines
> (Henderson) ; off Inchkeith, west side ; common (S.F.B.).

Monoculodes longimanus, B. & W.
> *Monoculodes longimanus*, B. & W., Brit. Sess.-eyed Crust., vol. ii.
> p. 507. One or two specimens among material dredged off
> Bo'ness (S.F.B.).

Pontocrates altimarinus (B. & W.).
> *Pontocrates altimarinus*, Norman, Mus. Nor., pt. iii. p. 15, 1886
> *Kroyera altimarinus*, B. & W., Brit. Sess.-eyed Crust., vol. i. p.
> 177. In the same material as the last (S.F.B.).

Paramphithoe bicuspis (Kröyer).
> *Paramphithoe bicuspis*, Norman, Mus. Nor., pt. iii. p. 15, 1886.
> *Pherusa bicuspis*, B. & W., Brit. Sess.-eyed Crust., vol. i. p. 253.
> Newhaven, from fishermen's lines (Henderson).

Paramphithoe fucicola (Leach).
    Paramphithoe fucicola, Norman, Mus. Nor., pt. iii. p. 15, 1886.
    Pherusa fucicola, B. & W., Brit. Sess.-eyed Crust., vol. i. p. 255,
        1863. Newhaven, from fishermen's lines (Henderson).
Iphimedia obesa, Rathke.
    Iphimedia obesa, B. & W., Brit. Sess.-eyed Crust., vol. i. p. 219,
        1863. Newhaven, from fishermen's lines, many specimens ;
        dredged to east of Inchkeith (Henderson) ; Granton Harbour,
        among Zoophytes attached to pier ; sometimes taken with
        surface-net (S.F.B.).
Dexamine spinosa (Mont) (Leach ?, L. & H.).
    Dexamine spinosa, Mont., B. & W., Brit. Sess. eyed Crust., vol.
        i. p. 237, 1863. Low water, Prestonpans (Cunningham,
        L. & H.). Gulland Bay ; a few specimens (S.F.B.).
Atylus Swammerdamii (Milne-Edwards).
    Atylus Swammerdamii, B. & W., Brit. Sess.-eyed Crust., vol. i.
        p. 246, 1863. Gulland Bay ; about a dozen specimens
        (S.F.B.).
Halirages bispinosus (Bate).
    Halirages bispinosus, Norman, Mus. Nor., pt. iii. p. 15, 1886.
    Atylus bispinosus, B. & W., Brit. Sess.-eyed Crust., vol. i. p. 250,
        1863. Gulland Bay ; scarce (S.F.B.).
Calliopius lœviusculus (Kröyer).
    Calliopius lœviusculus, Norman, Mus. Nor., pt. iii. p. 15, 1886.
    Calliope lœviuscula, B. & W., Brit. Sess.-eyed Crust., vol. i.
        p. 254, 1863. Not unfrequent between tide-marks near
        Joppa (S.F.B.).
Calliopius bidentatus, Norman.
    Calliopius bidentatus, Henderson, Proc. Roy. Phys. Soc. Edin., vol.
        viii. p. 310, 1884. Newhaven, many specimens from fisher-
        men's lines ; dredged off Fidra, 12 fathoms (Henderson).
Melita obtusata (Mont).
    Melita obtusata, B. & W., Brit. Sess.-eyed Crust., vol. i. p. 341,
        1863. Bass Rock, 24 fathoms ; off St Abb's Head, 40
        fathoms (Metzger ; L. & H.). Aberlady Bay (S.F.B.).
    Melita proxima, Bate, and *Megamœra Alderi*, Bate, are forms of
        this species (Norman, Mus. Nor., pt. iii. p.   ).
Melita gladiosa, Bate.
    Melita gladiosa, B. & W., Brit. Sess.-eyed Crust., vol. i. p. 346,
        1863. Largo Bay ; two specimens (S.F.B.).
Mœra grossimana, Mont.
    Mœra grossimana, B. & W., Brit. Sess.-eyed Crust., vol. i.
        p. 350, 1863. One specimen, among material dredged a little
        west of Inchkeith (11–11–87) (F.S.B.). This is an interest-
        ing addition to the fauna of the Forth. Bate and Westwood
        say : ' It would appear from the circumstance of its being
        ' very abundant on the south of England, whilst we have never
        ' received it from our numerous correspondents from the north
        ' (although it is recorded by Dr Johnston at Berwick, see Zool.
        ' Journ., iii. 180, where he records it as not very rare in
        ' Berwick Bay), to be essentially a species belonging to warmer
        ' latitudes.'
Mœra othonis (Milne-Edward).
    Mœra othonis, Norman, Mus. Nor., pt. iii. p. 16, 1886
    Megamœra othonis, B. & W., Brit. Sess.-eyed Crust., vol. i.
        p. 405, 1863. Gulland Bay ; one specimen (S.F.B.).

Gammarus locusta (Linné).
> *Gammarus locusta*, B. & W., Brit. Sess.-eyed Crust., vol. i.
> p. 378, 1863. Common among decaying sea-weed near high-
> water mark ; occasionally in the surface-net (S.F.B.).

Byblis Gaimardi, Kröyer.
> *Byblis Gaimardi*, L. & H., Invert. fauna Firth of Forth, p. 105,
> 1881. St Abb's Head, 40 fathoms (Metzger ; L. & H.).

Ampelisca typica (Bate).
> *Ampelisca typica*, Norman, Mus. Nor., pt. iii. p. 16, 1886.
> *Ampelisca Gaimardi*, B. & W., Brit. Sess.-eyed Crust., vol. i.
> p. 127, 1863. East of Fidra (S.F.B.).

Ampelisca tenuicornis, Lilljeborg.
> *Ampelisca tenuicornis*, Norman, Mus. Nor., pt. iii. p. 16, 1886.
> *Ampelisca lævigata*, B. & W., Brit. Sess-eyed Crust., vol. ii.
> p. 504, 1863. Bass Rock, 24 fathoms ; off St Abb's Head,
> 40 fathoms (Metzger ; L. & H.) ; Gulland Bay (S.F.B.).

Ampelisca macrocephala, Lilljeborg.
> *Ampelisca macrocephala*, L. & H. Invert. fauna Firth of Forth,
> p. 44, 1881. Firth of Forth, 24 fathoms (Metzger ; L. & H.).

Ampelisca æquicornis, Bruzelius.
> *Ampelisca æquicornis*, Henderson, Proc. Roy. Phys. Soc. Edin.,
> vol. viii. p. 310, 1884. West of May Island, 20 fathoms
> (Henderson).

Ptilocheirus hirsutimanus (Bate).
> *Ptilocheirus hirsutimanus*, Norman, Mus. Nor., pt. iii. p. 16, 1886.
> *Protomedia hirsutimana*, B. & W., Brit. Sess.-eyed Crust., vol. i.
> p. 298, 1863. East off Elie Ness, 15–17 fathoms (S.F.B.).

Aora gracilis, Bate.
> *Aora gracilis*, B. & W., Brit. Sess.-eyed Crust., vol. i. p. 281,
> 1863. Newhaven, many specimens from fishermen's lines,
> both sexes, off Fidra (Henderson) ; Gulland Bay, and deep
> water west of May Island (S.F.B.).

Protomedia fasciata, Kröyer.
> *Protomedia fasciata*, L. & H., Invert. fauna Firth of Forth, p. 44,
> 1881. St Abb's Head, 40 fathoms (Metzger ; L. and H.).

Gammaropsis erythrophthalmus (Lilljeborg).
> *Gammaropsis erythrophthalmus*, Norman, Mus. Nor., pt. iii.
> p. 17, 1886.
> *Eurystheus erythrophthalmus*, B. & W., Brit. Sess. eyed Crust.,
> vol. i. p. 354, 1863. One or two specimens, dredged a little
> west of Inchkeith.

COROPHIIDÆ.

Podoceropsis Sophiæ (Boeck).
> *Podoceropsis Sophiæ*, Norman, Mus. Nor., pt. iii. p 17, 1886.
> *Nænia tuberculosa*, B. & W., Brit. Sess.-eyed Crust., vol. i.
> p. 472, 1863. Dredged to the east of Inchmickery ; also
> south-west of Inchkeith (Henderson). Gulland Bay (S.F.B.).

Podoceropsis rimapalma (Bate).
> *Podoceropsis rimapalma*, Norman, Mus. Nor., pt. iii. p. 17, 1886.
> *Nænia rimapalma*, *excavata*, and *undata*, B. & W. Brit. Sess.-
> eyed Crust., vol. i. pp. 472 4 6, 1863. With the last in both
> localities (Henderson). Large Bay, 10 to 12 fathoms (S.F.B.).

Amphithoe podoceroides (Rathke).
> *Amphithoe podoceroides*, Norman, Mus. Nor., pt. iii. pt. 17, 1886

Amphithoe littorina, B. & W., Brit. Sess.-eyed Crust., vol. i.
    p. 422, 1863. Gulland Bay, and deep water west of May
    Island (S.F.B.).
Erichthonius difformis, Milne-Edwards.
    Erichthonius difformis, Norman, Mus. Nor., pt. iii. p. 17, 1886.
    Cerapus difformis, B. & W., Brit. Sess.-eyed Crust., vol. i.
    p. 457, 1863.
    Dercothoe punctatus (fem.), *loc. cit.*, p. 461. Gulland Bay (S.F.B.).
    Bass Rock (Metzger ; L. & H.).
Erichthonius abditus (Templeton).
    Erichthonius abditus, Norman, Mus. Nor., pt. iii. p. 17, 1886.
    Cerapus abditus, B. & W., Brit. Sess.-eyed Crust., vol. i. p. 455,
    1863. Gulland Bay ; a few specimens (S.F.B.).
Sunamphithoe hamulus, Bate.
    Sunamphithoe hamatus, B. & W., Brit. Sess.-eyed Crust., vol. i.
    p. 430, 1863. Off Pittenweem, among trawl refuse, and
    between tide-marks near Portobello (S.F.B.).
Podocerus falcatus (Mont).
    Podocerus falcatus, Norman, Mus. Nor., pt. iii. p. 17, 1886.
    Podocerus (Jassa) pelagicus, B. & W., Brit. Sess.-eyed Crust.,
    vol. i. pp. 436, 445, 447.
    Podocerus (Jassa) pulchellus, B. & W., *loc. cit.* Gulland Bay ; a
    few specimens (S.F.B.).
Janassa capillata (Rathke).
    Janassa capillata, Norman, Mus. Nor., pt. iii. p. 17, 1886.
    Podocerus capillatus, B. & W., Brit. Sess.-eyed Crust., vol. i.
    p. 442. 1863. 'We have dredged this species in 5 fathoms
    ' off Inchkeith ' (L. & H.).
Dryope crenatipalmata, Bate.
    Dryope crenatipalmata, B. & W., Brit. Sess.-eyed Crust., vol. i.
    p. 490, 1863. Among material dredged off Bo'ness (S.F.B.).
Corophium grossipes (Linné).
    Corophium grossipes, Norman, Mus. Nor., pt. iii. p. 17, 1886.
    Corophium longicorne, B. & W., Brit. Sess.-eyed Crust., vol. i.
    p. 493, 1863. Dunbar (Robertson ; L. & H.).
Corophium crassicorne, Bruzelius.
    Corophium crassicorne, Norman, Mus. Nor., pt. iii. p. 17, 1886.
    Corophium spinicorne, B. & W., Brit. Sess.-eyed Crust., vol. i.
    pp. 497-9.
    Corophium Bonellii, B. & W., *loc. cit.* Gulland Bay ; several
    specimens (S.F.B.).
Corophium tenuicorne, Norman.
    Corophium tenuicorne, Henderson, Proc. Roy. Phys. Soc. Edin.,
    vol. viii. p. 310, 1884. A single species dredged off Fidra
    in 12 fathoms (Henderson).

HYPERIIDÆ.

Hyperia medusarum (O. Fabricius).
    Hyperia medusarum, Norman, Mus. Nor., pt. iii. p. 13, 1863.
    Hyperia galba, B. & W., Brit. Sess.-eyed Crust., vol. ii. p. 12,
    1863.
    Lestrigonus Kinahani, B. & W., *loc. cit.*, p. 8. In the pouches
    of Medusæ (Cunningham ; L. & W.). Occasionally in the
    surface-net (S.F.B.).

Parathemisto oblivia (Kröyer).

> *Parathemisto oblivia*, Norman, Mus. Nor., pt. iii. p. 13, 1886.
> *Hyperia oblivia*, B. & W., Brit. Sess.-eyed Crust., vol. ii. p. 16,
> 1863. Several specimens taken in the tow-net off May
> Island and off Inchkeith ; occasionally in surface-net, Largo
> Bay, off North Berwick ; surface-net, Society bank, west of
> Queensferry ; one specimen (S.F.B.).

DULICHIIDÆ.

Dulichia falcata, Bate.

> *Dulichia falcata*, B. & W., Brit. Sess.-eyed Crust., vol. ii. p. 33,
> 1863. One specimen in surface-net in Granton Harbour
> (S.F.B.).

CAPRELLIDÆ.

Proto ventricosa (Müller).

> *Proto ventricosa*, Norman, Mus. Nor., pt. iii. p. 17, 1886.
> *Proto pedata*, B. & W., Brit. Sess.-eyed Crust., vol. ii. pp. 38, 42.
> *Proto Goodsiri*, B. & W., *loc. cit.* Newhaven, from the fisher-
> men's lines ; not uncommon (Henderson). Gulland Bay
> (S.F.B.).

Protella phasma, Mont.

> *Protella phasma*, B. & W., Brit. Sess.-eyed Crust., vol. ii. p. 45,
> 1863. Island of May (Brit. Mus.); Firth of Forth (H.D.S.
> Goodsir ; L. & H.). Two or three specimens, Gulland Bay,
> (S.F.B.).

Caprella linearis (Linné).

> *Caprella linearis*, Norman, Mus. Nor., pt. iii. p. 17, 1886.
> *Caprella linearis*, B. & W., Brit. Sess.-eyed Crust., vol. ii.
> pp. 52–57, 1863.
> *Caprella lobata*, B. & W., *loc. cit.* Plentiful in the upper
> laminarian zone ; we have also dredged it in a few fathoms
> (L. & H.). Gulland Bay ; frequent (S.F.B.).

Caprella acanthifera, Leach.

> *Caprella acanthifera*, B. & W., Brit. Sess.-eyed Crust, vol. ii.
> p. 65, 1863. Firth of Forth (Bell Collection, Oxford, Rev.
> J. Gordon ; L. & H.).

Caprella tuberculata, Guerin.

> *Caprella tuberculata*, B. & W., Brit. Sess.-eyed Crust., vol. ii.
> p. 68, 1863. Firth of Forth (Brit. Mus.; L. & H.). Gulland
> Bay (S.F.B.).

Podalirius typicus, Kröyer.

> *Podalirius typicus*, Norman, Mus. Nor., pt. iii. p. 17, 1886.
> *Caprella typica*, B. & W., Brit. Sess.-eyed Crust., vol. ii. p. 35,
> 1863. Firth of Forth (Bell Collection, Oxford ; L. & H.).
> In about 12 fathoms, east of Inchkeith and Largo Bay
> (S.F.B.).

Remarks.—It will be observed that by the investigations carried out
by instructions of the Fishery Board, several species of Amphipods are
here recorded for the first time from the estuary of the Forth, and
doubtless others will be discovered as these investigations are proceeded
with. Amphipods are frequently met with in the stomachs of fishes, but
so far as has been observed, those fishes which feed mostly near the

bottom, such as haddock and cod, appear to have no partiality for any particular species, but the stomachs of herring, especially of those taken on the East Coast, are found occasionally filled with species of Hyperiidæ to the exclusion of everything else, *Hyperia galba* and *Parathemisto oblivia* being the two generally observed. Of course the Hyperiidæ are exclusively pelagic in their habits, which may partly account for this, but the young and half-grown specimens of other species which are frequently met with in material collected with the surface-net have very seldom been noticed in herrings' stomachs, so that it is probable the species of Hyperiidæ are preferred by them.

ISOPODA.

The arrangement followed here is that adopted in Bate and Westwood's *Monograph of British Sessile-eyed Crustacea ;* the nomenclature that of Part iii. of *Museum Normanianum.*

TANAIDÆ.

Tanais vittatus (Rathke).
> *Tanais vittatus,* B. & W., Brit. Sess.-eyed Crust., vol. ii. p. 125 1863. May Island and Dunbar; about half-tide, living chiefly among mussels (Henderson).

Pseudotanais forcipatus (Lilljeborg).
> *Pseudotanais forcipatus,* Norman, Mus. Nor., pt. iii. p. 11, 1886.
> *Paratanais forcipatus,* B. & W., Brit. Sess.-eyed Crust., vol. ii. p. 138, 1863. Gulland Bay, frequent among dredged material; and from deep water west off Inchkeith (S.F.B.).

BOPYRIDÆ.

Phryxus abdominalis (Kröyer).
> *Phryxus abdominalis,* B. & W., Brit. Sess.-eyed Crust., vol. ii. p. 234, 1863. Off St Abb's Head, 40 fathoms (Metzger; L. & H.).

Athelges paguri (Rathke).
> *Athelges paguri,* Norman, Mus. Nor., pt. iii. p. 13, 1886.
> *Phryxus paguri,* B. & W., Brit. Sess.-eyed Crust., 1863. Firth of Forth (Dr Anderson).* On *Pagurus Bernhardus,* dredged in Gulland Bay (S.F.B.).

Liriopsis balani, Bate.
> *Liriope balani,* Bate, Brit. Assoc. Report, 1860, p. 225.
> *Cryptothiria balani,* Bate, B. & W., Brit. Sess.-eyed Crust., vol. ii. p. 267, 1863. Firth of Forth (H. Goodsir, as male of *Balanus balanoides,* in *Edin. New Phil. Journ.,* 1843; B. & W.).

ÆGIDÆ.

Eurydice pulchra, Leach.
> *Eurydice pulchra,* B. & W., Brit. Sess.-eyed Crust., vol. ii. p. 310, 1863. Taken in surface-net, off Preston Island, a few miles west of Queensferry ; one specimen (28.11.87, S.F.B.).

ASELLIDÆ.

Jæra albifrons (Mont).
> *Jæra albifrons,* B. & W., Brit. Sess.-eyed Crust., vol. ii. p. 317, 1863. Common at the May Island and Granton Quarry (Henderson).

* *Proc. Roy. Phys. Soc. Edin.,* vol. vii.

Jæra Nordmanni (Rathke).
　　Jæra Nordmanni, B. & W., Brit. Sess.-eyed Crust., vol. ii. p. 320,
　　　　1863. Among dredged material, off Inchkeith. Between
　　　　tide-marks near Joppa (S.F.B.).
Munna Kröyeri, Goodsir.
　　Munna Kröyeri, Goodsir, Edin. New Phil. Journ., 1842.
　　Munna Kröyeri, B. & W., Brit. Sess.-eyed Crust., vol. ii. p. 326,
　　　　1863. Firth of Forth (Goodsir). Among material dredged
　　　　a little west of Inchkeith ; several specimens (S.F.B.).
Janira maculosa, Leach.
　　Janira maculosa, B. & W., Brit. Sess.-eyed Crust., vol. ii. p. 338,
　　　　1863. Among dredged materials off Bo'ness (S.F.B.).
Limnoria lignorum (Rathke).
　　Limnoria lignorum, B. & W., Brit. Sess.-eyed Crust., vol. ii.
　　　　p. 351, 1863. 'We obtained it at Elie' (L. & H.).

ARCTURIDÆ.

Astacilla longicornis (Sowerby).
　　Astacilla longicornis, Norman, Mus. Nor., pt. iii. p. 12, 1886.
　　Arcturus longicornis, B. & W., Brit. Sess.-eyed Crust., vol. ii.
　　　　p. 365, 1863. Firth of Forth ; not unfrequent throughout
　　　　the estuary.
Astacilla intermedia (Goodsir).
　　Leachia intermedia, Goodsir, Edin. New Phil. Journ., 1842.
　　Astacilla intermedia, Norman, Mus. Nor., pt. iii. p. 12, 1886.
　　Arcturus intermedia, B. & W., Brit. Sess.-eyed Crust., vol. ii.
　　　　p. 371, 1863. Off Anstruther (Goodsir).
Astacilla gracilis (Goodsir).
　　Leachia gracilis, Goodsir, Edin. New Phil. Journ., 1842.
　　Astacilla gracilis, Norman, Mus. Nor., pt. iii. p. 12, 1886.
　　Arcturus gracilis, B. & W., Brit. Sess.-eyed Crust., vol. ii. p. 373,
　　　　1863. Off Anstruther (Goodsir).

IDOTEIDÆ

Idotea marina, Linné. (Fabr. ?).
　　Idotea marina, Norman, Mus. Nor., pt. iii. p. 12, 1886.
　　Idotea tricuspidata, B. & W., Brit. Sess.-eyed Crust., vol. ii. pp. 379,
　　　　384, 1863.
　　Idotea pelagica, B. & W., *loc. cit.* Common throughout the Forth.
Idotea linearis (Linné) (Penn. ?).
　　Idotea linearis, B. & W., Brit. Sess.-eyed Crust., vol. ii. p. 388,
　　　　1863. Occasionally at Newhaven, from the fishermen's lines
　　　　(Henderson). Off Crail ; several specimens (S.F.B.).
Idotea emarginata (Fabricius).
　　Idotea emarginata, B. & W., Brit. Sess.-eyed Crust., vol. ii.
　　　　p. 386, 1863. Several specimens taken with the dredge in
　　　　Aberlady Bay.

ONISCIDÆ.

Ligia oceanica (Linné).
　　Ligia oceanica, B. & W., Brit. Sess.-eyed Crust., vol. ii. p. 444,
　　　　1863. Under stones at and above high-water mark ; gene-
　　　　rally distributed.

Remarks.—Isopods, especially the commoner Idoteidæ, are frequently found among the contents of the stomachs of fishes such as cod, haddock, saith, and other ground-feeders, but seldom in the stomachs of pelagic fishes.

CUMACEA.*

DIASTYLIDÆ.

Iphinoe gracilis, Bate.
    Iphinoe gracilis, L. & H., Invert. Fauna Firth of Forth, p. 106.
    Venilia gracilis, White, Pop. Hist. Brit. Crust., p. 153, 1857.
        Bass Rock, 24 fathoms (Metzger; L. & H.).
Cumopsis Goodsiri, Van Beneden.
    Between tide-marks near Joppa (S.F.B.).
Lamprops fasciata, G. O. Sars.
    Frequent between tide-marks near Joppa, where the beach is composed of sand; also taken with surface-net in Granton Harbour (S.F.B.).
Leucon nassica, Kröyer.
    Leucon nassica, L. & H., Invert. Fauna Firth of Forth, p. 106, 1881. St Abb's Head, 40 fathoms (Metzger; L. & H.).
Diastylis Rathkii, Kröyer.
    Alauna rostrata, Bell, Brit. Stalk-eyed Crust., p. 331, 1853.
    Diastylis Rathkii, Sp. Bate, Ann. and Mag. Nat. Hist., vol. xvii. p. 451, 1856.
    Diastylis Rathkii, White's Pop. Hist. Brit. Crust., p. 150, 1857. Firth of Forth (Goodsir; L. & H.). A considerable number of specimens among material dredged a little west of Inchkeith (S.F.B.).
Diastylis lucifera, Kröyer.
    One specimen among material dredged off the west side of Inchkeith (S.F.B.).
Diastylis lævis, Norman.
    Off Fidra, 12 fathoms (Henderson).
Pseudocuma cercaria, Van Beneden.
    Several specimens between tide-marks near Joppa; taken with surface-net in Granton Harbour; also taken with surface-net, inshore, near Charleston (S.F.B.).
(?) *Cuma Edwardsii,* Goodsir.
    Cuma Edwardsii, White, Pop. Hist. Brit. Crust., p. 151, 1857. Firth of Forth (H. Goodsir). Largo Bay (L. & H.).
Cuma scorpioides, Mont.
    Cuma scorpioides, White, Pop. Hist. Brit. Crust., p. 150, 1857. Firth of Forth (H. Goodsir; L. & H.).
(?) *Cuma trispinosa,* Goodsir.
    Cuma trispinosa, Bell, Brit. Stalk-eyed Crust., p. 329, 1853.
    Halia trispinosa, White, Pop. Hist. Brit. Crust., p. 152, 1857.
    Halia trispinosa, Bate, Ann. and Mag. Nat. Hist., vol. xvii. p. 459, 1856. Firth of Forth (H. Goodsir; L. & H.).
(?) *Bodotria arenosa,* Goodsir.
    Bodotria arenosa, Bell, Brit. Stalk-eyed Crust., p. 332, 1853. Firth of Forth (H. Goodsir; L. & H.).

Remarks.—Some of Goodsir's *Cumas* cannot satisfactorily be ascribed to known species. Four of the Cumacea here recorded are additions to the Forth fauna.

* The arrangement followed here and in the list of the Schizopoda is that of part ii. of Museum Normandanum.

SCHIZOPODA.

EUPHAUSIDÆ.

Boreophausia raschii (M. Sars).

    Thysanopoda Raschii, M. Sars, Vidensk. Selsk. Förhandl., 1863, p. 14.

    Boreophausia Raschii, G. O. Sars, Prelim. Notices Challenger Schizopoda, Vidensk. Selsk. Förhandl., No. 7, p. 11, 1883. ' We took specimens in the Firth of Forth in November 1884.' (Henderson).

Nyctiphanes norvegica (M. Sars).

    Thysanopoda norvegica, M. Sars, Vidensk. Selsk. Förhandl., 1863, p. 2.

    Nyctophanes norvegica, G. O. Sars, Prelim. Notices Challenger Schizopoda, 1883, No. 7, p. 23. Many specimens taken in the surface-net north-east of Inchkeith (Henderson). This species is of frequent occurrence in the Firth of Forth (S.F.B.).

MYSIDÆ.

Mysidopsis gibbosa, G. O. Sars.

    Mysidopsis gibbosa, G. O. Sars, Carcinologisk Bidrag til Norges Fauna I. Monogr. over de ved norges kyster forekommende Mysider Hefte ii. p. 23, pl. xxviii. figs. 1–13, 1872. One specimen taken with surface-net in Granton Harbour, November 1887 (S.F.B.).

Leptomysis lingvura, G. O. Sars.

    Leptomysis lingvura, G. O. Sars, *op. cit.*, Hefte. iii. p. 35, pl. xxi., 1879. Several specimens were secured near extreme low-water mark, a little east of Joppa, September 1887 (S.F.B.).

Siriella crassipes, G. O. Sars.

    Siriella crassipes, G. O. Sars. One specimen taken at extreme low water, a little east of Joppa, September 1887 (S.F.B.).

Gastrosaccus spinifer, Göes.

    Gastrosaccus spinifer, Stebbing, Ann. and Mag. Nat. Hist., ser. 5 vol. vi., 1880. Frequent in material collected in surface-net off Bo'ness.

Macropsis slabberi, Van Beneden.

    Macropsis slabberi, Henderson, Decapod and Schizopod. Crust. Clyde, p. 11, 1886.

    Podopsis slabberi, Henderson, Proc. Roy. Phys. Soc. Edin., vol. viii. p. 311, 1884. Taken by Mr Cunninghame and Dr Henderson with surface-net, below Grangemouth, in October 1884, and subsequently near Granton Quarry, Inchmickery, and Inchkeith (Henderson). Off Bo'ness, a very common species in the upper waters of the Forth, but does not yet appear to have been met with anywhere else in Britain (S.F.B.).

Mysis flexuosa, Müller.

    Mysis flexuosus, G. O. Sars, *op. cit.*, Hefte iii. p. 45, pls. xxiv, xxv., 1879.

    Mysis chamæleon, Bell, Brit. Stalk-eyed Crust., p. 336, 1853. Rock pools, Seafield (M'B.) ; Firth of Forth (Leach; L. and H.). West of Granton at low water ; frequent, and in pools left by the ebbing tide (S.F.B.).

    [Exceedingly plentiful in Cromarty Firth, near Invergordon, September 1887 (S.F.B.).]

Mysis inermis, Rathke.
> *Mysis inermis,* G. O. Sars, *op. cit.,* Hefte iii. p. 54, pl. xxvii., 1879. Taken by surface-net off Bo'ness, November 1887 (S.F.B.).

Mysis spiritus, Norman.
> *Mysis spiritus,* G. O. Sars, *op. cit.,* Hefte iii. p. 58, pl. xxviii., 1879. Taken in the same locality as the last; several specimens (S.F.B.).
> [Also taken by tow-net at 14 fathoms near Aberdeen, October 1886 (S.F.B.).]

Mysis ornata, G. O. Sars.
> *Mysis ornata,* G. O. Sars, *op. cit.,* Hefte iii. p. 62, pl. xxix., 1879. Taken with the last two (S.F.B.).

Mysis lamornæ, R. Q. Couch.
> *Mysis lamornæ,* White, Pop. Hist. Brit. Crust., p. 143, 1857.
> *Mysis aurantia,* G. O. Sars, Beretrung om en i Sommeren, 1863, foretagen Zoologisk Reise, p. 30. Also taken in surface-net off Bo'ness ; two or three specimens (S.F.B.).

Mysis vulgaris, J. Vaughan Thompson.
> *Mysis vulgaris,* Bell, Brit. Stalk-eyed Crust., p. 339, 1853. A common species in surface-net, material in various parts of the Forth west of Queensferry, and occasionally in Granton Harbour (S.F.B.).

Remarks.—With the exception of the last, the species of *Mysis* recorded here have the telson more or less bifurcate, but the telson of *Mysis vulgaris* terminates in a blunt point, and the antennal scales also differ from those of the other species of the genus. *Cynthia flemingii* (Goodsir), *Themisto longispinosa* (Goodsir), and *Themisto brevispinosa* (Goodsir), recorded from the Firth of Forth by Mr H. Goodsir, are doubtful species. Dr Norman says that it is impossible to identify Goodsir's Mysidæ, but suggests that his *Cynthia flemingii* may be the *Siriella crassipes* of G. O. Sars, and *Themisto longispinosa* and *Themisto brevispinosa* the males of species of *Mysis.*

Through the investigations carried on by the Fishery Board, nine species of Schizopoda have been added to the fauna of the Firth of Forth. It may be worth noting that though a considerable number of species of Schizopods are now included in the British fauna, only two (*Nyctiphanes norvegica* and *Boreophausia raschii*) have been observed among the contents of fishes' stomachs, *e.g.,* those of the herring and mackerel.

DECAPODA.

BRACHYURA.

INACHIDÆ.

Inachus dorsettensis (Pennant).
> *Inachus dorsettensis,* Bell, Brit. Stalk-eyed Crust., p. 13, 1853, 'Deep sea lines' (Howden ; L. & H.).

MAIIDÆ.

Hyas araneus (Linné).
> *Hyas araneus,* Bell, Brit. Stalk-eyed Crust., p. 31, 1853. Common between tide-marks, and dredged in all parts of the estuary. Fishermen have a great dislike to this crab, and

generally kill every one they get ; it is blamed for eating the bait from the hooks of the long lines, to the great annoyance and loss of the fishermen.

Hyas coarctatus Leach.

> *Hyas coarctatus*, Bell, Brit. Stalk-eyed Crust., p. 35, 1853. A fairly common species, from the laminarian zone outwards.

LEPTOPODIDÆ.

Stenorhynchus rostratus, Linné.

> *Stenorhynchus rostratus*, Norman, Mus. Nor., pt. iii. p. 6, 1886.
> *Stenorhynchus phalangium*, Bell, Brit. Stalk-eyed Crust., p. 2, 1853. Firth of Forth, on mud and sand, generally distributed (Howden ; L. & H.) ; common in the vicinity of Inchkeith (S.F.B.).

PARTHENOPIDÆ.

Eurynome aspera (Pennant).

> *Eurynome aspera*, Bell, Brit. Stalk-eyed Crust., p. 46, 1853. Off Prestonpans and Portseaton (Howden ; L. & H.).

CANCRIDÆ.

Cancer pagurus, Linné.

> *Cancer pagurus*, Bell, Brit. Stalk-eyed Crust., p. 59, 1853. Common in the laminarian and littoral zones (L. & H.). Occasionally in the trawl-net (S.F.B.).

PORTUNIDÆ.

Portunus puber (Linné).

> *Portunus puber*, Bell, Brit. Stalk-eyed Crust., p. 90, 1853. One specimen on the deep-sea lines, from the mouth of the Forth (L. & H.)

Portunus depurator (Linné).

> *Portunus depurator*, Bell, Brit. Stalk-eyed Crust., p. 101, 1853. A common species throughout the estuary.

Portunus marmoreus, Leach.

> *Portunus marmoreus*, Bell, Brit. Stalk-eyed Crust., p. 105, 1853. Portobello and Musselburgh beaches (Howden ; L. & H.). Taken occasionally in the dredge, in the vicinity of Inchkeith, and in the trawl west of May Island (S.F.B.).

Portunus holsatus, Fabricius.

> *Portunus holsatus*, Bell, Brit. Stalk-eyed Crust., p. 109, 1853. Dr Leach found one amongst a number of specimens of *Portunus depurator* at Newhaven (L. & H.). One of the species commonly met with on the ' Oyster Banks ' (Henderson).[*]

Portunus pusillus, Leach.

> *Portunus pusillus*, Bell, Brit. Stalk-eyed Crust, p. 112, 1853. Off Prestonpans (Howden). We have frequently dredged it near Inchkeith, &c. (L. & H.). Largo Bay (S.F.B.).

---

[*] *Decapod and Schizopod Crustacea of the Clyde*, p. 10.

PLATYONYCHIDÆ.

Carcinus mænas, Linné.
　　Carcinus mænas, Bell, Brit. Stalk-eyed Crust., p. 76, 1853.
　　　　Between tide-marks, and in the laminarian zone ; everywhere
　　　　common.
Portumnus latipes, Pennant.
　　Portumnus latipes, White, Pop. Hist. Brit. Crust., p. 43.
　　Portumnus variegatus, Bell, Brit. Stalk-eyed Crust., p. 85, 1853.
　　　　Prestonpans and Portseaton (Howden). We have taken this
　　　　species at Portobello (L. & H.). The Rev. A. M. Norman,
　　　　in Part III. of *Mus. Nov.*, retains this species in Milne-
　　　　Edwards' genus *Platyonychus*.

CORYSTIDÆ.

Atelecyclus septemdentatus (Mont).
　　Atelecyclus septemdentatus, White, Pop. Hist. Brit. Crust., p. 64,
　　　　1857.
　　Atelecyclus heterodon, Bell, Brit. Stalk-eyed Crust., p. 153, 1853.
　　　　Firth of Forth, rare (Goodsir). Portobello beach (M'Bain ;
　　　　L. & H.). One specimen taken by trawl-net west of May
　　　　Island (S.F.B.).
Clyroess cassivelaunus (Pennant).
　　Corystes cassivelaunus, Bell, Brit. Stalk-eyed Crust., p. 159, 1853.
　　　　Off Inchkeith (M'Bain). Bass Rock, 24 fathoms (Metzger).
　　　　Newhaven (C. W. Peach). ' We have dredged it in Aberlady
　　　　Bay' (L. & H.). Occasionally in the trawl-net west of May
　　　　Island (S.F.B.).

PINNOTHERIDÆ.

Pinnotheres pisum (Linné).
　　Pinnotheres pisum, Bell, Brit. Stalk-eyed Crust., p. 121, 1853.
　　　　Taken with the dredge associated with *Mytilus modiolus*, in
　　　　the vicinity of Inchkeith (S.F.B.). Off Longniddry in
　　　　14 fathoms (L. & H.).

LEUCOSIDÆ.

Ebalia tuberosa (Pennant).
　　Ebalia Pennantii, Leach, Bell, Brit. Stalk-eyed Crust., p. 141,
　　　　1853. Firth of Forth, rare (H. Goodsir; L. & H.). One
　　　　specimen among trawl material, west of May Island (S.F.B.).
Ebalia Cranchii, Leach.
　　Ebalia cranchi, Bell, Brit. Stalk-eyed Crust., p. 148, 1853.
　　　　2½ miles off Dunbar, in 25 fathoms (F. M. Balfour ; L. & H.).

ANOMURA.

PORCELLANIDÆ.

Porcellana platycheles (Pennant).
　　Porcellana platycheles, Bell, Brit. Stalk-eyed Crust., p. 190, 1853.
　　　　Crail and Fifeness at low water (Howden) ; at Elie, and on
　　　　the shore near N. Berwick (L. & H.).

Porcellana longicornis (Pennant).
> *Porcellana longicornis*, Bell, Brit. Stalk-eyed Crust., p. 193, 1853.
> Upper part of the Firth (Howden). Bass Rock, 24 fathoms
> (Metzger). Off May Island, in 8 fathoms, near Elie, and near
> Inchkeith (L. & H.). Not unfrequent among dredged and
> trawled material from between Inchkeith and May Island
> (S.F.B.).

LITHODIDÆ.

Litho les maia (Linné).
> *Lithodes maia*, Bell, Brit. Stalk-eyed Crust., p. 165, 1853.
> This species is not uncommon near the mouth of the Firth.
> It is often obtained by fishermen near the Island of May
> (L. & H.). One specimen in deep water, west of May Island
> (S.F.B.).

PAGURIDÆ.

Eupagurus bernhardus (Linné).
> *Eupagurus bernhardus*, Brandt, Middend. Sibir. Reise, Zool. i.
> p. 105.
> *Pagurus Bernhardus*, Bell, Brit. Stalk-eyed Crust., p. 171.
> *Pagurus Ulidianus*, W. Thomp. Rep. Brit. Assoc., p. 267, 1843 ;
> Bell, Brit. Stalk-eyed Crust., p. 180, 1853 (a dwarf variety).
> Very common within the littoral and laminarian zones.

Eupagurus pubescens (Kröyer).
> *Eupagurus pubescens*, Stimpson, Proc. Acad. Nat. Sci. Philad.
> 1858, p. 75.
> *Pagurus Thomsoni*, Bell, Brit. Stalk-eyed Crust., p. 372, 1853.
> West of May Island, 20 fathoms (Henderson). East of Inch-
> keith, 9 to 10 fathoms (S.F.B.).

Eupagurus sculptimanus (Lucas).
> *Eupagurus sculptimanus*, Norman, Mus. Nor., pt. iii. p. 7, 1886.
> *Pagurus Forbesii*, Bell, Brit. Stalk-eyed Crust., p. 186, 1853.
> Firth of Forth (Howden).

Eupagurus cuanensis (Thompson).
> *Eupagurus cuanensis*, Stimpson, Proc. Acad. Nat. Sci. Philad.,
> 1858, p. 75.
> *Pagurus cuanensis*, Bell, Brit. Stalk-eyed Crust., p. 178, 1853.
> In *Turretella*, Firth of Forth (F. M. Balfour ; L. & H.).

Spiropagurus Hyndmanni (Thompson).
> *Spiropagurus Hyndmanni*, Norman, Mus. Nor., pt. iii. p. 7, 1886.
> *Anapagurus Hyndmanni*, Henderson, Decapod and Schiz. Crust.
> Firth of Clyde, p. 27.
> *Pagurus Hyndmanni*, Bell, Brit. Stalk-eyed Crust., p. 182, 1853.
> Off Musselburgh and Prestonpans (Howden ; L. & H.).
> Three species dredged east of Inchkeith (S.F.B.).

Spiropagurus lævis (Thompson).
> *Spiropagurus lævis*, Norman, Mus. Nor., pt. iii. p. 7, 1886.
> *Anapagurus lævis*, Henderson, Decapod and Schiz. Crust. Firth
> of Clyde, p. 28, 1886.
> *Pagurus lævis*, Bell, Brit. Stalk-eyed Crust., p. 184, 1853. Firth
> of Forth (Howden ; L. & H.). Not unfrequent in the vicinity
> of Inchkeith (S.F.B.).

GALATHEIDÆ.

Munida rondeletii, Bell.
> *Munida rondeletii*, Bell, Brit. Stalk-eyed Crust., p. 208, 1853.
> *Munida rugosa*, Leach, Dict. des Sci. Nat., tom. xviii. p. 52.
> *Munida Bamffica*, White, Pop. Hist. Brit. Crust., p. 89, 1857.
> Not uncommon at Dunbar (R. Gray; L. & H.).

Galathea squamifera, Leach.
> *Galathea squamifera*, White, Pop. Hist. Brit. Crust., p. 87, 1857.
> *Galathea squamifera*, Bell., Brit. Stalk-eyed Crust., p. 197,
> 1853. Common in the littoral and laminarian zones.

Galathea strigosa (Linné).
> *Galathea strigosa*, Bell, Brit. Stalk-eyed Crust., p. 200, 1853.
> Off the Bass Rock (Howden). Plentiful near Dunbar (R.
> Gray; L. & H.).

Galathea nexa, Embleton.
> *Galathea nexa*, Bell, Brit. Stalk-eyed Crust., p 240, 1853.
> Off Portseaton (Howden; L. & H.).

Galathea intermedia, Lilljeborg.
> *Galathea intermedia*, Lilljeborg, Ofvers. Vet. Acad. Förhandl.,
> 1851, p. 21.
> *Galathea Andrewsii*, Kinahan, Nat. Hist. Rev., vol. iv. pt. 2,
> p. 228; Trans. Irish Acad., 1871, p. 95. Firth of Forth
> (Dr Anderson; * L. & H.).

Galathea dispersa, Bate.
> *Galathea dispersa*, Bate, Proc. Linn. Soc. (Zool.), vol. iii. p. 3.
> Commonly met with on the so-called 'Oyster banks'
> (Henderson). †

MACRURA.

ASTACIDÆ.

Homarus vulgaris, Milne-Edwards.
> *Homarus vulgaris*, Bell, Brit. Stalk-eyed Crust., p. 242, 1853.
> *Homarus gammarus*, Linné-Henderson, Decapod and Schiz. Crust.
> Firth of Clyde, p. 31, 1886. Firth of Forth, at many places
> at low water (Howden). Caught in considerable numbers for
> the markets on all the rocky coasts near the mouth of the
> estuary (L. & H.).

Nephrops norvegicus (Linné).
> *Nephrops norvegicus*, Bell, Brit. Stalk-eyed Crust., p. 251, 1853.
> This is a very common species in the lower parts of the Firth,
> and it is equally common in Rothesay Bay. It is frequently
> observed in the stomachs of large cod.

CARIDA.

CRANGONIDÆ.

Crangon vulgaris (Fabricius).
> *Crangon vulgaris*, Bell, Brit. Stalk-eyed Crust., p. 256, 1853.
> Common within the littoral zone where the shore is sandy,
> and occasionally taken with the dredge in moderately deep
> water.

---

* *Proc. Roy. Phys. Soc. Edin.*, vol. ii.
† *Decapod and Schizopod Crustacea of the Clyde*, p. 10.

Crangon allmanni, Kinahan.

  Crangon allmanni, Kinahan, Proc. Dublin Nat. Hist. Soc., vol. iv.
  p. 80, 1857. Bass Rock, 24 fathoms (Metzger ; L. & H.).
  Taken with the trawl in deep water, west of May Island.
  This species is more frequently observed in the stomachs of
  cod and haddock than any other Crangon (S.F.B.).

Crangon nanus, Kröyer.

  Crangon nanus, Kröyer, Nat. Hist. Tidsskr., iv. p. 231.
  Crangon bispinosus, Bell, Brit. Stalk-eyed Crust., p. 268, 1853.
  Bass Rock, 24 fathoms (Metzger ; L. & H.). One specimen,
  taken with the trawl, off Prestonpans (S.F.B.). This appears
  to be the first record of the occurrence of this species so far
  up the estuary.

PALÆMONIDÆ.

Hippolyte spinus (Sowerby).

  Hippolyte spinus, Bell, Brit. Stalk-eyed Crust., p. 284, 1853.
  Newhaven (Leach). The species is rather common in the
  laminarian and littoral zones (L. & H.). One specimen,
  dredged near Inchkeith (S.F.B.). This species does not seem
  to have been yet recorded from the Clyde, the next species
  having been usually mistaken for it.

Hippolyte securifrons, Norman.

  Hippolyte securifrons, Norman, Trans. Tyneside Nat. Field Club,
  vol. iv. p. 267, 1863. Off St Abb's Head, 40 fathoms
  (Metzger ; L. & H.).

Hippolyte pusiola, Kroyer.

  Hippolyte pusiola, Kröyer, Monogr. Fremstilling af Hippol. Nord.
  Arter., p. iii.
  Hippolyte Barleei, White, Pop. Hist. Brit. Crust., pp. 124, 335.
  Hippolyte Andrewsii, White, *loc. cit.* Newhaven, from the fisher-
  mens' lines (Henderson). Several specimens were taken with
  the dredge in the vicinity of Inchkeith. This seems to be
  the species referred to by Dr James Howden as being common
  at Crail.[*]

Hippolyte cranchii, Leach.

  Hippolyte cranchii, Bell, Brit. Stalk-eyed Crust., p. 288, 1853.
  Rocks off Broxmouth, near Dunbar (F. M. Balfour ; L. & H.).

Virbius varians (Leach).

  Virbius varians, Norman, Mus. Nor., pt. iii. p. 8, 1886.
  Hippolyte varians, Bell, Brit. Stalk-eyed Crust., p. 286, 1853.
  Firth of Forth in pools (Howden). Rocks off Broxmouth,
  near Dunbar (F. M. Balfour; L. & H.). Frequent in pools
  left by the receding tide, shore above Granton (S.F.B.).

Virbius fasciger (Gosse).

  Hippolyte fascigera, Gosse, Ann. and Mag. Nat. Hist., p. 153,
  1853.
    "   "   White, Pop. Hist. Brit. Crust., p. 119, 1857.
  Virbius fasciger, Norman, Mus. Nor., pt. iii. p. 8, 1886. Several
  specimens of this species were taken with the hand-net
  amongst the weed at the edge of low water, at Cramond
  Island. It is easily distinguished when alive from all other
  British species of *Hippolyte* by the peculiar arrangement of

[*] *Trans. Roy. Phys. Soc. Edin.*, 1853.

dark blotches and streaks on its integument, which is otherwise pellucid and almost transparent. There does not seem to be any previous record of its occurrence in the Forth (S.F.B.).

Pandalus annulicornis, Leach.

> *Pandalus annulicornis,* Bell, Brit. Stalk-eyed Crust., p. 297, 1853.
> A common species throughout the Forth ; frequently observed in the stomachs of the haddock and cod.

Pandalus brevirostris, Rathke.

> *Pandalus brevirostris,* Norman, Mus. Nor., pt. iii. p. 8, 1886.
> *Hippolyte Thomsoni,* Bell, Brit. Stalk-eyed Crust., p. 290, 1853.
> Firth of Forth (F. M. Balfour ; L. & H.). One specimen dredged in deep water a little west of Inchkeith, October 1887 (S.F.B.). This appears to be a rare species in the Forth.

(?) *Palæmon squilla* (Linné).

> *Palæmon squilla,* Bell, Brit. Stalk-eyed Crust., p. 305, 1853.
> Frequent in rock pools near the mouth of the Firth (L. & H.).

Remarks.—The Decapod Crustacea here recorded as having been observed in the Firth of Forth, amount in number to nearly a half of the whole British species. Scarcely a third of them, however, are of frequent or common occurrence, and of several only a few specimens have as yet been noticed in the estuary. The common species of the Decapods are frequently observed among the contents of fishes' stomachs, and form a considerable part of the food of those fishes that feed at or near the bottom. It does not appear that the species of Crustacea belonging to this group are to any appreciable extent more abundant at one season than another, although some are found to frequent the littoral zone during the spring and summer months more than at other times, and *Hyas araneus* may be cited as an example of this. It is rather curious that this spider crab, though not confined so exclusively to deep water as the other species, *Hyas coarctatus* is yet frequently captured by the dredge in water of 10 or 15 fathoms depth, it is nevertheless very rarely found in the stomachs of fishes, even of that omniverous feeder the cod ; whereas *Hyas coarctatus* is of common occurrence, especially in cods' stomachs, six, eight, and sometimes a dozen or more specimens being found in a single fish.

That *Hyas araneus* is more a littoral species cannot be given as a satisfactory reason for its practical exemption ; that it is larger is a reason hardly more tenable, for cuttle-fish, Norwegian lobsters, and even seafowl are devoured by cod ; nor are they so active in their movements as to be able to escape by that means more readily than other crabs. Whatever the reason is, it seems evident that fishes, as well as fishermen, have a decided dislike to *Hyas araneus.*

The common food fishes, throughout all stages of their growth, feed very generally and largely on species belonging to nearly all the orders of Crustacea, both in their young and mature conditions. Where Crustacea are abundant it may reasonably be expected that fishes will be more or less numerous. It goes without saying, then, that the study of the Crustacea,—their distribution, habits, and development,—forms by no means an unimportant part of fishery investigations.

With comparatively few exceptions, the various species of Decapods frequent the bottom, seeking shelter under stones, among sea-weed, Loophytes, or, as is the case with some, burrowing in the mud or sand, and sometimes to a considerable depth. The Cumacea seem also to frequent the bottom. The Schizopoda, or at least very many of them are,

x 2

on the other hand, like the Calanidæ among the Copepoda, and the Hyperiidæ among the Amphipoda, pelagic in their habits. The phosphorescence of the sea is also apparently to some extent due to the power they have—notably the species of the Euphausidæ—of emitting light from various parts of their bodies; and it is a curious circumstance that it is those species which have been observed to have preeminently the power of becoming luminous that are most frequently found in the stomachs of herrings, namely, _Nyctiphanes_ and _Boreophausia_. Whether the property of emitting light which these Schizopods possess has anything to do with their being so commonly selected as food by the herring, cannot be easily answered, though it is probable that their luminosity may have some connection with it.

The somewhat singular auditory organs observed in many of the Mysidæ are very interesting. They are conspicuous owing to their glistening transparency; they appear as clear circular vesicles near the base of each of the inner caudal lamellæ, which are at this part suddenly enlarged to afford space for the vesicles. Viewed with a low power, each vesicle appears to be formed of concentric zones or laminæ, which are alternately more and less clearly transparent; with a moderately high power, the auditory ossicle may be observed. The position of the auditory organs here referred to seems confined to the Mysidæ. The Cumaceæ according to Claus do not possess auditory organs.

The distribution of the Ostracoda, as might be expected from their frequenting the mud, sea-weed, and zoophytes at the bottom, is, like that of the Decapods, little influenced by the various seasons. The free-swimming Copepoda are decidedly different in this respect. During the later months of spring, and in summer and early autumn, they are at times captured in great abundance by the surface-net; whereas during the colder months, comparatively few are to be met with, even though the net be sunk to a considerable depth. In April and May I have also found the larval or free-swimming forms of _Balani_ exceedingly numerous— much more so than at any other season. If adult _Balani_ be collected about this time, and left in sea-water for an hour or two, swarms of these larval forms may be observed swimming about. It is probable that during this time when the sea around our coasts is swarming with these very minute organisms, the herring, and possibly other fishes also, may make use of them as food, by drawing them into their mouths along with the water of respiration, and retaining them while the water passes to through the opercula, as Dr Möbius suggested; but there is no doubt that herring can, and do, discriminate between one form of food and another, and purposely capture the organisms on which they generally feed, whether Copepods, Amphipods, or Schizopods.

The importance of this extensive class of organisms—perhaps the most extensive among the Invertebrata—is a sufficient reason, if only from a commercial point of view, for its being carefully and thoroughly studied.

In the list here given, over 230 species of Crustacea are recorded as occurring in the Firth of Forth, including 41 species of Ostracoda, 42 of Copepoda, and 13 Schizopods.

In the list prepared by Leslie and Herdman, 99 species are enumerated; and Dr Henderson in his paper added other 21 species, which, including a few doubtful forms, brought the total up to 120 species. The number of species in the present list, exclusive of the two additional groups, Ostracoda and Copepoda, and leaving out doubtful forms, is about 150, which shows that the investigations carried out during the past year under the directions of the Fishery Board have been fairly successful in adding to our knowledge of the distribution of this important class of organisms.

An Extra Large 'Squid.'

Ommatostrephes todarus,	D'Orbigny.,
Ommastrephes todarus,	Forbes and Hanley.
,, ,,	Gosse.
Loligo sagittata,	Alder.
,, ,,	Fleming.

This specimen, which was captured by hand in shallow water, a mile or
so to the west of Granton, is of unusual size, as the following measure-
ments show.

Length, from posterior end of body to tip of longest arms,	4 feet 1 inch.
Length from posterior end of body to edge of mantle,	$21\frac{1}{2}$ inches.
Length of head from edge of mantle,	$4\frac{1}{2}$,,
Length of tentacles,	23 ,,
Length of arms,	16 ,,
Breadth of body,	$5\frac{3}{4}$,,
Breadth from tip to tip of fins,	$14\frac{1}{2}$,,
Length of fin,	$10\frac{1}{2}$,,
Length of shell (or 'bone'),	$20\frac{3}{8}$,,
Breadth of shell (or 'bone'),	$\frac{7}{8}$,,
Diameter of large sucker,	$\frac{1}{2}$,,

The usual size of this species is very much less, being as a rule about
12 to 15 inches long, exclusive of the tentacles, with a breadth of 3 or $3\frac{1}{2}$.

Its distribution is European, and it has been recorded from several
localities on the Scottish coasts.

O. *todarus,* D'Orbigny, differs from O. *sagittatus,* Lam., in the shell
having a rib along the middle and one along each edge, instead of two to
four ribs on each side of the mid rib, as in the last.

The suckers are also each 'encircled by a horny denticulated ring,'
which does not seem to be the case with those of O. *sagittatus.*

A Large Haddock.

This fish which was caught off Ailsa Craig, in February 1888, was sent from Girvan, and had the following dimensions.

Whole length,	. 31 inches.	Length of caudal fin,	
Height, .	7⅝ ,,	from commence-	
Circumference,	. 20 ,,	ment of fin, .	5½ inches.
Length of head,	. 7¼ ,,	Length of body, .	20⅜ ,,
Length of caudal fin,		Weight of fish, .	13½ lbs.
from termination of		Weight of ovaries	
body, .	. 3 ,,	(notfullydeveloped),	21 ozs.

The internal organs were normal. The stomach contained only a little pulpy matter, while the intestine was fairly well filled; among the objects recognisable in the latter being—a shell of *Bulla utriculus* (half-grown); parts of a *Portunus pusillus*; *Calocaris macandrei*, nearly whole, but soft and very much discoloured; and plates of starfishes (Ophiuroids).

Judging from the presence of *Calocaris* in the intestine, the fish had been recently feeding in moderately deep water (30 fathoms or more). It appeared to be in a healthy condition, as far as could be observed; no nematode parasites were present, either in the stomach or intestine.

III

6

8 AR.

1840

DESCRIPTION OF PLATES.

The figures, except figs. 16 and 17, were drawn by means of the camera lucida.

Reference Letters.

a.c.	alimentary canal.	*ms.*	somites.
a.	anus.	*n.c.*	neurenteric canal.
au.	ear.	*no.*	notochord.
b.	blood.	*o.*	eye.
br.	brain.	*ol.o.*	olfactory organ.
b.o.	branchial aorta.	*p.b.*	pineal body.
b.v.	blood-vessel.	*p.r.*	pigment layer of retina.
c.	claspers.	*r.*	retina.
e.	epiblast.	*s.t.*	sensory tubes.
epi.	epithelium.	*sp.*	spiracle.
e.g.	external gill.	*t.*	tail.
g.c.	gill clefts.	*u.c.*	umbilical cord.
ht.	heart.	*u.f.*	unpaired fins.
hy.	hypophysis.	*v.*	trigeminus nerve.
l.	lens.	*vii.*	facial nerve.
m.	site of mouth.	*viii.*	auditory nerve.
m.c.	medullary canal.	*ix.*	glossopharyngeal nerve.
ms.c.	mesoblast cells.	*x.*	vagus nerve.
me.	merocytes.		

Fig. 1. Superficial segmentation.
Fig. 2. Deep segmentation.
Fig. 3. Shows segmentation cavity in section.
Fig. 4. Entire blastoderm of preceding stage, aged thirty-one days.
Fig. 5. First embryonic thickening, forty days.
Fig. 6. Embryo of fifty-four days (winter).
Fig. 7. Embryo of ten weeks (winter).
Fig. 8. Embryo of eleven and a half weeks (winter).
Fig. 9. Embryo of twelve weeks (winter).
Fig. 10. Embryo of about twelve weeks (winter).
Fig. 11. Embryo of nearly fourteen weeks (winter).
Fig. 12. Embryo of fourteen weeks (winter).
Fig. 13. Embryo of thirteen and a half weeks (spring), length 5·8 mm.
Fig. 14. Embryo of eleven weeks (early summer), length 7·5 mm.
Fig. 15. Head end of embryo of *R. clavata.*
Fig. 16. Dorsal view of embryo of six months, life size.
Fig. 17. The same as seen from the ventral aspect.
Fig. 18. End of an external gill-filament of an embryo twenty-three weeks
old.   × 60 diameters.

All the figures, except fig. 15, refer to *R. batis.*

VII.—ADDITIONS TO THE FAUNA OF THE FIRTH OF FORTH.

By Thomas Scott, F.L.S.   (Plates XII., XIII.)

In the present paper there are recorded over 90 species not previously recognised as belonging to the fauna of the Firth of Forth. A few of these are now recorded for the first time for the east of Scotland, one or two are additions to the British fauna, and one or two new to science.

Most of them were obtained last year, during the investigations carried on on board the 'Garland,' since the publication of the Seventh Annual Report, only a few having been obtained earlier, but not determined in time to be included in either of the two previous papers on the Forth fauna.*

The forms here recorded belong exclusively to the Invertebrata, and comprise 23 species of Foraminifera, 61 species of Crustacea, and 7 species of Mollusca. Other invertebrate groups are being studied, and information as to their distribution, &c., collected with a view to publication later on.

As was pointed out in a previous paper, the study of the marine Invertebrata, from a fishery point of view, is of considerable interest and importance. As regards the Crustacea, the Rev. A. M. Norman says (*Museum Normanianum*, part 3) :—'I venture to prophecy that when the ' Crustacean Fauna of the Arctic and Temperate regions shall have been ' thoroughly investigated, it will hereafter be found to embrace not less ' than 5000 species. It was little suspected a generation ago that the ' Crustacea is the class which undoubtedly embraces more forms than any ' other outside the Insecta.' The scientific investigations carried out under the directions of the Fishery Board have helped very much to prove that the Crustacea is also one of the most important groups—if not the most important of the Invertebrata—that constitute the food supply of fishes. The movements of fishes are also undoubtedly partly influenced by the prevalence in particular localities of invertebrate forms which they, for the time being, may be partial to as a source of food. The study therefore of the distribution, habits, and life-histories of the Invertebrata should hold a place next in importance to the study of the food, distribution, habits, spawning, and development of the fishes themselves. This study has been and is being carried on from year to year on board the 'Garland,' along with the other and more important fishery investigations, as opportunity offers, and the present and previous similar papers are the results of an attempt to collect all the information within reach bearing on the distribution and habits of these lower forms of life, especially within the area of the Firth of Forth.

The information contained in these papers, especially in the present one and in the one published last year, is mainly the outcome of a personal examination of the various organisms referred to in them ; and though the restricted area to which the information principally applies imparts to it a value which is perhaps chiefly of local importance, yet the more thorough and accurate the information relating to the fauna of separate areas becomes, its value will increase tenfold, because reliable comparisons of various kinds will become possible, and from these comparisons theories and principles of great importance may be worked out. There has also resulted a wider acquaintance with, and greater certainty

---

* *Sixth Ann. Report*, Part iii. p. 235, 1888 ; *Seventh Ann. Report*, Part iii. p. 311, 1889.

in identifying, the different objects observed in the stomachs of fishes ; and therefore more satisfactory and reliable information respecting the ̄ood of fishes is now being collected.

In preparing this paper the following among other works have been consulted :—

1850. Baird, *British Entomostraca.*
1868. Williamson, *Recent Foraminifera of Great Britain.*
1870. H. B. Brady, " The Foraminifera of Tidal Rivers, *Annals and Magazine of Natural History.*
1884.   „   *Foraminifera of the Challenger Expedition.*
  „   G. S. Brady, *Monograph of the British Copepoda.*
1868.   „   *Monograph of Recent British Ostracoda.*
1870.   „   and David Robertson, *The Ostracoda of Tidal Rivers.*
1889.   „   and A. M. Norman, *Monograph of the Marine and Fresh Water Ostracoda of the North Atlantic and North-Western Europe.*
1863. Bate and Westwood, *British Sessile-eyed Crustacea.*
1872. G. O. Sars, *Monograph of the Norwegian Mysidæ.*
1876–79.   „   *Monograph of the Mediterranean Mysidæ and Cumacea.*
1862–69. J. G. Jeffreys, *British Conchology.*

I have also to acknowledge the kindness of Professor G. S. Brady, F.R.S., Dr H. B. Brady, F.R.S., the Rev. A. M. Norman, D.C.L., F.L.S., Rev. T. R. R. Stebbing, F.L.S., and Mr David Robertson, F.L.S., F.G.S., in naming obscure and difficult species. Indeed, but for the help of these gentlemen, this paper could not possibly have been so full or so valuable.

FORAMINIFERA.

I am indebted to Mr Robertson for indentifying a few of the species in this group mentioned below ; and also for notes of the occurrence of others which have not as yet come under my own observation.

MILIOLIDÆ.

Miliolina tricarinata (d'Orbigny).
  Triloculina tricarinata, d'Orb., Ann. Sci. Nat., tome vii. p. 277, No. 7 *a* ; Modelé, No. 94 (1826).
  Miliolina tricarinata, H. B. Brady, Foram. Chall. Exped., p. 165, pl. iii. fig. 17, *a–b* (1884).

Habitat.—Vicinity of Bass Rock. This species is easily distinguished from *M. trigonula* by the three sharp keel-like ridges extending from end to end and about equidistant from each other. It is much rarer in the Forth then *M. trigonula.*

Miliolina fusca, Brady.
  Miliolina fusca, Brady, Ann. and Mag. Nat. Hist., ser. iv., vol. vi. p. 286, pl. xi. fig. 2, *a–c* (1870).
  Miliolina fusca, Robertson, Fauna and Flora of the W. of Scotland, p. 51 (1876).

Habitat.—Brackish water pools by the shore near Aberlady, common. This is an arenaceous species. It is much smaller than *M. agglutinans,* and frequently dark brownish in colour. It is considered to be a somewhat rare species, and seems confined to water more or less brackish.

ASTRORHIZEDÆ.

Psammosphæra fusca, F. E. Schulze.
> *Psammosphæra fusca*, F. E. Schulze, II. Jahresberichte d. Komm.
> Untersucht d. deutsch. Meere., p. 113, pl. ii. fig. 8, *a–f*
> (1874).
> *Psammosphæra fusca*, H. B. Brady, Foram. of the Chall. Exp., p.
> 249, pl. xviii. figs. 1–8 (1884).

Habitat.—East of Inchkeith, not common. This species has been obtained off Loch Scavaig, Skye, in 45 to 60 fathoms. It has also been found in seven of the Challenger stations in the South Atlantic in depths of from 150 to 2800 fathoms, and in the North Atlantic from 440 to 2750 fathoms.

LITUOLIDÆ.

Reophax fusiformis (Williamson).
> *Proteonina fusiformis*, Williamson, Rec. Foram. Gt. Brit., p. 1,
> pl. i. fig. 1 (1858).
> *Reophax fusiformis*, H. B. Brady, *op. cit.*, p. 290, pl. xxx. figs.
> 7–11 (1884).

Habitat.—Largo Bay, off St Monance, and other parts of the Forth, but nowhere very common—a much more robust species than *R. scorpiurus*.

Reophax nodulosa (?) H. B. Brady.
> *Reophax nodulosa*, Brady,[*] Quart. Jour. Micr. Sci., vol. xix.
> N. S., p. 52, pl. iv., figs. 7, 8 (1879); Brady, Foram. Chall.
> Exped., p. 394, pl. xxii. figs. 1–9 (1884).

Habitat.—Off St Monance, Largo Bay, and other parts of the Forth, frequent. This is a very variable species as regards size. The Forth specimens are very small, but H. B. Brady says (*op. cit.*) that there are specimens of this species which are amongst the very largest of recent arenaceous Foraminifera.

Reophax findens (Parker).
> *Lituola findens*, Parker (in Dawson's paper) Canad. Nat. vol. v.
> N.S., p. 177, pl. 180 fig. 1 (1870).
> *Reophax findens*, H. B. Brady, Foram. of the Chall. Exp., p. 299,
> pl. xxxi. figs, 10–11 (1884).

Habitat.—East of Inchkeith, not very common. The only other British examples are from the estuary of the Dee, N. Wales (J. D. Siddal). There appears to be no other authenticated British locality for this species.

Ammodiscus gordialis (Jones and Parker).
> *Trochamina squamata gordialis*, Jones and Parker, Quart. Jour.
> Geol. Soc., vol. xvi. p. 304 (1860).
> *Ammodiscus gordialis*, H. B. Brady, *op. cit.*, p. 333, pl. xxxviii.
> figs. 7–9.

Habitat.—Aberlady Bay, rather rare, structure arenaceous, colour brownish; the test consists of a tube coiled upon itself in an irregular manner, inclining to complanate.

---

[*] There is some diversity of opinion as to whether the organisms here referred to *R. nodulosa* are foraminiferal; it i therefore with some hesitation that they are included in the present list.

Trochamina inflata (Montagu).

  Nautilus inflata, Montagu, Test. Brit., Suppl., p. 81, pl. xviii. fig. 3 (1808).

  Rotalina inflata, Williamson, Rec. Foram. Gt. Brit., p. 50, pl. iv. figs. 93, 94 (1858).

  Trochamina inflata, H. B. Brady, *op. cit.*, p. 338, pl. xli. fig. 4, *a–c.*

Habitat.—Brackish water pools by the shore near Aberlady; vicinity of Inchkeith, and others parts of the Forth; rare, except at Aberlady, where it is comparatively common. This seems to be an inshore, rather than a deep water species. Colour brownish.

Trochamina macrescens (?) Brady.

  Trochamina inflata, var. *macrescens*, Brady, Ann. and Mag. Nat. Hist., ser. iv. vol. vi. p. 290, pl. xi. fig. 5, *a–c.* (1870).

  Trochamina macrescens, Robertson, Fauna and Flora of the W. of Scotland, p. 51 (1876).

Habitat.—With the last, but not so common; the cells of this species are concave above and below, and look as if the sides had been partially crushed in. The cells are not all equally concave, those in the centre being frequently only flattened—probably a form of *H. canariense.*

Trochamina ochracea (Williamson).

  Rotalina ochracea, Williamson, Rec. For. Gt. Brit., p. 55, pl. iv. fig. 112; pl. v. fig. 113 (1858).

  Trochamina ochracea, H. B. Brady, Foram. of the Chall. Exp., p. 338 (1884).

Habitat.—Off St Monance, common; other parts of the Forth rather rare. This is a very small species, 'and,' as Brady remarks, 'composed of a large number of segments.' 'On the inferior side the septal lines are arcuate, 'flexuose, and very prominent.' It is not uncommon in the British seas, though not previously recorded for the Forth.

TEXTULARIDÆ.

Textularia gramen, d'Orbigny.

  Textularia gramen, d'Orbigny, For. Foss. Vien., p. 248, pl. xv., figs. 4–6 (1846).

  Textularia gramen, H. B. Brady, *op. cit.*, p. 365, pl. xliii. figs. 9–10.

Habitat.—Off St Monance, and from other parts of the Forth. This seems to be a generally distributed species. The test is of an elongated tapering form, is not so compressed as *T. sagittula*, and the lateral edges are rounded instead of being sharply keeled.

Textularia variabilis, Williamson.

  Textularia variabilis, Williamson, Rec. Foram. Gt. Brit., p. pl. figs.

Habitat.—Granton Harbour (David Robertson).

Gaudryina filiformis, Berthelin.

  Gaudryina filiformis, Berthelin, Mém. Soc. géol. France, ser. 3, vol. i. No. 5, p. 25, pl. i., fig. 8 (1880).

  Gaudryina filiformis, H. B. Brady, Foram. of the Chall. Exp. p. 380, pl. xlvi. fig. 12, *a b c* (1884).

Habitat.—East of Inchkeith, not very common.

Bulimina elegans, d'Orbigny.

  Bulimina elegans, d'Orbigny, Ann. Sci. Nat., vol. vii. p. 270, No. 10; Modelé, No. 9 (1826).

  Bulimina elegans, Brady, *op. cit.*, p. 398, pl. l. figs. 1–4.

Habitat.—Bo'ness (David Robertson). East of Inchkeith, not uncommon. In this species the cells are arranged in a triserial manner, and
the shell tapers gradually towards the apex, it is thus distinctly different
from *B. elegantissima*, d'Orb.

Bulimina fusiformis, Williamson.
>   *Bulimina pupoides*, var. *fusiformis*, Williamson, Rec. Foram. Gt.
>   Brit., p. 63, pl. v. figs. 129–130 (1868).
>   *Habitat.*—East of Inchkeith, not common.

LAGENIDÆ.

Lagena pulchella, Brady.
>   *Lagena pulchella*, H. B. Brady, Ann. and Mag. Nat. Hist., ser. iv.,
>   vol. vi. p. 294 ; pl. xii. fig. 1, *a. b.* (1870).
>   *Habitat.*—Granton Harbour (David Robertson). This species has a
tricarinate form, and the "convex" faces are ornamented with irregular
longitudinal, branching costæ.

Lagena melo (d'Orbigny).
>   *Oolina melo*, d'Orbigny, Form. Amer. Merid., p. 20, pl. v. fig. 9
>   (1839).
>   *Habitat.*—Granton Harbour (David Robertson).

Polymorphina compressa, d'Orbigny (fistulose variety).
>   *Polymorphina compressa*, d'Orbigny, Foram. Foss. Vien., p. 233,
>   pl. xii. figs. 32–34 (1846).
>   *Polymorphina compressa*, var. *fistulosa*, Williamson, Rec. Foram.
>   Gt. Brit., p. 72, pl. vi. fig. 150 (1858).
>   *Polymorphina compressa* (fistulose form), H. B. Brady, *op. cit.*, p.
>   566, pl. lxxiii. fig. 17.
>   *Habitat.*—Near Phidra, rather rare.

Uvigerina angulosa, Williamson.
>   *Uvigerina angulosa*, Williamson, Rec. Foram. Gt. Brit., p. 67, pl.
>   v. fig. 140 (1868).
>   *Uvigerina angulosa*, H. B. Brady, Foram. of the Chall. Exp., p.
>   576, pl. lxxiv. figs. 15–18 (1884).
>   *Habitat.*—East of Inchkeith, not very common.

ROTALIDÆ.

Rotalia nitida (Williamson).
>   *Rotalia nitida*, Will., Rec. Foram. Gt. Brit., p. 54, pl. iv. figs.
>   106–108 (1858).
>   *Rotalia nitida*, Robertson, Fauna and Flora of the West of
>   Scotland, p. 52 (1876).
>   *Habitat.*—Various parts of the estuary of the Forth, frequent ; a much
smaller and more delicate species than *R. beccarii*.

Spirillina vivipara, Ehrenberg.
>   *Spirillina vivipara*, Ehrenberg, Abhandl. k. Akad. Wiss. Berlin,
>   p. 442, pl. iii. fig. 41 (1841).
>   *Spirillina vivipara perforata*, Williamson, Rec. Foram. Gt. Brit.,
>   p. 92, pl. vii. fig. 202 (1858).
>   *Spirillina vivipara*, H. B. Brady, *op. cit.*, p. 630, pl. lxxxv. figs.
>   1–5.
>   *Habitat.*—Aberlady Bay, rare. This is a widely distributed species,
but does not seem to be very common ; I have found it also in the Clyde.

Patellina corrugata, Williamson.
>> *Patellina corrugata,* Williamson, Rec. Foram. Gt. Brit., p. 46, pl.
>> iii. figs. 86–89 (1858).
>> *Patellina corrugata,* H. B. Brady, *op. cit.,* p. 634, pl. lxxxvi. figs.
>> 1–7.

Habitat.—Largo Bay, rare; a small but pretty species. It has been found at a depth of 620 fathoms in the South Pacific.

NUMMULINIDÆ.

Operculina ammonoides (Gronovius).
>> *Nautilus ammonoides,* Gronovius, Zooph. Gron., p. 282, No. 1220,
>> and pl. v. (1781).
>> *Nonionina elegans,* Williamson, Rec. Foram. Gt. Brit., p. 35, pl
>> iii. figs. 74, 75 (1858).
>> *Operculina ammonoides,* H. B. Brady, *op. cit.,* p. 745, pl. cxii.
>> figs. 1–2.

Habitat.—Largo Bay, not very common.

Note.—The curious Rhizopods *Dendrophrya erecta,* Str. Wright, and *Dendrophrya radiata,* Str. Wright, discovered by Dr Wright in low-water pools in the Old Quarry at Granton, and described by him in the Annals and Magazine of Natural History in 1861, seem to have been overlooked by the authors of the "Invertebrate Fauna of the Firth of Forth." I am indebted to my friend Mr David Robertson for drawing my attention to these species; he informs me that he also has found *D. erecta* in Granton Old Quarry; he has found both forms in low-tide pools at Cumbrae. So far as I can learn, there does not seem to be any known British habitat for these curious organisms other than the localities here referred to.

CRUSTACEA.
COPEPODA.

16 Specie

CALANIDÆ.

Candace pectinata, Brady.
>> *Candace pectinata,* Brady, Mon. Brit. Copep., vol. i. p. 49, pl. viii.
>> figs. 14, 15; pl. x. figs. 1–12 (1878).

Habitat.—In surface and bottom tow-net gatherings from various parts of the Forth between Inchkeith and May Island, moderately frequent, and easily distinguished from the other and commoner Copepoda by the dark-coloured plumes and terminal spines of the swimming feet. The only place where this species was obtained by Dr Brady and Mr Robertson, as stated in the monograph referred to above, was 'on very hard ground, and in a 'depth of about 40 fathoms south-west of the Island of St Agnes, Scilly,' where a very few specimens were dredged. The dark-coloured strongly-toothed crest on the joint next to and above the hinge of the right antennæ of the male is a peculiar and striking object. I have also obtained this species in St Andrews Bay, and off Montrose, 20 to 30 miles S.E.

MISOPHRIIDÆ.

Pseudocyclops obtusatus, Brady and Robertson.
>> *Pseudocyclops obtusatus,* Brady and Robertson, Ann. and Mag.
>> Nat. Hist., ser. iv., vol. xii. p. 12; pl. viii. figs. 4–7 (1873).
>> *Pseudocyclops obtusatus,* Brady, *op. cit.,* vol. i. p. 84; pl. xii. figs.
>> 1–13 (1878).

Habitat.—Off St Monance, where it was taken with the dredge, but somewhat sparingly. The body is robust, and the dorsal aspect is boldly arched, the abdomen is slender, the antennæ moderately short and stout.

CYCLOPIDÆ.

Thorellia brunnea, Boeck.

> *Thorellia brunnea*, Boeck, Oversigt over de ved Norges Kyster iagt. Copep., p. 26 (1864).
> *Thorellia brunnea*, Brady, *op. cit.*, vol. i. p. 95, pl. xvi. figs. 1–10.

Habitat.—Near Oxcar, 30 to 40 fathoms, and Largo Bay, rather rare. I have found this species frequently at Tarbert, Loch Fyne, and at Rothesay amongst weeds in shallow water.

HARPACTICIDÆ.

Ectinosoma melaniceps, Boeck.

> *Ectinosoma melaniceps*, Boeck, Oversigt Norges Copepoder, p. 30 (1864).
> *Ectinosoma melaniceps*, Brady, *op. cit.*, vol. ii. p. 11, pl. xi. figs. 17–20.

Habitat.—Largo Bay and other parts of the Forth, moderately frequent. This is a smaller species than *E. spinipes*, which it somewhat resembles, but from which it may be distinguished by a small, more or less distinct, black patch near the base of the rostrum.

Ectinosoma erythrops, Brady.

> *Ectinosoma erythrops*, Brady, *op. cit.*, vol. ii. p. 12, pl. xxxvi. figs. 11–17.

Habitat.—Off St Monance, 10 to 15 fathoms, bottom clean sand or gravel; and Largo Bay, rather rare. This species is readily distinguished by its having two brilliant red eye-spots, one on each side close to the anterior margin of the cephalic segment. The eye-spots appear to lose their colour when the specimens are kept a while in spirit.

Ameira longipes, Boeck.

> *Ameira longipes*, Boeck, Oversigt Norges Copepoder, p. 49 (1864).
> *Ameira longipes*, Brady, *op. cit.*, vol. ii. p. 37, pl. liii. figs. 1–10.

Habitat.—Largo Bay and off St Monance, as well as other parts of the Forth. This is not a very satisfactory species, and great care is required in discriminating between it and *Stenhelia ima*.

Laophonte serrata (Claus).

> *Cleta serrata*, Claus, Die frei-lebenden Copepoden, p. 123, t. xv. figs. 13–20 (1863).
> *Laophonte serrata*, Brady, *op. cit.*, vol. ii. p. 71, pl. xxxiii. figs. 1–14.

Habitat.—Off St Monance, rare.

Laophonte longicaudata, Boeck.

> *Laophonte longicaudata*, Boeck, Oversigt Norges Copepoder, p. 55 (1864).
> *Laophonte longicaudata*, Brady, *op. cit.*, vol. ii. p. 82, pl. lxxiv., figs. 12–15; pl. lxxvi. figs. 10–15.

Habitat.—Off St Monance, rather scarce.

Laophonte hispida (Brady and Robertson).

> *Asellopsis hispida*, B. and R., Ann. and Mag. Nat. Hist., vol. xii. p. 137, pl. ix. figs. 6–10 (1873).
> *Laophonte hispida*, Brady, *op. cit.*, vol. ii. p. 85, pl. lxxxi. figs. 1–11.

Habitat.—Largo Bay, frequent. This species is rather robust, with short caudal segments; these and one or two of the last abdominal segments are more or less covered with close-set short hairs.

Cletodes limicola, var. *gracilis,* Brady.

    Cletodes limicola, var. *gracilis,* Brady, *op. cit.,* vol. ii. p. 96.

Habitat.—Largo Bay, off St Monance, and other parts of the Forth in company with the type. The caudal segments in this form are long and slender, and have a prominent jointed (?) spine arising nearly at right angles from the upper surface and near the middle of each segment. I have observed both male and female, the latter with ova, in material dredged off St Monance in from 12 to 14 fathoms. With the exception of the long caudal segments (which are fully two-thirds the length of those of *C. longicaudata*), very little difference can be observed between this variety and the typical *C. limicola.*

Cletodes longicaudata, Brady and Robertson.

    Cletodes longicaudata, B. & R., Brit. Assoc. Report, p. 196 (1875).
    Cletodes longicaudata, Brady, *op. cit.,* vol. ii. p. 92, pl. lxxix. figs. 13–19.

Habitat.—Off St Monance, rare. This species has long, slender, caudal segments, and differs from the *C. limicola,* var. *gracilis,* by the form of the fifth feet and anterior antennæ; the caudal segments are also longer.

Enhydrosoma curvatum (Brady and Robertson).

    Rhizothrix curvata, B. & R., Brit. Assoc. Report, p. 197 (1875).
    Enhydrosoma curvatum, Brady, *op. cit.,* vol. ii. p. 98, pl. lxxxi. figs. 12–15 , pl. lxxxii. figs. 11–19.

Habitat.—Largo Bay, not uncommon; the extremities of both branches of the first feet are furnished with two long slender setæ, at the ends of which are a few fine flagellum-like hairs.

Thalestris serrulata, Brady.

    Thalestris serrulata, Brady, Mon. Brit. Cop., vol. ii., p. 133, pl. lix., figs. 2–11 (1880).

Habitat.—East of Inchkeith, several specimens taken with surface net. This species was described by Dr Brady from a single specimen—a male —dredged on a bottom of muddy sand in New Grimsbay Harbour, Scilly. Last year (1889) another specimen—a female—was observed by I. C. Thompson in a tow net gathering from Puffin Island.* The Forth specimens comprised both male and female, and were of a dark brick-red colour, which made them very conspicuous in the tow-netting. Some of the coloured copepoda, as *Alteutha,* retain their colour for a considerable time after being in spirit ; but in the case of this *Thalestris* not a trace of colour remained after a few hours immersion. I have obtained this species also in Dornoch Firth. This seems to be the first record of it for Scotland.

Harpacticus flexus, Brady and Robertson.

    Harpacticus flexus, B. & R., Ann. and Mag. Nat. Hist., ser. iv., vol. xii. p. 134, pl. ix. figs. 17–21 (1873).
    Harpacticus flexus, Brady, *op. cit.,* vol. ii. p. 152, pl. lxiv. figs. 12–18.

Habitat.—Off St Monance, scarce.

Zaus goodsiri, Brady.

    Zaus ovalis, Claus, Die frei-lebenden Copepoden, p. 146, tab. xxii. fig. 18; tab. xxiii. figs. 11–18 (1863).
    Zaus goodsiri, Brady, *op. cit.,* vol. ii. p. 156, pl. lxvi. figs. 10–13.

---

* *Proc. Biol. Soc., Liverpool,* iii., p. 188 (1889).

Habitat.—Off St Monance, frequent.   Dr Brady says that this species
' must be looked upon as one of the rarest, as it is certainly one of the
' finest of the British Harpacticidæ.'   It has, somewhat like *Alteutha
depressa*, a broad reddish purple band across the thorax.

(Family uncertain.)

Cylindropsyllus lævis, Brady.
    Cylindropsyllus lævis, Brady, *op. cit.,* vol. iii. p. 30, pl. lxxxiv.
        figs. 1–8.

Habitat.—Off St Monance, frequent.   This species, which does not seem
to have been previously recorded for Scotland, might be easily passed over
as belonging to some other group than the *Copepoda*.   Its comparatively
long and cylindrical form and short swimming feet impart to it a somewhat
close resemblance to a young *Pseudatanais*—a kind of Isopod.   Both the
genus and species were described from a single specimen dredged off
Hartlepool, and from the structure of the mouth it was conjectured to be
of parasitic or semiparasitic habits.   All the specimens found by me have,
however, been unattached to any other organism.   I also found this species
in East Loch Tarbert (Loch Fyne) in 1885, but it was not recorded.   It
has been observed by the Rev. A. M. Norman at Plymouth.

Fifteen species of Copepoda are recorded above, which brings up the
number observed within the area of the Firth of Forth to sixty.   I
expect that this number will be yet further increased.   There are several
forms that are doubtful, or that have not yet been identified with
described species, which will be recorded later on.   I am greatly indebted
to Dr G. S. Brady for the trouble he has taken in examining and identifying doubtful species, not only belonging to this, but also to the following
group, the Ostracoda.

OSTRACODA.

Thirty-two species of Ostracoda are here added to those recorded in my
two previous papers.   Four of these have not as yet been identified with
known species, and are for the present provisionally named and described.
I am also indebted to Mr David Robertson for notes of a few species not
as yet observed by me in the Firth of Forth.

PODOCOPA.

CYPRIDIDÆ.

Aglaia complanata, Brady and Robertson.
    Aglaia complanata, Brady and Robertson, Ann. and Mag. Nat.
        Hist., ser. iv., vol. iii. p. 66, pl. xx. figs. 4, 5 (1869).
    Aglaia complanata, Brady and Norman, Mon. of the M. and Fw.
        Ostrac. of the N. Atlantic and N.-W. Europe, p. 94 (1889).

Habitat.— Bo'ness (David Robertson).   A note of the occurrence of this
rare and interesting species was communicated to me by Mr Robertson,
who observed it among some material he had collected at Bo'ness some
years ago.   The only localities where it had previously been recorded from
are Westport Bay, Roundstone Bay, and Birterbuy Bay, Ireland.

Pontocypris acupunctata, Brady.
    Pontocypris acupunctata, Brady, Mon. Rec. Brit. Ostrac., p. 386,
        pl. xxiv. figs. 53–56 (1868).
Pontocypris acupunctata, Brady and Norman, *op. cit.,* p. 109.

Habitat.—Off St Monance, several specimens, and one or two from other parts of the Forth. This species seems to have been previously recorded from only two places in Scotland—St Magnus Bay, Shetland, and the Minch (see Monograph by Brady and Norman). I have, however, also observed it among some material dredged last year (1889) among the Orkney Islands.

Pontocypris trigonella, G. O. Sars.

 Pontocypris trigonella, Brady, *op. cit.,* p. 387, pl. xxv. figs. 31–34; pl. xxxviii. fig. 3.

 Pontocypris trigonella, Brady and Norman, *op. cit.,* p. 109 pl xxii. figs. 18–25; pl. xxiii. fig. 6.

Habitat.—Largo Bay and other parts of the Estuary, but not very common.

BAIRDIIDÆ.

Bairdia inflata, Norman.

 Bairdia inflata, Brady, *op. cit.,* p. 388, pl. xxvii. figs. 9–17; pl xxxviii., fig. 5.

 Bairdia inflata, Brady and Norman, *op. cit.,* p. 112.

Habitat.—Off St Monance, rare.

CYTHERIDÆ.

Loxoconcha viridis (Müller).

 Cythere viridis, Müller, Entom., p. 64, pl. vii. figs. 1, 2 (1785), non Brady.

 Loxoconcha elliptica, Brady, *op. cit.,* p. 435, pl. xxvii. figs. 38, 39; 45–48; pl. xl. fig. 3.

 Loxoconcha viridis, Brady and Norman, *op. cit.,* p. 185.

Habitat.—Granton Harbour (David Robertson). This is a brackish-water species, and may have accidentally got into the harbour.

Loxoconcha multifora (Norman).

 Cytheropteron multiforum, Brady, *op. cit.,* p. 449, pl. xxix. figs. 38–42.

 Loxoconcha multifora, Brady and Norman, *op. cit.,* p. 185.

Habitat.—Granton Harbour (David Robertson). These two species of *Loxoconcha* were observed by Mr Robertson in material collected by him in Granton Harbour twenty years ago.

Cythere finmarchica (G. O. Sars), ♂ ♀.

 Cythere finmarchica, Brady, *op. cit.,* p. 410, pl. xxxi. figs. 9–13.

 Cythere finmarchica, Brady and Norman, *op. cit.,* p. 163.

Habitat.—Off St Monance, frequent.

Cythere whitei (Baird).

 Cythereis whitei, Baird, Brit. Entom., p. 175, t. xx. figs. 3, 3a (1850).

 Cythere whitei, Brady, *op. cit.,* p. 416, pl. xxx. figs. 21–24.

 Cythere whitei, Brady and Norman, *op. cit.,* p. 169.

Habitat.—Largo Bay, rather rare.

Cythere (?) *semiovata,* n. s. (Pl. XII. figs. 1–2).

Shell seen from the side semiovate, dorsal and ventral margins nearly parallel; dorsal margin a flattened curve sloping downwards posteriorly, and forming with the nearly straight ventral margin a somewhat bluntly angular extremity; anterior end sharply rounded below, then curving obliquely upwards and backwards till it merges in the dorsal margin.

Seen from above, the width is greatest near the anterior end, but varies little for about three quarters of the length, when the sides converge and form posteriorly a somewhat wedge-shaped extremity. The anterior end is broadly rounded, inclining to angular in the middle, where the valves meet; greatest breadth equal to height; height about ½ the length. Surface of the valves smooth, but having a slightly resinous appearance. Length, ·35 mm.

Habitat.—Off St Monance, not very rare. Specimens of this form have been dredged on several occasions at this place, depth 12 to 14 fathoms, bottom clean gravel and sand. The animal has not yet been made out, the species is therefore for the present doubtfully referred to *Cythere*.

Cytheridea torosa (Jones).

> *Cytheridea torosa*, Brady, *op. cit.*, p. 425, pl. xxviii. figs. 7-12; pl. xxxix. fig. 5.
> *Cytheridea torosa*, Brady and Norman, *op. cit.*, p. 175.

Habitat.—Brackish water pools by the shore at Aberlady Bay, common. Associated with *Cytherura gibba* (Müller), *Candona candida* (Müller), *Trochamina inflata* (Mont.), *Haplophragmium canariense*, &c., Granton Harbour (Robertson). I have this species also from Montrose Basin and from Orkney; it is a brakish-water species. The above are the only records of its occurrence on the east of Scotland.

Krithe bartonensis (Jones).

> *Krithe bartonensis*, Brady, *op. cit.*, p. 432, pl. xxxiv. figs. 11-14; pl. xl. fig. 5.
> *Krithe bartonensis*, Brady and Norman, *op. cit.*, p. 179.

Habitat.—Near the mouth of the Estuary, moderately common. This species is new to the east of Scotland.

Cytherura gibba (Müller).

> *Cythere gibba*, Müller, Entomostraca, p. 66, pl. vii. figs. 7-9, ♀ (1785).
> *Cytherura robertsoni*, Brady, *op. cit.*, p. 444, pl. xxxii. figs. 16-18, ♀.
> *Cytherura gibba*, Brady and Norman, *op. cit.*, p. 190 (non *Cytherura gibba*, Brady, Mon. Rec. Brit. Ostrac.)

Habitat.—Largo Bay, rare (dead), frequent in brackish-water pools at Aberlady Bay (living); Granton Harbour (Robertson); it occurs also in Montrose Basin. This is a brackish-water species, and is sometimes observed in moderate abundance where the water is only slightly saline. Its occurrence in Largo Bay and in Granton Harbour is probably accidental.

Cytherura cornuta, Brady.

> *Cytherura cornuta*, Brady, *op. cit.*, p. 445, pl. xxxii. figs. 12-15.
> *Cytherura gibba*, idem ibidem, p. 444, pl. xxxii. figs. 68-70, ♀ (non *Cytherura gibba*, Müller).
> *Cytherura affinis*, idem ibidem, p. 443, pl. xxxii. figs. 17-21, ♀ var. (non *Cytherus affinis*, G. O. Sars).
> *Cytherura lineata*, idem ibidem, p. 443, pl. xxxii. figs. 30-34 (jun.).
> *Cytherura cornuta*, Brady and Norman, *op. cit.*, p. 192, pl. xviii. figs. 21, 22.

Habitat.—Vicinity of Phidra, off Musselburgh, and Burntisland, but not common. Though of frequent occurrence on the west coast, I do not find any previous record of it from the east coast of Scotland.

Cytherura bodotria,* n. s.   (Pl. XII. figs. 6, 7).

Shell seen from the side of nearly equal height throughout ; dorsal and

---

* Bodotria, the ancient name of the Forth.

ventral margins nearly straight, the former is slightly convex towards the anterior extremity ; anterior margin evenly rounded, posterior extremity with a short beak situated about the middle, its termination narrow, truncate. Seen from above, ovate, slightly constricted in front, where the valves meet. At the posterior end, the middle is bluntly mucronate, and the sides are produced to an acute angle, so as to impart to it a somewhat tridentate appearance ; dorsal ridge prominent, where it bends downwards in front. Surface sculptured with flexuous longitudinal riblets, crossed by a few indistinct ones arranged irregularly. Length, ·5 mm. ; breadth, ⅔ length ; height, fully ½ the length.

Habitat.—Off St Monance, in 12 to 14 fathoms, bottom sand and gravel, rare.

This species somewhat resembles *Cytherura acuticostata*, but differs in being not so stout, and in having the valves produced backwards, so that the posterior extremity of the shell has a tridentate form.

Cytherura mucronata, n. s.   (Pl. XII. figs. 3, 5).

Shell seen from the side, elongate, narrow ; height about equal at both ends, length two and a half times the height ; dorsal margin nearly straight, ventral margin slightly and evenly concave, posterior end much produced and wedge-shaped, forming a ' beak,' which is situated below the middle ; anterior margin broadly rounded, somewhat produced in the middle. Seen from above, oval, with the ends acuminate ; the margin at each end, especially the anterior margin, is produced, so as to form a distinct ' mucro.' The surface is marked with indistinct raised lines, which are somewhat irregularly distributed ; the breadth is equal to the height ; length, ·33 mm.

Habitat.—Off St Monance, not very rare.

Cytherura simplex, Brady and Norman.

   Cytherura simpex (name only), Brady and Robertson, Ann. and
      Mag. Nat. Hist., ser. iv., vol. xi. p. 66 (1872).
   Cytherura sarsii (" local variety "), idem ibidem, vol. xiii. p. 117,
      pl. iv. figs. 6, 7 (1874).
   Cytherura simplex, Brady and Norman, *op. cit.*, p. 200, pl. xviii.
      figs. 1, 2.

Habitat.—Off St Monance, frequent, depth 12 to 15 fathoms ; bottom clean sand, part gravel. Viewed laterally, the shell of this species differs somewhat from the usual form of *Cytherura*, which has a more or less distinct ' beak ' at the posterior end, whereas this has no posterior beak. New to the east of Scotland.

Cytherura fulva, Brady and Robertson.

   Cytherura fulva, Brady and Robertson, Ann. and Mag. Nat. Hist.,
      ser. iv., vol. xiii. p. 116, pl. iv. figs. 1–5 (1874).
   Cytherura fulva, Brady and Norman, *op. cit.*, p. 205, pl. xix. figs.
      9–11.

Habitat.—Largo Bay and other parts of the Estuary, but not common. New to the east of Scotland.

Cytheropteron punctatum, Brady.

   Cytheropteron punctatum, Brady, *op. cit.*, p. 449, pl. xxxiv. figs.
      45–48.
   Cytheropteron punctatum, Brady and Norman, *op. cit.*, p. 211.

Habitat.—Off St Monance, rather rare. I do not find any previous record of this species for the east of Scotland.

Bythocythere turgida, G. O. Sars.
> *Bythocythere turgida*, Brady, *op. cit.*, p. 452, pl. xxxiv. figs. 35–38.
> *Bythocythere turgida*, Brady and Norman, *op. cit.*, p. 221.

Habitat.—Off Musselburgh and other parts of the Estuary, but not common. The only other Scotch localities where this species has been observed are the Clyde, Orkney, and Shetland.

Bythocythere recta (Brady).
> *Cytheropteron rectum*, Brady, *op. cit.*, p. 476.
> *Bythocythere recta*, Brady and Norman, *op. cit.*, p. 222, pl. xix. figs. 13–14.

Habitat.—Largo Bay, rare. This species has also been recorded from Lerwick and St Magnus Bays, Shetland, which appear to be the only records of it for Scotland.

Cytherois fischeri (G. O. Sars).
> *Paradoxostoma fischeri*, Brady, Nat. Hist. Trans. Northumb. and Durham, vol. iii. p. 362, pl. xii. figs. 1–3 (1870).
> *Cytherois fischeri*, Brady and Norman, *op. cit.*, p. 228, pl. xxi. figs. 20–22.

Habitat.—Generally distributed throughout the Estuary; common in brackish pools by the shore at Aberlady, where it is more or less of a dark bluish colour : those dredged off St Monance are nearly white.

PARADOXOSTOMATIDÆ.

Paradoxostoma variabile (Baird).
> *Paradoxostoma variabile*, Brady, *op. cit.*, p. 459, pl. xxxv. figs. 1–7, 12–17; pl. xli. fig. 8.
> *Paradoxostoma variabile*, Brady and Norman, *op. cit.*, p. 229, pl. xxiii. fig. 10.

Habitat —Largo Bay and other places, frequent

Paradoxostoma obliquum, G. O. Sars.
> *Paradoxostoma obliquum*, Brady, *op. cit.*, p. 459, pl. xxxv. figs. 18–21.
> *Paradoxostoma obliquum*, Brady and Norman, *op. cit.*, p. 230.

Habitat.—Off Phidra, Musselburgh, and Burntisland, rare.

Paradoxostoma hibernicum, Brady.
> *Paradoxostoma hibernicum*, Brady, *op. cit.*, p. 460, pl. xxxv. figs. 35, 36; pl. xl. fig. 7.
> *Paradoxostoma sarniense*, idem ibidem, p. 460, pl. xxxv. figs. 26–29; pl. xl. fig. 9.
> *Paradoxostoma hibernicum*, Brady and Norman, *op. cit.*, p. 232, pl. xxi. figs. 15–17.

Habitat.—Largo Bay, rare. Neither this nor the previous species appear to have been recorded before for the east of Scotland.

Paradoxostoma arcuatum, Brady.
> *Paradoxostoma* (?) *arcuatum*, Brady, *op. cit.*, p. 461, pl. xxv. figs 37–38.
> *Paradoxostoma arcuatum*, Brady and Norman, *op. cit.*, p. 234, pl. xxi. figs. 5, 6.

Habitat.—Off St Monance, Largo Bay, and near Inchkeith; several specimens Granton Harbour (Robertson).

Paradoxostoma hodgei, Brady
> *Paradoxostoma hodgei*, Brady, Nat. Hist. Trans. Northumberland
> and Durham, vol. iii. p. 371, pl. xii. figs. 12, 13 (1870).
> *Paradoxostoma hodgei*, Brady and Norman, *op. cit.*, p. 235, pl. xxi.
> figs. 7, 8.

Habitat.—Off St Monance and Phidra, frequent. New to the east of Scotland.

Paradoxostoma (?) *affine*, provisional name. (Pl. XII. figs. 8–9).

Shell seen from the side elongate, subovate, highest a little behind the middle; dorsal margin evenly but not boldly arched, inferior nearly straight, slightly sinuate towards the anterior extremity; anterior extremity rather higher than the posterior, and the margins of both evenly rounded; surface smooth, with a few irregular scratched lines. Outline seen from above compressed, ovate, the posterior half of nearly equal breadth, with the extremity obtusely pointed; anteriorly the shell is more compressed, the extremity being somewhat acuminate; breadth about equal to height and a third of the length; length, ·42 mm.

This form resembles a small *P. arcuatum*, but is not so narrow posteriorly, and the greatest breadth is nearer the posterior extremity.

Habitat.—Off St Monance, not common.

MYODOCOPA.

CYPRIDINIDÆ.

Asterope marie (Baird).
> *Cypridina mariæ*, Baird, Proc. Zool. Soc. Lond., part xviii. (1850),
> p. 257, pl. xvii. figs. 5–7.
> *Cylindroleberis mariæ*, Brady, Mon. Rec. Brit. Ostrac., p. 465, pl.
> xxxiii. figs. 18–22; pl. xli. fig. 1 (1868).
> *Asterope mariæ*, Robertson, Fauna and Flora of the West of Scotland,
> p. 39 (1876).

Habitat.—Bass Rock, but not common. This is a generally distributed, though not an abundant species. I have specimens from the Moray Firth and from Orkney: it is not uncommon in the Clyde.

CLADOCOPA.

POLYCOPIDÆ.

Polycope orbicularis, G. O. Sars.
> *Polycope orbicularis*, G. O. Sars, Oversigt af Norges Marine
> Ostracoder, p. 122.
> *Polycope orbicularis*, Brady, *op. cit.*, p. 471, pl. xxxv. figs. 53–57.

Habitat.—Off Phidra, rare. There is no previous record of this species for the east of Scotland.

Note.—In the Monograph by Brady and Norman, recently published by the Royal Dublin Society, the following species are recorded from the Firth of Forth:—*Loxoconcha fragilis*, G. O. Sars; *Loxoconcha pusilla*, Brady and Robertson; and *Cythere pulchella*, Brady, which, with the exception of the first, I also have observed in different parts of the Estuary.

AMPHIPODA.

GAMMARIDÆ.

Gitana sarsi, Boeck.
> *Gitana sarsi*, Boeck, De Skand. Arkt. Amphip., p. 439, pl. xi.
> fig. 2 (1876).

Amphilochus sabrinæ, Stebbing, Ann. and Mag. Nat. Hist., p. 364, pl. xv. (1878).

Habitat.—Off Inchkeith (Nov. 1889) rare. This is a small species, and easily missed when mixed up among a lot of other things.

Guernia coalita (Norman).

Helleria coalita, Norman, Ann. and Mag. Nat. Hist., p. 418, pl. xxii. fig. 8 ; pl. xxiii. figs. 1–6 (1868).

Guernia coalita, Chevreux, Cat. Amphip. du Sud-ouest de la Bretagne (1889).

Habitat.—Off St Monance. A few specimens only of this curious little species were observed in material dredged off St Monance, depth from 12 to 14 fathoms, bottom sand and gravel.

Hippomedon holbölli (Kröyer).

Anonyx holbölli, Kröyer, Natur. Hist. Tidsskr., 2 R, 2 B, p. 8 (1846).

Anonyx denticulatus, Sp. Bate, Cat. Amphip. Crust. Brit. Mus., p. 75 (1862).

Hippomedon holbölli, A. Boeck, De Skand. Arkt. Amphip., p. 136, pl. v. fig. 6 ; pl. vi. fig. 7 (1876).

Habitat.—A little north-west of May Island (1888), rare.

Megaluropus agilis, Norman.

Megaluropus agilis, Norman, Ann. and Mag. Nat. Hist. (1889), p. 446, pl. xviii. figs. 1–10.

Habitat.—Largo Bay, frequent. 'The most remarkable characters in the genus,' to which this species belongs, 'are the eye, which is situated on a greatly projected lobe, and the expanded foliaceous branches of the last uropods.'[*] The peculiar form of these uropods is even more striking than the prominent eye on its curious stalk-like lobe, which projects forward between the peduncles of the antennules and antennæ. In Scotland this species has been observed at Cumbrae, Firth of Clyde (D. Robertson), and 25 miles off May Island, Firth of Forth (John Murray). This last station is considerably beyond the limits of the Forth, and the present is there-fore the first record of the occurrence within the Estuary.

Monoculodes carinatus, Bate.

Westwoodia carinata, Bate, Brit. Assoc. Rep. (1855), p. 58.

Monoculodes carinata, Bate and Westw., Brit. Sess.-eyed Crust., vol. i. p. 165.

Monoculodes stimpsoni, ibid ibidem, p. 160, ♂ (jun.)

Monoculodes affinis, Boeck, Crust. Amphip., bor. et arct., p. 84 (1870).

Monoculodes carinatus, Norman, Ann. and Mag. Nat. Hist. (1889), p. 447, pl. xix. figs. 1–5.

Habitat.—Off St Monance, near Phidra, and in Largo Bay, but not common. In Largo Bay, *M. longimanus*, Bate (a species I have already recorded for the Forth), is of frequent occurrence ; females with ova are occasionally observed. This species is not so large nor so robust as the other, being scarcely half the size. *M. carinatus* has been taken '25 miles 'off May Island,' which is considerably beyond the limits of the Firth of Forth. This is the first record of its occurrence within the Estuary. Mr Robertson records it from several places in the Firth of Clyde, and T. Edward at Banff.

---

[*] Norman, Ann. and Mag. Nat. Hist. (1889), p. 446.

Urothoe elegans, Spence Bate.

> *Gammarus elegans,* Spence Bate, Brit. Assoc. Rep. (1855).
> *Urothoe elegans,* Spence Bate. Sess.-eyed Crust., vol. i. p. 200 (1863).

Habitat.—Largo Bay, not uncommon. A small but robust species, which does not appear to have been previously recorded for the Forth.

Leucothoe spinicarpa (Abildgaard).

> *Gammarus spinicarpos,* Abildgaard, Zool. Dan., vol. iii. p. 66, pl. cxxix. figs. 1–4.
> *Leucothoe spinicarpa,* A. Boeck, Crust. Amph., bor. et arct., p. 78 (1870).
> *Leucothoa spinicarpa,* Bate and Westwood, Brit. Sess.-eyed Crust., vol. i. p. (1863).

Habitat.—Largo Bay, rare. *Leucothoe* is readily distinguished by the peculiar form of the hands of the first pair of gnathopods, which somewhat resemble the blades of a pair of scissors with curved points. I have frequently taken this species, but usually in the branchial cavities of large Ascidians, and very seldom otherwise. I have observed it in such situations at East Loch Tarbert (Loch Fyne), at Scapa Flow, Orkney, and in the Moray Firth. It is of a delicate reddish or pink colour, and moderately active. It is curious that this somewhat semiparasitic habit of *L. spinicarpa* has been so seldom referred to by authors.

Phoxocephalus fultoni,[*] n. s. (Pl. XII. figs. 10–12), and Pl. XIII. figs. 13–19.

Rostrum (fig. 12) extending to about the end of the second joint of the peduncle of the antennules. Antennules short, not longer than the peduncle of the antennæ; joints of peduncle stout, sparsely furnished with hairs, the last rather more than half the length of the penultimate joint; flagellum shorter than the peduncle, 4-jointed joints sub-equal; secondary appendage 3-jointed, extending to the end of the second joint of the flagellum. Antennæ short, stout, furnished with a few hairs, especially on the upper distal margin of the joints. There is no very marked difference between the peduncle and flagellum; second and third joints of peduncle about equal in length; flagellum 3-jointed, rather longer than the last joint of the peduncle. The thigh of the first gnathopods is long, the anterior distal angle of the short stout meros is produced into a small rounded process; the adjacent parts of meros and wrist are correspondingly hollowed out, and thus a kind of ball and socket joint is formed (fig. 15, *a*); hand (fig. 15) subquadrate, the length about twice the breadth; sides nearly straight and parallel; palm slightly convex, and produced forward at an obtuse angle from the joint of the finger; finger slightly curved, the point reaching nearly to the extremity of the palm, and fitting into a small notch. Second gnathopods very like the first, but the hand is to some extent proportionally broader; the hands of both first and second gnathopods have a fringe of short hairs along each side of the palm. The first, second, and third pereiopods are short and stout; the fourth are longer, the fifth are also short and stout. The outer branch of posterior pleiopods is 2-jointed, the terminal joint being very much shorter than the other; the inner branch is 1-jointed, and small, being scarcely more than half the length of the first joint of the outer branch (fig. 19).

I obtained two forms of this species; they resemble each other closely.

[*] It gives me much pleasure to have the opportunity to name this species after my friend, Dr T. Wemyss Fulton, Secretary to the Scientific Department of the Fishery Board.

The one that seems to be the female differs from that now described chiefly in the following points :—The flagellum of the antennules is 5-jointed, the first and second joints rather shorter than the others (figs. 10–11). The flagellum of the antennæ is 10-jointed ; the first joint is moderately long—longer than the next two together, which are short, and about equal in length, fourth joint rather longer than the preceding ; the remaining joints gradually increase in length, and become more slender (fig. 11). The inner joint of the posterior pleiopods, which is also 1-jointed, is rather longer than, and as stout as the first joint of the outer branch; the two forms are very much alike otherwise.

Habitat —Off St Monance, in 12 to 15 fathoms, not very common.

Amphithopsis latipes (M. Sars).

    Calliope ossiani, Bate and Westwood, Brit. Sess.-eyed Crust., vol. i. p. 261 (1868).

    Calliope fingalli, idem ibidem, vol. i. p. 263.

    Amphithopsis latipes, Norman, Mus. Norm., part iii. p. 15 (1868).

Habitat.—Several specimens attached to a Zoophyte (*Antennularia*) brought up in the trawl-net a few miles east of Inchkeith; they were, with one or two exceptions, all prettily marked by brown bands extending from the side along the posterior edge of each segment of the posterior pleon; the coxæ were also of the same colour. In the form of the antennules and antennæ, and of the gnathopods and in the coloration, they agreed with the form described by Spence Bate as *Calliope ossiani*. The Rev. T. R. R. Stebbing, to whom I submitted specimens, and who corrobated my diagnosis, informs me that Boeck and Norman identify *Calliope ossiani* and *C. fingalli*, Bate and Westwood with *Amphithopsis latipes* (M. Sars). I have therefore followed them in ascribing my specimens to Sars's species.

Epimeria cornigera (Fabricius).

    Gammarus cornigera, Fab., Reisenach Norwegen (1779), p. 383.

    Acanthonotus testudo, White, Cat. Crust. Brit. Mus. (1847), p. 57.

    Acanthonotus owenii, Bate and Westwood, Brit. Sess.-eyed Crust., vol. i. p. (1863).

    Epimeria cornigera, A. Boeck, Crust. Amphip., bor. et arct. (1870), p. 105.

Habitat.—East of Inchkeith, about 3 miles. The colour of this pretty species is white, somewhat pellucid, beautifully variegated with bright red; the postero-lateral margins of each segment is of this colour, which is also more or less diffused over the dorsal surface. I have also got this species in the Moray Firth ; and Mr Robertson records it from various places in the Clyde district.

COROPHIIDÆ.

Siphonœcetus colletti (?), Boeck.

    Siphonœcetus colletti, Boeck, Crust. Amphip., bor. et arct. (1870), p. 178.

    Siphonœcetus colletti, idem, De Skand. og Arkt. Amph. (1876), p. 633, pl. xxviii. fig. 9.

Habitat.—Largo Bay, frequent. New to Britain. It is quite possible that this may have been passed over as a *Corophium*, otherwise it is difficult to account for its not being previously recorded. The specimen comes very near to *S. typicus*, and it may ultimately be found to belong to that species. I prefer therefore to consider it for the present as doubtful.

ISOPODA.

SPHÆROMIDÆ.

Sphæroma rugicauda, Leach.

    Sphæroma rugicauda, Leach, Edin. Enc., vol. vii. pp. 405, 408.

    Sphæroma rugicauda, Bate & Westwood, Brit. Sess.-eyed Crust., vol ii. p. 408 (1863).

    Habitat.—In brackish-water pools on the shore at Aberlady Bay, Common. They appeared mostly to creep upon or through the surface layer of the soft oozy mud forming the bottom of the pools; and only when the mud was stirred would they rise and swim very rapidly through the water for a short distance, then drop down again and burrow among the mud. When prevented from swimming, or when taken out of the water, they rolled themselves into a ball. Though observed at Berwick-on-Tweed by Dr Johnston, they do not appear to have been previously recorded for the east of Scotland. Mr Robertson found them plentiful in a weedy brackish pool with a soft muddy bottom at Hunterston, Ayrshire.

CUMACEA.

Only four species are added to the Forth Cumacea in this Report, viz. :—

CUMIDÆ.

Cuma pulchella, G. O. Sars, ♂ ♀.

    Cuma pulchella, G. O. Sars, Nye Bidrag til Kundakaben om Middelhavets Invert-fauna, part ii., Cumacear, p. 24, tab. vi. and tab. lx. (1879).

    Habitat.—Off St Monance, and in the vicinity of Phidra; Largo Bay, common. This is a small species, and easily overlooked. Dr Norman says that 'a good point for distinguishing the species is the first joint of 'the second foot, which is furnished with a series of backward directed 'tooth-like processes,' which is well shown in tab. lx. fig. 7, of Sars' Monograph referred to above. This seems to be the first time that *C. pulchella* has been observed in Britain; previously it has been noticed at Naples by G. O. Sars, and Bayonne by Marquis de Folin. The integument is ornamented with numerous microscopic circular depressions arranged in irregular oblique rows; the anterior part of the cephalon is dorsally of a dusky colour, and is darkest in the vicinity of the rostrum.

Eudorellopsis deformis (Kröyer).

    Leucon deformis, Kröyer, Voyage en Skand., pl. vi. fig. 3.

    Eudorella? deformis, G. O. Sars, Beskrivelse af de paa Fregatten Josephines Exped., fundne Cumaceer, p. 50, figs. 118–121 (1871).

    Habitat.—Off St Monance and Aberlady Bay, not common. Dr Norman states in reference to this species, 'not yet recorded as British, but I have 'had specimens in my collection, determined, since 1866, when I found 'them in a gathering from Bridlington, sent me by G. S. Brady.' It does not seem to have been observed anywhere else in Britain, and thus forms an interesting addition to the Forth Fauna. In 1882 G. O. Sars described this under the generic name *Eudorellopsis.*

Diastylus rugosa, G. O. Sars.

    Diastylus rugosa, G. O. Sars, Om den aberrante Krebsdyrgruppe Cumacea og dens nordiske Arter, p. 41.

    Diastylus strigata, Norman, Ann. Nat. Hist., ser. 5, vol. iii. p. 62 (male).

Diastylus rugosa, G. O. Sars, Middelhavets Cumaceer, p. 98,
    Tab. 34–38 (1879).
    Habitat.—Largo Bay and other parts of the Forth. This seems to be
a well-marked species, though it has not previously been recorded for the
Estuary.

Campylaspis affinis, G. O. Sars, ♂ ♀.
    Campylaspis affinis, G. O. Sars, Nye Dybrands Crustaceer fra
        Lofoten, p. 160 (1870); Extract, p. 16.
    Habitat.—Vicinity of the Bass Rock, rare. The cephalic shield is
thickly sprinkled with purple spots, which impart to it a somewhat uni-
form purplish colour. The Rev. T. R. R. Stebbing, to whom I submitted
the specimen, points out that it comes very close to *C. rubricunda*
(Liljeborg) in the form of the tail appendages ; it differs in the coloration.

SCHIZOPODA.

MYSIDÆ.

Erythrops serrata, G. O. Sars.
    Nematopus serratus, G. O. Sars, Beretning om en Sommeren
        (1862), foretagen Zoologisk Reise i Christianias og Trondhjems
        Stifter, p. 43.
    Nematopus serratus, Norman, Last Report on Dredging among
        the Shetland Isles : Report Brit. Assoc. (1868), p. 270.
    Erythrops serrata, G. O. Sars, Mon. over de ved Norges Kyster
        Forkommende Mysider., Frste Hefte, p. 27, tab. ii. figs. 1–2
        (1870).
    Habitat.—South-east of the Bass Rock 4 or 5 miles, rather rare. In
this species the outside edges of the antennal scales are deeply toothed,
with the teeth pointing forwards, and thus differs from the other two
species of *Erythrops* recorded for Britain. There appears to have been
some confusion in previous records of the distribution of *Nyctiphanes
norvegica* and *Boreophausia raschii*, the first being understood to be a more
common species. I find, on the contrary, that the latter is comparatively
abundant, especially in the outward part of the Estuary, while the other is
rather rare.

MOLLUSCA.

Miss J. E. Carphin kindly placed her extensive collections of Forth
Mollusca at my service, which has enabled me to include a few interesting
additions to the local list of species belonging to this group.

LAMELLIBRANCHIATA.

LUCINIDÆ.

Diplodonta rotundata (Montagu).
    Tellen rotundata, Mont., Test. Brit., p. 74, t. ii. fig. 3.
    Diplodonta rotundata, Jeffreys, Brit. Conch., vol. ii. p. 254 ; vol.
        v. pl. xxxiii. fig. 4.
    One living specimen of this pretty bivalve was found at Newhaven Pier
by Miss J. E. Carphin. It had been brought in from the outer part of
the Estuary on the fishermen's lines.

CARDIIDÆ.

Cardium nodosum, Turton.
    Cardium nodosum, Turt., Conch. Dith., p. 186, t. xiii. fig. 8.
    Cardium nodosum, Jeffreys, Brit. Conch., vol. ii. p. 285 ; vol. v.
        pl. xxxv. fig. 4.

Habitat.—Between Inchkeith and May Island, several specimens of this species were dredged, a few being alive. Though widely distributed, it does not appear to be a common species. My friend Mr J. T. Marshall, M.C.S., Torquay, kindly examined one of the specimens for me, and confirmed my identification. He says 'it is a rare species in Scotland, 'and I have it thence from only two localities.' It may be remarked that some experience is necessary to enable one to discriminate the smaller species of Cardium.

CYPRINIDÆ.

Circe minima (Montagu).

> *Venus minima*, Mont., Test. Brit., p. 121, t. iii. fig. 3.
> *Circe minima*, Jeff., Brit. Conch., vol. ii. p. 322, pl. vi. fig. 4 ; vol. v. pl. xxxvii. fig. 6.

Procured from the fishermen's lines at Newhaven Pier by Miss J. E. Carphin.

GASTEROPODA.

TROCHIDÆ.

Trochus montacuti, W. Wood.

> *Trochus montagui*, Wood, Ind. Test., Suppl., pl. vi. fig. 43.
> *Trochus montacuti*, Jeff., Brit. Conch., vol. iii. p. 320 ; vol. lxiii. fig. 1.

This species was found at Newhaven Pier by Miss J. E. Carphin, having been brought in from the outward part of the Estuary on the fishermen's lines.

Trochus zizyphinus, L., var *lyonsii* (Leach).

> *Trochus lyonsii*, Flem., Brit. Anim., p. 323.
> *Trochus zizyphinus*, var. *lyonsii*, Jeff., Brit. Conch., vol. iii. p. 331.

Two fine and living specimens of this pretty variety of *Trochus zizyphinus* were found by Miss J. E. Carphin at Newhaven Pier ; they had been brought in attached to the fisherman's lines from the outer part of the Estuary. I have also obtained two living specimens of this variety and one typical specimen among trawl refuse a few miles west of May Island, while trawling the Forth stations.

EULIMIDÆ.

Eulima polita (Linné).

> *Turbo politus*, Linn, S. N. p. 1241.
> *Eulima polita*, Jeff., Brit. Conch., vol. iv. p. 201 ; vol. v. pl. lxxvii. fig. 3.

Habitat.—Off St Monance, rare. Two adult living specimens and one or two young ones were dredged at this locality

NUDIBRANCHIATA.

HERMÆIDÆ, A. & H.

Alderia modesta (Loven).

> *Stiliger modestus*, Loven, Trans. Royal Swedish Academy.
> *Alderia modesta*, idem, Index Molluscorum Scandinaviæ.
> *Alderia modesta*, A. & H., Brit. Nud. Moll., fam. 3, pl. xli. figs. 1–5.
> *Alderia modesta*, J. G. Jeffreys, Brit. Conch., vol. v. p. 33 (1869).

Habitat.—Brackish-water pools between tide marks, Aberlady Bay, frequent, but easily overlooked. Not before recorded for the Forth. Jeffreys says*—'This curious animal is almost amphibious, being only

---

* *Brit. Conch.*, vol. v. p. 33.

' found in very shallow brackish-water barely within the reach of the
' tide, and occasionally crawling on the moist weed beyond. It is a rare
' [or local] species, but generally plentiful where it does occur.' I obtained
one specimen of this species in the vicinity of Skeirvuie—a small island near
the head of East Loch Tarbert (Loch Fyne)—where *Zostera marina* grows
in considerable abundance ; the specimen was kept alive for some time,
and carefully examined by myself and others, so that though the conditions of the locality mentioned are different from those of the *habitat*
which this species is said to be restricted to as stated above, there was no
doubt as to the correct identification of the specimen.

Note.—In the course of our examination of the stomachs of fishes, taken
in the Firth of Forth by the 'Garland's' trawl-net, the Annelids *Priapulus
caudatus* and *Echiurus oxyurus*, and the Tunicate *Pelonaia corrugata*
have been occasionally observed, and in some instances so little injured as
to indicate that they had been quite recently captured by the fish. It is
in the stomach of the haddock and cod that these organisms are usually
observed. *Priapulus* and *Pelonaia* have been recorded from the Forth ;
but so far as I know, *Echiurus* has not been hitherto observed in the
Estuary. In St Andrews Bay, however, it is occasionally met with.
Macropsis slabberi, which, as a British species, was considered to be confined to the upper part of the Firth of Forth, has been taken by me
during the last year in the vicinity of the Bass, in St Andrews Bay, and
and in the Estuary of the Tay opposite Tayport ; this would indicate that
its distribution is not so restricted as was supposed, or that it is spreading gradually to other parts of our coast. I have also obtained the
somewhat rare *Isocardia cor* and *Palmipes membranacea* (*placenta*) in the
Moray Firth. Of the first, two large specimens—one living and one dead
—were brought up by the trawl of the 'Southesk' last year during the time
I was on board ; a specimen of the other was brought up by the 'Southesk's'
trawl on one or two occasions while I was on board in the early part of
this year (1890).

CORRIGENDA.

In my paper 'Some Additions to the Fauna of the Firth of Forth,' in
last year's Report, the Amphipods referred by me to *Gammarus edwardsi*,
Spence Bate (p. 321), I am now satisfied do not belong to that species,
but are a form of *G. locusta*, L.

Note on *Cymbasoma rigidum* (Thompson), Scott, 'Some Additions to the
' Fauna of the Firth of Forth,' *Seventh Annual Report*, pt. iii. p. 316
(1889).

In a paper by G. C. Bourne, M.A., F.L.S., Director of the Plymouth
Laboratory of the Marine Biological Association, on the genus *Monstrilla*,
Dana, in the *Quarterly Journal of Microscopical Science*,† this genus is
fully and carefully described ; short descriptions are also added of various
species belonging to it which have been more or less satisfactorily
determined. In this paper Mr Bourne identifies *Cymbasoma*, Thompson,
with *Monstrilla*, Dana, and refers the form recorded by me for the Firth
of Forth in the *Seventh Annual Report* as *Cymbasoma rigidum* to *Monstrilla helgolandica*, Claus (of which there is no previous record for

† Vol. xxx. pt. iv. pp. 565–578, pl. xxxvii. (February 1890).

Britain). Its distinctive characters as described by Bourne —who examined the dissections of my specimens—are six on each furcal member, four abdominal segments, antennæ four-jointed, the two last joints elongate subequal. The protopodite of the swimming feet bears a spine on its interior lower angle.

Habitat.—Heligoland and Firth of Forth.

Explanation of Plates XII, XIII.

Plate XII

Fig. 1. Cythere semiovata, seen from above.
Fig. 2.   "   " seen from the right side.
Fig. 3. Cytherura cuneata, seen from the right side.
Fig. 4.   "   " end view.
Fig. 5.   "   " seen from above.
Fig. 6. Cytherura bollotria, seen from the left side.
Fig. 7.   "   " seen from above.
Fig. 8. Paradoxostoma affine, seen from above.
Fig. 9.   "   " seen from right side.
Fig 10. Phoxocephalus fultoni ♀ ?
Fig. 11.   "   " ♀ ? a rostrum, b antennules, c antennæ.
Fig. 12.   "   " ♂ ? a rostrum, b antennules, c antennæ.

Plate XIII

Fig. 13. Phoxocephalus fultoni, 2nd perieopods.
Fig. 14.   "   " 1st Gnathopods.
Fig. 15.   "   " 2nd "
Fig. 16.   "   " 3rd perieopods.
Fig. 17.   "   " 1st pleiopods.
Fig. 18.   "   " 1st pleiopods.
Fig. 19.   "   " posterior pleiopods.

PLATE XII.

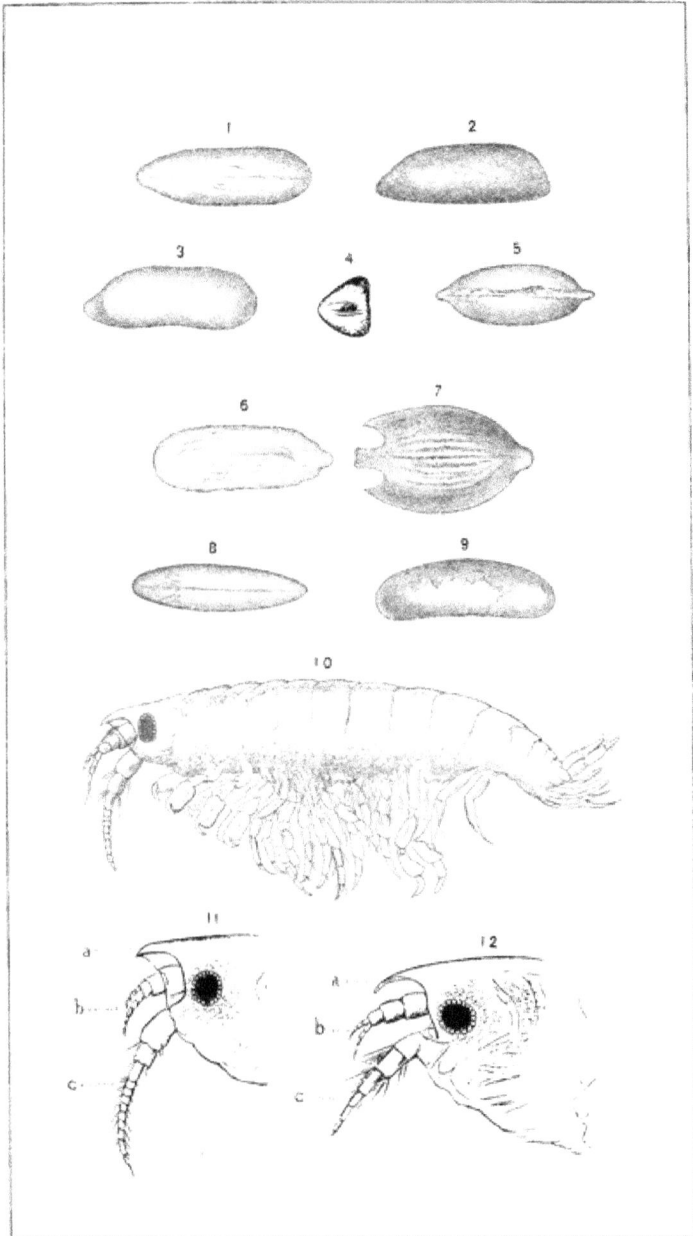

PLATE XIII.

13

14

15

16

17

18

19

No. VIII.—THE INVERTEBRATE FAUNA OF INLAND WATERS.
—I. REPORT ON LOCH COULTER AND THE COULTER
BURN, STIRLINGSHIRE.   By Thomas Scott, F.L.S.

The systematic investigation of the fauna of fresh water lakes,
especially in those of great depths, has in recent years been largely undertaken on the Continent, by Forel, Pavesi, Fol, and a number of other
investigators; but in this country comparatively little has yet been done
on this subject. It is now proposed to make, from time to time as
opportunity allows, a biological and physical investigation of the great
Scottish lochs and inland waters, which cannot fail to be of interest in
regard to the food-fishes which inhabit them, and will also incidentally throw
light upon many other questions of scientific interest. The physical, and to
some extent the biological, conditions of some of the West Coast lochs have
been inquired into by Dr John Murray, the Director of the Challenger
Commission; and Mr J. S. Grant Wilson a year or two ago made a
physical examination of the lochs in Perthshire. A comparison will be
made of the fauna of lochs, which have been in recent geological times
cut off from the sea, with the fauna of typical inland lochs, and with that
of those where there now occurs an admixture of sea water. The nature
and distribution of the invertebrate organisms in the waters of very deep
lochs will also be investigated. It is well known that marked differences
exist between the trout and other edible fishes of many lochs. This
inquiry by determining the main food of these fishes may lead to useful
measures being recommended.

In compliance with instructions received, I began in June 1889 an
investigation of the invertebrate fauna—especially the Crustacea and
Mollusca—of Loch Coulter, and its effluent, the Coulter Burn.

Loch Coulter is situated in a natural hollow about 300 yards to the
east of the Stirling and Kilsyth Road, and nearly midway between these
two places. It lies almost due east and west; its greatest length is about
1100 to 1200 yards, its greatest breadth 600 to 700 yards, and it has a
somewhat quadrangular outline. From the peculiar physical conditions
of the district in which this loch is situated, only a few ditches drain
into it, but one or more springs are reported to exist somewhere within
its area. The Coulter Burn is the only outlet for the water of the loch.
It takes its rise from the north-west corner, and flows east and north by
a rather circuitous route, passing on its way through the well-known
Howietoun Fisheries and Goldenhoof Dam, and joins the Bannock
Burn a little to the south of the site of the historical battle of that name.

Sir J. Ramsay-Gibson-Maitland, Bart., kindly gave me information
and advice, which were of the greatest value to me in making arrangements as to the manner in which the investigation should be made.

As I had to return to Edinburgh, in order to continue my investigations on the 'Garland,' it was decided, that as soon as arrangements
could be made for proceeding with the work, my son, Mr Andrew Scott,
should take my place in carrying on the proposed investigations, which
he did on 12th June. In order to enable him to carry out the examination as carefully and accurately as possible, I drew up for his guidance
a plan of work, dividing the district into sections, and instructing him to
examine each separately and consecutively, and to take notes of the
organisms observed and collect samples of material. My son completed
his investigations on the 22nd of June, having examined Loch Coulter

and traversed and examined the Coulter Burn from its source to its confluence with the Bannock, and thence to the Forth, a distance of between
10 and 12 miles,—and it is chiefly from his notes and the material he
collected that this Report is prepared.

The scheme of work to which I have referred, and which was adhered
to as closely as possible, was as follows :—

1. Loch Coulter.
2. The Coulter Burn from the Loch to Craigquarter Wood.
3. The Coulter Burn from Craigquarter Wood to the Hatching
   House (connected with the Howietoun Fisheries).
4. The Coulter Burn from the Hatching House to the Ponds at
   Howietoun, including as far as possible an examination of the
   Ponds.
5. The Coulter Burn from the Ponds to Goldenhoof Dam, including
   an examination of the Dam.
6. The Coulter Burn from Goldenhoof Dam to its confluence with
   the Bannock.
7. The Bannock Burn from thence to the Forth.

In stating the results of the examination of the loch and its effluent
it will perhaps be better to refer to each section separately, as this will
to some extent simplify and localise the information secured.

1. LOCH COULTER.

In our examination of Loch Coulter, the tow-net, hand-net, and dredge
were used. We first of all used the tow-net, towing it for a time
just under the surface of the water, when we soon ascertained that the
water all over the loch was teeming with Entomostracan organisms.
The tow-net was then fixed to the dredge rope, the dredge being
used as a 'sinker,' and towed close to the bottom. Various parts of
the loch were examined in this way, and Entomostraca were again
observed to be abundant ; in fact, on looking over the side of the boat,
they could easily be observed in great numbers swimming about near the
surface. The bottom was next examined by means of the dredge, and
the results showed that the depth of the loch did not much exceed 5
fathoms at the deepest part, which was near the east end. At this end
the bottom was hard and stony, and appeared to be unsuitable for the
existence of Mollusca or other non-pelagic organisms, as very few were
obtained at this part. Towards the north end the bottom was found to
consist, more or less, of fine vegetable mud, on which several species of
Molluscs appeared to live, a few of them being common, while others
were more sparingly distributed. The examination of the mud also yielded
a number of species of Ostracoda ; the individuals of this group were,
however, not very plentiful, the macrospores of *Isoetes lacustris*—an
aquatic plant allied to the Clubmosses—were very common in the mud.
As Entomostraca were observed to be so abundant in the water of Loch
Coulter, we endeavoured to ascertain whether the fish in the loch were
feeding on them. For this purpose efforts were made, by means of hook
and line, to capture some of the fish, but a few perch only were obtained ;
and though the stomachs of these were carefully examined, no Entomostraca were observed, a few insects being the only objects discernible.
Though trout were noticed swimming about in the water, they appeared
to be very shy, and none were caught ; I am, therefore, unable to say
whether they were feeding on the Entomostraca or not. The following
is a list of the Mollusca and Crustacea observed in and round the sides of
the loch :—

MOLLUSCA.

Sphærium corneum (Linné).   Not common.
Pisidium amnicum.   Rare.
    „   *fontinale* (Draparnaud).   Frequent.
    „   *pusillum* (Gmelin).   Frequent.
    „   *nitidum* (Jenyns).   Common.
Valvata piscinalis (Müller).   Common.
Planorbis albus, Müller.   Rare.
    „   *contortus* (Linné).   Common.
Limnæa peregra (Müller).   Not common.
    „   *truncatula* (Müller).   Not Common.

CRUSTACEA.

DAPHNIADÆ.

Daphnia pulex,	Very Common.	Ephippia also
„ *vetula*,	Frequent.	frequent.

COPEPODA.

Diaptomus castor,	Very common.
Cyclops pulchellus,	Common.
„ *strenuus*,	Frequent.
„ *gigas*,	Very common (several stages).

OSTRACODA.

Cypria ophthalmica (Jurine).   Frequent.
Cypria serena (Koch).   Frequent.
Cyclocypris globosa (G. O. Sars).   Not very common.
Erpetocypris strigata (O. F. Müller).   Not common.
    „   *tumefacta*, Brady and Robertson.   Not common.
Cypridopsis villosa (Jurine).   Not very common.
Candona candida (O. F. Müller).   Frequent.
    „   *rostrata*, Brady and Norman.   Scarce.
    „   *kingsleii*, Brady and Robertson.   Scarce.

Insect larvæ, aquatic Coleoptera, Diatoms, and Confervæ were also observed to be more or less common both in the loch and around its margins.

2. THE COULTER BURN FROM THE LOCH TO CRAIGQUARTER WOOD.

After the loch had been carefully examined we next proceeded to examine this section of the Coulter Burn. Its course is over open moorland, and there are comparatively few places along its banks which form suitable habitats for aquatic organisms. No Mollusca* nor Crustacea were observed in this part of the burn, the only things noticed being insect larvæ—chiefly of the Phryganeidæ—but in the few marshy places and pools along its sides several species of Entomostraca and one or two of Mollusca were obtained. The water of the burn was very pure.

The following is a list of the Mollusca and Crustacea observed :—

_____
* *Sphærium corneum* is, however, very abundant in the covered passages through which the burn runs from the sluice of Loch Coulter for several yards.

MOLLUSCA.

Pisidium pusillum (Jurine). Frequent.
Limnæa truncatula (Müller). Frequent.

CRUSTACEA.

OSTRACODA.

Cypria ophthalmica (Jurine). Frequent.
Cyclocypris globosa (G. O. Sars). Rare.
Erpetocypris tumefacta (Brady and Robertson). Frequent.
Cypridopsis villosa (Jurine). Not common.
Candona candida (O. F. Müller). Frequent.
    ,,   *rostrata*, Brady and Norman. Not very common.
    ,,   *kingsleii*, Brady and Robertson. Not very common.

3. THE COULTER BURN FROM CRAIGQUARTER WOOD TO THE HATCHING HOUSE.

Part of the course of the burn in this section is alongside the Kilsyth Road, so that its banks presented conditions even less favourable to the existence of aquatic organisms than in the previous section. The burn itself was also unproductive of anything noteworthy. The species observed and identified in this section were—

MOLLUSCA.

Pisidium pusillum (Gmelin). Few.
Limnæa peregra (Müller). Few.

CRUSTACEA.

OSTRACODA.

Cypria ophthalmica (Jurine). Few.
    ,,   *serena* (Koch). Not very common.
Cyclocypris globosa (G. O. Sars). Few.
Erpetocypris strigata (O. F. Müller). Not common.
    ,,   *tumefacta* (Brady and Robertson). Not common.
Cypridopsis villosa (Jurine). Not common.
Candona candida (Müller). Frequent.
    ,,   *kingsleii*, Brady and Robertson. Few.

4. THE COULTER BURN FROM THE HATCHING HOUSE TO THE PONDS AT HOWIETOUN.

This section included the examination of the burn from the Hatching House to the ponds at Howietoun and a few of the ponds. In describing our examination of Loch Coulter, reference was made to the immense number of Entomostraca in the water there, and the question of what became of these organisms suggested itself as one of the first points requiring consideration. An explanation, that seemed a fairly reasonable one, was that a great many of them would be carried down the burn with the overflow water, especially when during wet weather a larger quantity of water than usual passed down the burn. At the time we visited the

y

loch, the level of the water, we were informed, was about 4 feet higher than it usually is at that season, and consequently the overflow was greater than usual. In order, therefore, to ascertain whether Entomostraca were being carried down the stream, a tow-net was fixed in midchannel a short distance up from where the water is led off into the rearing ponds at Howietoun; and, further, the net was so arranged that a large proportion of the water would pass through it, and thus give the experiment a fair trial. The net was fixed in position at 7.15 P.M. of the 13th, and removed about 8.15 A.M. of the 14th. Thus the water was allowed to pass through the net for fully 13 hours, but the result was not what had been expected, only a few *Cyclops, Gammarus,* Ostracods, and the larvæ of insects being captured. There was also a small quantity of mud in the net. The *Gammarus,* Ostracods, and insect larvæ were very likely carried into the net from some place in the vicinity of where the net was fixed, while the few *Cyclops* were probably the only organisms carried down from the loch. It was thus fairly evident that though Entomostraca were abundant in Loch Coulter, very few found their way down the stream.

On the stones in the burn a few of the common *Limnæa peregra* were observed, and a somewhat rare Ostracod—*Candona acuminata*—was obtained in a marshy place at the side, otherwise nothing requiring special notice was observed between the Hatching House and the ponds. In the ponds nothing of special interest was observed except that in one of them *Candona acuminata* was again noticed, and *Cypria exculpta* in another, in the 'Botanical pond' *Conferva* was moderately common.

The following are the species of Mollusca and Crustacea observed in this section:—

MOLLUSCA.

Sphærium lacustre (Müller).   Plentiful in one of the ponds.
Pisidium fontinale (Draparnaud).   Not common.
     ,,   *pusillum* (Gmelin).   Not common.
     ,,   *nitidum* (Jenyns).   Not common.
Ancylus fluviatilis (Müller).   Common.
Limnæa peregra (Müller).   Abundant.

CRUSTACEA.

AMPHIPODA.

Gammarus pulex (Linné).   Frequent in the stream.

COPEPODA.

Cyclops pulchellus, Koch.   A few in the net fixed in the stream.
     ,,   *serrulatus,* Fischer.   A few in material collected by handnet in one of the ponds.
     ,,   *crassicornis,* Müller.   A few in the same material with the last.
Canthocamptus minutus (Müller).   Frequent in one of the ponds.

OSTRACODA.

Cypria exculpta (S. Fischer).   Frequent in one of the ponds.
     ,,   *ophthalmica* (Jurine).   Frequent in one of the ponds.

Erpetocypris tumefacta (Brady and Robertson). A few in the burn and in the ponds.

Cypridopsis vidua (O. F. Müller). A few in one of the ponds.

"   *villosa* (Jurine). A few in the burn and in one of the ponds.

Candona candida (Müller). Frequent in the burn and ponds.

"   *lactea*, Baird. In one of the ponds, rather rare.

"   *acuminata* (Fischer). In the burn and one of the ponds, rather rare.

"   *kingsleii*, Brady and Robertson. In marshy ground by the side of the burn, not common.

Aquatic *Acari* and *Coleoptera*, the larvæ of *Coleoptera* and Phryganeidæ, were also observed both in the burn and in the ponds.

5. THE COULTER BURN FROM HOWIETOUN TO GOLDENHOOF.

This section includes the Howietoun Ponds to Goldenhoof Dam as well as the dam itself. Between the ponds and Goldenhoof the burn is locally known by the name of the Stockbridge Burn. Very few organisms were observed in this part of the burn. It flows too rapidly to permit of anything, except perhaps insect larvæ or *Ancylus*, getting a foothold. There were, however, a few marshy places by the side of the burn that yielded a number of *Ostracoda*, among which was *Cypris reticulata*—a species which, though generally distributed, is not very common.

The dam is of comparatively small area ; part of its margin is densely fringed with reeds and other aquatic plants, which afford shelter to numerous organisms. Among these *Mollusca* and various groups of Entomostraca were common, as well as Diatoms, Confervæ, and the larvæ of insects. The water is not very deep, the deepest part being scarcely over 6 feet ; the bottom is formed of fine mud, evidently consisting very much of vegetable débris, for when it was disturbed bubbles of marsh-gas rose to the surface of the water. Trout were moderately common in the dam, and it was ascertained, by the examination of the stomachs of some of them, that they were feeding on insects and *Entomostraca*, especially Ostracoda. The following are the species of Mollusca and Crustacea observed and identified in this section :—

MOLLUSCA.

Pisidium pusillum (Gmelin). Frequent in the dam.

Planorbis spirorbis, Müller. Frequent in the dam.

Limnæa peregra (Müller). Frequent in the dam and burn.

CRUSTACEA.

DAPHNIADÆ.

Eurycercus lamellatus (Müller). Not common in the dam.

Chydorus sphæricus (Müller). Not common in the dam.

Acroperus harpæ, Baird. In the dam not common.

COPEPODA.

Cyclops serrulatus, Fischer. Frequent in the dam.

Canthocamptus minutus (Müller). Frequent in the dam.

OSTRACODA.

Cypria exsculpta (Fischer)   In the dam moderately common.
    ,,   *ophthalmica* (Jurine).   In the dam frequent.
    ,,   *serena* (Koch).   Marshy ground by the side of the burn, and
        in the dam frequent.
Cyclocypris globosa (G. O. Sars).   In the dam not common.
Cypris reticulata, (Zaddach).   Marshy ground by the side of the
        burn, rare.
Erpetocypris reptans (Baird).   In the dam not common.
    ,,   *strigata* (Müller).   Marshy ground by the side of the
            burn, rare.
    ,,   *tumefacta* (Brady and Robertson).   In the dam not
            common.
Cypridopsis villosa (Jurine).   In the dam frequent.
    ,,   *vidua* (Müller).   In the dam frequent.
Candona candida (Müller).   By the side of the burn and in the
        dam frequent.
    ,,   *lactea*, Baird.   In the dam not common.
    ,,   *rostrata*, Brady and Norman.   In the dam not common.
    ,,   *acuminata* (Fischer).   In the dam not common.
    ,,   *kingsleii*, Brady and Robertson.   In the dam not
            common.

6 & 7. THE COULTER BURN FROM GOLDENHOOF TO THE BANNOCK AND THENCE TO THE FORTH.

The Coulter Burn after leaving Goldenhoof Dam runs eastward for 600 to 700 yards, then northward for a few hundred yards more, and joins the Bannock not very far from the site of the famous battle of Bannockburn. Along the part of its course the water flows with considerable rapidity except in a few places where quiet pools are formed. The freshwater limpet *Ancylus fluviatilis*, and the larvae of Caddis flies, were observed under or attached to the stones in the stream; while in the quiet pools, as well as in the marshy places along its banks, a number of Entomostraca and a few Mollusca were obtained. The Bannock Burn, from where it is joined by the Coulter down to the village of Bannockburn, also yielded a number of Mollusca and Entomostraca, but from that village to the Forth, very few such organisms were observed, probably owing to the water being more or less contaminated by the refuse from the public works on its banks.   Between the village of Bannockburn and the Forth the Bannock flows between steep banks through a comparatively level tract of country and its course is very tortuous.   The distance as the crow flies from where the Bannock is joined by the Coulter Burn to its union with the Forth is scarcely 3½ miles, while the course of the burn measures fully 6 miles.   Though the water appeared to be contaminated, both trout and sticklebacks were observed to be moderately frequent; the impurity of the water did not seem to have so much effect on these as on the Entomostraca.

The following is the list of the Mollusca and Crustacea observed and identified with Sections VI. and VII :—

MOLLUSCA.

Pisidium pusillum (Gmelin).   Moderately common in Section VI.
        and upper part of Section VII.

Pisidium nitidum, Jenyns.   Moderately common in upper part of
Section VII.
   „   *roseum,* Sholtz.   Rare, lower part of Section VII.
Planorbis contortus (Linné).   Frequent in upper part of Section
   VII.
Ancylus fluviatilis (Müller).   Frequent in Section VI.
Limnæa peregra (Müller).   Not very common in both sections.
   „   *truncatula* (Müller).   Not very common in both sections.

CRUSTACEA.

DAPHNIADÆ.

Chydorus sphæricus Müller.   Not very common, Section VI.

COPEPODA.

Cyclops serrulatus, Fischer.   Frequent in Section VI.
Canthocamptus minutus (Müller).   Frequent in Section VI.

OSTRACODA.

Cypria ophthalmica (Jurine).   Frequent in Section VI. and upper
   part of Section VII.
   „   *serena* (Koch).   Frequent in upper part of Section VII.
Cyclocypris globosa (G. O. Sars).   Not common in upper part of
   Section VII.
Erpetocypris reptans (Baird).   Not common in upper part of
   Section VII.
   „   *strigata* (Müller).   Rare in Section VI.
Cypridopsis villosa (Jurine).   Not common in Section VI.
   „   *vidua* (Müller).   Not common in Section VI.
Candona candida (Müller).   Frequent in both sections.
   „   *rostrata,* Brady and Norman.   Rare in Section VI.
   „   *kingsleii,* Brady and Robertson.   Rare in Section VII.
   „   *fabæformis* (Fischer).   Rare in Section VI.
   „   *acuminata* (Fischer).   Rare in Section VI.
Ilyocypris gibba (Ramdohr).   Not common in lower part of Section
   VII.

As considerable changes have recently been made in the terminology of
the fresh-water Ostracoda, I propose now to give a list of the species
observed throughout the district examined, adding to each a synonymy
sufficiently full to allow of the older works on this group of Crustacea
being referred to with greater facility, together with notes on the distribution of the rarer species.

OSTRACODA.

Cypria exsculpta (S. Fischer).

1854. *Cypris exsculpta,* Fischer, Beitrag zur Kenntniss der Ostrac., p. 18,
   pl. xix. figs. 36–38.
1868. *Cypris striolata,* Brady, Mon. rec. Brit. Ostrac, p. 372, pl. xxiv.
   figs. 6–10.
1880. *Cypris granulosa,* Robertson, Fresh and Brackish water Ostrac.
   of Clydesdale, p. 18, (jun.)
1889. *Cypria exsculpta,* Brady and Norman, Mon. M. and Fw. Ostrac. of
   the N. Atlantic and N. W. Europe, p. 68, pl. xi. figs. 1–4.

This is a widely distributed species, though not previously recorded for Stirlingshire.

Cypria ophthalmica (Jurine).

1820. *Monoculus ophthalmicus*, Jurine, Hist. des Monocles, p. 178, pl. xix. figs. 16–17

1835. *Cypris compressa*, Baird, Trans. Berw. Nat. Club, vol. i. p. 100, pl. iii. fig. 16.

1868. *Cypris compressa*, Brady, *op. cit.*, p. 372, pl. xxiv. figs. 1–5; pl. xxxvi. fig. 6.

1889. *Cypria ophthalmica*, Brady and Norman, *op cit.*, p. 69, pl. xi. figs. 5–9.

One of the commonest of the British species.

Cypria serena (Koch).

1838. *Cypris serena*, Koch, Deutschlands Crustaceen, H. xxi., 22.

1868. *Cypris lævis*, Brady, *op. cit.*, p. 374, pl. xxiv. figs. 6–8.

1889. *Cypria serena*, Brady and Norman, *op cit.*, p. 70.

A common species in Britain.

Cyclocypris globosa (G. O. Sars).

1863. *Cypris globosa*, G. O. Sars, Om en i Sommeren 1862 foretagen Zoologisk Reise i Christianias og Trondhjems Stifter, p. 27.

1868. *Cypris cinerea*, Brady, *op. cit.*, p. 374, pl. xxiv. figs. 39–42; pl. xxxvi. fig. 7.

1889. *Cyclocypris globosa*, Brady and Norman, *op. cit.*, p. 71, pl. xiv. figs. 1–2; pl. xi. figs. 10–18.

The distribution of this species seems to be more restricted. In Scotland it has been observed in the islands of Lewis and Bute, at West Loch Tarbert (Loch Fyne); in Kirkcudbrightshire; Loch Fitty, Loch Dow, and Black Loch, in Fife. Not previously recorded for Stirlingshire.

Cypris reticulata, Zaddach.

1844. *Cypris reticulata*, Zaddach, Synops. Crust. Pruss. Prodr., p. 24 (jun.)

1868. *Cypris tessellata* (in part), Brady, *op. cit.*, p. 336, pl. xxiii. figs. 39–45.

1883. *Cypris affinis*, Lilljeborg, International Fisheries Exhib. London. Sweden Cat., p. 146.

1889. *Cypris reticulata*, Brady and Norman, *op. cit.*, p. 76, pl. viii., figs. 1–2; pl. xi. figs. 5–7.

This does not seem to be a commonly distributed species. The following are the Scotch localities where it has been observed :—Johnston Loch; Possil Marsh; Bishop Loch; side of Paisley Canal; side of Loch Ascog, Bute; Hairmyres, near East Kilbride; Mill Loch, Lochmaben; and Barron Loch, Peebles.

Erpetocypris reptans (Baird).

1850. *Candona reptans* and *similis*, Baird, Brit. Entom., pp. 162, 167.

1868. *Cypris reptans*, Brady, *op. cit.*, p. 370, pl. xxv. figs. 10–14; pl. xxxvi. fig. 4.

1889. *Erpetocypris reptans*, Brady and Norman, *op. cit.*, p. 84 pl xiii. fig. 27.

A common British species.

Erpetocypris strigata (O. F. Müller).

1785. *Cypris strigata,* O. F. Müller, Entomostraca, p. 54, pl. iv. figs. 4–6.
1844. *Cypris jurinii,* Zaddach, Synops. Crust. Pruss. Prodr., p. 36.
1870. *Cypris ornata,* Brady (non Müller), Nat. Hist. Trans. Northumb. and Durham, vol. iii. p. 364, pl. xiv. figs. 1–3.
1889. *Erpetocypris strigata,* Brady and Norman, *op. cit.,* p. 85, pl. viii. figs. 14, 15.

This is not so commonly distributed as the last. The following are some Scotch localities :—Duddingston Loch ; Ponds near Taymouth Castle ; Isle of Cumbrae ; Hayston Dam, Peebles. Not previously recorded for Stirlingshire.

Erpetocypris tumefacta (Brady and Robertson).

1870. *Cypris tumefacta,* Brady and Robertson, Ostracoda and Foraminifera of Tidal Rivers, Ann. Nat. Hist., ser. iv., vol. vi. p. 13, pl. iv. figs. 4–6.
1889. *Erpetocypris tumefacta,* Brady and Norman, *op. cit.,* p. 87, pl. viii. figs. 5–7 ; pl. xiii. fig. 18.

This seems to be one of the less common species. It has not been previously recorded for Stirlingshire.

Cypridopsis vidua (Müller).

1785. *Cypris vidua* (Müller), Entomostraca, p. 55.
1850. *Cypris sella,* Baird, British Entom., p.
1868. *Cypridopsis vidua,* Brady, *op. cit.,* p. 375, pl. xxiv. figs. 27–36, 46.
1869. *Cypridopsis obesa,* Brady and Robertson, Ann. Nat. Hist., ser. iv., vol. iii. p. 364, pl. xviii. figs. 5–7.
1889. *Cypridopsis vidua,* Brady and Norman, *op. cit.,* p. 89.

This is a widely distributed species.

Cypridopsis villosa (Jurine).

1820. *Monoculus villosa,* Jurine, Hist. des Monocles, p. 178.
1850. *Cypris westwoodii* and *elongata,* Baird, Brit. Entom., p. 156.
1868. *Cypridopsis villosa,* Brady, *op. cit.,* p. 377, pl. xxiv. figs. 11–15 ; pl. xxxvi. fig. 9.
1889. *Cypridopsis villosa,* Brady and Norman, *op. cit.,* p. 90.

This is moderately common in Scotland.

Candona candida (Müller).

1785. *Cypris candida,* Müller, Entom., p. 62, tab. vi. figs. 7–9.
1850. *Candona lucens,* Baird, Brit. Entom., p. 160, tab. xix. fig. 1.
1889. *Candona candida,* Brady and Norman, *op. cit.,* p. 98, pl. x. figs. 1, 2, 14–23.

Common everywhere—very variable.

Candona lactea, Baird.

Candona lactea, Baird, Proc. Zool. Soc. Lond., p. 255, pl. xviii. figs. 25–27.
1868. *Candona detecta,* Brady (var.), *op. cit.,* p. 384, pl. xxiv. figs. 35–38 ; pl. xxxvii. fig. 2.
1889. *Candona lactea,* Brady and Norman, *op. cit.,* p. 100.

Generally distributed, but not so common as the last.

Candona rostrata, Brady and Norman.

1857. *Cypris compressa*, Fischer, Ueber das genus Cypris, p. 144, pl. ii. figs. 7–12 ; pl. iii. figs. 1–5.
1889. *Candona rostrata*, Brady and Norman, *op. cit.*, p. 101, pl. ix. figs. 11, 12, 12 *a* and *b*; pl. xii. figs. 22–31.

This is one of the less common *Candona*, though it may be more widely distributed than we know of at present. It has been observed in Duddingston Loch ; in Loch Fitty and Lurg Loch, in Fife ; Loch Fad, in Bute, &c.; but it does not appear to have been recorded before for Stirling-shire.

Candona kingsleii, Brady and Robertson.

1870. *Candona kingsleii*, Brady and Robertson, Ann. and Mag. Nat. Hist., ser. ix., vol. vi. p. 17, pl. ix. figs. 9–12.
1889. *Candona kingsleii*, Brady and Norman, *op. cit.*, p. 102, pl. ix. figs. 19–22 ; pl. xiiii. fig. 19.

A widely distributed species, but not previously recorded for Stirling-shire.

Candona fabæformis (Fischer).

1851. *Cypris fabæformis*, Fischer, Ueber das Genus Cypris, p. 146, pl. iii. figs. 6–16.
1870. *Candona diaphana*, Brady and Robertson, Ann. and Mag. Nat. Hist., ser. iv., vol. vi. pl. v. figs. 1–3.
1889. *Candona fabæformis*, Brady and Norman, *op. cit.*, p. 103, pl. ix. figs. 1–4.

This species is somewhat restricted in its distribution, and does not appear to have been previously recorded for Stirlingshire. It is found at Corstorphine and Luffness Links.

Candona acuminata (Fischer).

1857. *Cypris acuminata*, Fischer, Ueber das Genus Cypris, p. 148, pl. iv. figs. 12–16.
1889. *Candona acuminata*, Brady and Norman, *op. cit.*, p. 104, pl. ix. figs. 9–10 ; pl. x. figs. 5, 6.

This is considered to be a rare species in Britain, though it is probable that its distribution is more extensive than is known at present. It has not been previously recorded for Stirlingshire.

Ilyocypris gibba (Ramdohr).

(?) *Cypris gibba*, Ramdohr, Mag. und Geselesch. Naturforsch. Freunde zu Berlin, ii. p. 91.
1868. *Cypris gibba*, Brady, *op. cit.*, p. 369, pl. xxiv. figs, 47–54 ; pl. xxxvi. fig. 2.
1889. *Ilyocypris gibba*, Brady and Norman, *op. cit.*, p. 107, pl. xxi. figs. 1–5.

A common British species.

TABLE I.—MOLLUSCA.

List of the Mollusca obtained, showing their Distribution in the District, in June 1889, as well as their Distribution in Britain.

× signifies 'rare,' × × 'frequent,' and × × × 'common.'

Name of Species.	Distribution in the District examined.							Distribution in Britain.		
	Section I. Loch Coulter.	Section II. Coulter Burn.	Section III. Coulter Burn.	Section IV. Coulter Burn and Ponds.	Section V. Coulter Burn and Coltenhave Dam.	Section VI. Coulter Burn.	Section VII. Bannock Burn.	England.	Scotland.	Ireland.
Sphaerium corneum,	×							× × ×	× × ×	× × ×
„ lacustre,								× × ×	× × ×	× ×
Pisidium amnicum,	× ×	× ×						× × ×	× × ×	× × ×
„ fontinale,	× ×		×	× × ×				× × ×	× × ×	× × ×
„ pusillum,	× × ×			× × ×				× × ×	× × ×	× × ×
„ nitidum,	× × ×							× × ×	× ×	× × ×
„ roseum?,	× ×							× × ×	×	×
Valvata piscinalis,	× × ×				× ×	× ×	× ×	× × ×	× × ×	× × ×
Planorbis albus,	× ×							× × ×	× × ×	× × ×
„ spirorbis,	× × ×							× × ×	× × ×	× × ×
„ contortus,	× ×	× ×	×	×	× ×	× ×	× ×	× × ×	× × ×	× × ×
Limnaea peregra,	×							× × ×	× × ×	× × ×
„ truncatula,						× ×	× ×	× × ×	× × ×	× × ×
Ancylus fluviatilis,								× × ×	× × ×	× × ×

TABLE II.—DAPHNIADÆ AND COPEPODA.

List of Daphniadæ and Copepoda obtained, showing their Distribution in the District, in June 1889, and also their Distribution in Britain.

× signifies 'rare,' × × 'frequent,' and × × × 'common.'

Name of Species.	Distribution in the District examined.							Distribution in Britain.		
	Section I. Loch Coulter.	Section II. Coulter Burn.	Section III. Coulter Burn.	Section IV. Coulter Burn and Ponds.	Section V. Coulter Burn and Dam.	Section VI. Coulter Burn.	Section VII. Bannock Burn.	England.	Scotland.	Ireland.
DAPHNIADÆ										
Daphnia pulex,	× × ×							× ×	× ×	×
" vitula,	× ×							× ×	× ×	× ·-·-·-·
Acroperus harpæ,								× ×	× ×	×
Chydorus sphæricus,					× × ×	×		× × ×	× × ×	
Eurycercus lamellatus,								×	× × ×	
COPEPODA.										
Diaptomus castor,	× × ×			×				× × ×	× × ×	× ×
Cyclops gigas,	× × ×				×	× ×		× × ×	× × ×	× ×
" serrulatus,	× ×			×				× ×	×	×
" strenuus,	× × ×			× ×	×			×	× ×	
" pulchellus,				× ×				×	×	× ×
" coronicornis,				× ×				× ×	× ×	
Canthocamptus minutus,				× ×	× ×	×		×	×	×

TABLE III.—OSTRACODA.

List of the Ostracoda obtained, showing their Distribution in the District, in June 1889, and also their Distribution in Britain.

× signifies 'rare,' × × 'frequent,' and × × × 'common.'

Name of Species.	Distribution in the District examined.							Distribution in Britain.		
	Section I. Loch Coulter.	Section II. Coulter Burn.	Section III. Coulter Burn.	Section IV. Coulter Burn and Ponds.	Section V. Coulter Burn and Dam.	Section VI. Coulter Burn.	Section VII. Bannock Burn.	England.	Scotland.	Ireland.
Cypria exsculpta,	× ×		× ×	× ×	× × ×	×	× ×	× × ×	× × ×	× × ×
" ophthalmica,	× ×	×	× ×	× ×	× ×	×	× ×	× × ×	× × ×	× × ×
" serena,	×	×			× ×			× ×	×	× ×
Cyclocypris globosa,					×			×	× × ×	
Cypris reticulata,	× ×	×	× ×	× × ×	× × ×	×	×	× × ×	× × ×	× × ×
Erpetocypris reptans,	×	×		× ×	× ×	× × ×	× × ×	× × ×	× × ×	× × ×
" strigata,					× ×			× × ×	× × ×	
" tumefacta	× ×	×	× ×	× ×	× ×	× ×	×	× × ×	× × ×	× × ×
Cypridopsis vidua,	×	×	×	× ×	× ×	×	×	× × ×	× × ×	× × ×
" villosa,								×	×	
Candona candida,	×	× ×	×	×	× × ×	× ×		× × ×	× × ×	× × ×
" lactea,	× ×	× ×		×	×	×		× × ×	× ×	× ×
" rostrata,						× ×		× × ×	× × ×	
" kingslii,								×	×	×
" fabaeformis,	× ×	× ×	× ×	×	× × ×	× ×	×	× × ×	× × ×	
" acuminata,								× × ×	× × ×	
Ilyocypris gibba,	× ×		× ×	×	×	×	×	×	×	× ×

.

FW

1X APR

III.—THE INVERTEBRATE FAUNA OF THE INLAND WATERS OF SCOTLAND. Part II. By Thomas Scott, F.L.S. (Plates V., VI.)

In this, the second report on the Invertebrate Fauna of Scottish Inland Waters, reference is made to the following lochs:—

Loch Leven, Kinross-shire.	Loch Ness, ⎫ the Caledonian
Raith Lake, near Kirk- ⎫	Loch Oich, ⎬ Canal Lochs,
caldy ⎪ Fife-	Loch Lochy, ⎭ Inverness-shire.
Camilla Loch, ⎬ shire.	Loch Balnagowan, ⎫ Lismore
Loch Gelly, ⎭	Loch Kilcheran, ⎬ Island,
Loch Strathbeg, Aberdeenshire.	Loch Fiart, ⎭ Argyleshire.
Loch Achnacloich, near Inver-	Loch Hempriggs, ⎫ near Wick,
gordon, Ross-shire.	Loch Wester, ⎬ Caithness.
	Loch Harray, ⎫ Orkney.
	Loch Stenness, ⎬

Only a short visit, however, was made to all, except Loch Leven, and even the examination of Loch Leven cannot be considered as being in any way exhaustive; to arrive at a fairly accurate knowledge of its rich invertebrate fauna, several visits would require to be made. The results of our examination of these lochs has nevertheless proved highly interesting.

I propose in the following report to describe my examination of the lochs named in the order in which they are arranged. In preparing this report numerous works have been consulted, some of the more important of which are as follows:—

1819-20. H. E. Strauss, Mem. sur les Daphnia (Ann. du Mus. d'hist. Nat., vols. v., vi.).
1828. Fleming, British Animals.
1844. Zaddach, Synop. Crust. Pruss. Prod.
1850. Baird, Brit. Entom.
1851. Zenker, Phys. Bem. Kun. über die Daphnoid (Mull. Achiv. für Phys).
1861. Pritchard, Infusoria.
1865. G. O. Sars, Norges. Ferskvand. Branch. Clad. Cten.
1866. Schœdler, Die Clad. des Frisc. Halfs (Archiv. für Nat., vol. xxxii.).
1867. Norman and Brady, Mon. Brit. Entom. (Nat. Hist. Trans. of Northumb.).
1868. Brady, Mon. recent Brit. Ostrac. (Trans. Lin. Soc., vol. xxvi.).
1871. Stavely, British Insects.
1875. Claus, Die Schaleudruse der Daphn. (Zeit. für Wiss. Zool., vol. xxv.).
1878. Brady, Monograph of the British Copepoda.
1884. Herrick, Crustacea of Minnesota, U.S.A.
1889. Brady and Norman, Mon. Marine and Fresh-Water Ostrac. of the N. Atlantic and N.W. Europe.

To my ever kind friend, Professor G. S. Brady, I am greatly indebted for much valuable help. I have also to acknowledge the kind encouragement accorded to me in many ways by Dr T. Wemyss Fulton and Mr W. Anderson Smith while carrying out the investigation of the inland waters of Scotland. Professor W. C. M'Intosh has favoured me with the names of some of the Loch Leven Annelides. J. Rae, M.D., and Mr R. Kidston, F.R.S.E., F.G.S., have kindly undertaken the examination of the Loch Leven Diatomacea, and have prepared a list of species. I have also to acknowledge the active and kindly co-operation of Captain R. E. Simpson, master of the 'Garland,' both in connection with the investigations here referred to and in our regular marine fishery investigations. My son, Mr Andrew Scott, has prepared the drawings which accompany this paper.

1. Loch Leven, Kinross-shire.

Loch Leven, which belongs to Sir G. Graham-Montgomery, Bart., has been long famous for the peculiar delicacy in the colour and flavour of its trout. In the Old Statistical Account of Scotland, they are referred to as follows :—" The high flavour and bright red colour of the trout seem evidently to arise from the food which Nature has provided for them in the loch. . . . . What appears to contribute most to the redness and rich taste of the Loch Leven trout is the vast quantity of a small shell-fish, red in its colour, which abounds all over the bottom of the loch, especially among the aquatic weeds; the trout when caught have often their stomachs full of them." *

Reference is also made to the Loch Leven trout in the New Statistical Account.†

Though Loch Leven has thus long been noted for the superiority of its trout, it is only within the last thirty-five years or so that it has become such a famous resort for anglers. Previous to 1856 fishing with rod and line appears to have been so disappointing in its results that few anglers cared to give the loch a second trial. From some cause that does not appear to have been satisfactorily explained the fish were observed about or shortly after the time stated to rise to the bait more freely than in previous years, and consequently angling became more successful and encouraging, and the result was that Loch Leven ere long formed a rendezvous for anglers from all parts of the country.‡

The management of the Loch Leven Fishery was for a long time in the hands of a tacksman, but some years ago it was undertaken by a limited liability company, called the Loch Leven Angling Association, who pay a rental of £1000 a year.

As an example of the large number of fish taken during the season it may be stated that in 1888 over 23,500 trout, weighing 21,000 lbs.—being on an average nearly 1 lb. each—were captured. Perch also abound in the loch, and are frequently fished ; pike are not uncommon, but as they are destructive to the other fish, their number is being rapidly reduced. Charr (*Salmo alpinus*, Linné) used also to be frequent, but I can find no record of any of them being caught in recent years.

The following extract from the Old Statistical Account of Scotland may be of interest as showing the present greatly increased value of the Loch Leven Fishery, compared with what it was last century : 'The fish of Loch Leven only a few years ago sold here at 1d. each, great and small, for the trout, and the perch at a 1d. per dozen, and about 25 years ago at half that price ; the fishing was then let at 200 merks Scotch. The trout are now raised to 4d. per lb., the perch to 2d. per dozen, and the pike for 2d. per lb.; the present rent of the fishing is £80 sterling, and for next year it is fixed at £100.§ In 1845 the rent had been increased to £204 per annum and " 2 boats and 4 boatmen were employed during part of the fishing season."‖ There are now (1891) 22 boats on the loch for the use of anglers, and the rent, as stated above, is £1000 per annum.

During the first half of the present century extensive operations were carried out for the partial draining of the loch ; these were completed about 1845 at a cost of £40,000. By these operations the level of the loch was lowered 4½ feet and its area, which previously extended to 4,638

* Vol. vi. pp. 166, 168 (1793).
† Vol. ix. (Kinross), p. 7 (1845).
‡ " Sportsman's Guide," September 1890, p. 235.
§ Old Stat. Acc. of Scot., vol. vi. (Kinross), pp. 166–168 (1793).
‖ New Stat. Acc. of Scot., vol. ix. (Kinross), p. 6 (1845).

imperial acres was reduced to 3,238 acres.* Its present area is now somewhat greater, being about 3,406 acres.†

As so much interest centres round this famous loch, it was considered desirable that its invertebrate fauna should be carefully investigated, in order to ascertain if the marked superiority of the trout was in any way owing to a difference in the kind of organisms that form their food supply.

I made an examination of the loch during the month of June last (1890) by means of dredge and tow-net, worked from a rowing-boat, kindly placed at my service by Mr Hall, the manager of the Fishery, and also by hand-net from the shore. Unfortunately, the weather was rather stormy at the time, and interfered somewhat with the satisfactory carrying out of the investigations ; nevertheless, the results of the examination of the loch were of considerable interest. The fauna was found to be abundant and varied—Mollusca, Arthropoda, Annelida, and Protozoa being more or less common all over the loch. The present report deals chiefly with the first two classes named, because they are in several respects the more important of the invertebrata ; the others, while they also are referred to here, may be more fully worked out in a future report.

I propose to refer to the various classes of organisms in the order in which they are arranged above.

Mollusca.

Mollusca were common and generally distributed, except at that part of the loch called the 'Shallows,' the bottom of which consists of little else than fine sand, and is therefore not so suitable as a habitat for these organisms as where the bottom consists of mud or vegetable debris. Fourteen species of Mollusca were obtained, comprising five of Lamellibranchs, and nine of Gasteropods. The more common forms were *Sphærium corneum*, *Valvata piscinalis*, and *Planorbis contortus*. The swan mussel (*Anodonta cygnæa*) appeared also to be frequent,—that is, if the number of dead shells observed at several places along the shore can be relied on as evidence of the presence of this shell-fish. I was only able, however, to obtain *living*, one adult, and a few young specimens ; the latter were found burrowing in the sand in the shallow water at the north-east side of the loch. The following is a list of the species of Mollusca obtained :—

MOLLUSCA.

1. LAMELLIBRACHIATA.

Sphærium corneum (Linné). Generally distributed, common.
Pisidium fontinale (Draparnaud). Generally distributed, common.
,,   *pusillum* (Gmelin). Generally distributed, frequent.
,,   *nitidum* (Jenyns). Generally distributed, frequent.
Anodonta cygnæa (Linné). [Frequent (?), 1 adult and several young living.

2. GASTROPODA.

Valvata piscinalis (Müller). Generally distributed, common.
,,   *cristata*, Müller. Generally distributed, but scarce.
Planorbis albus, Müller. Generally distributed, rather scarce.
,,   *nautileus* (Linné). West end of the loch, but not common.
,,   *contortus* (Linné). Generally distributed, common.

---

* I understand the mill-owners have still the power of diminishing the lake by 2½ feet, a total of 7 feet lower than it was in the beginning of the century.
† Fifth Annual Report, Fishery Board for Scotland, p. 367, 1887.

Limnæa peregra (Müller).   Generally distributed, but not very common.
   ,,   *palustris* (Müller).   South from the landing-stage, not common.
   ,,   *truncatula* (Müller).   North from the landing-stage, not common.
Physa fontinalis (Linné).   North-west end of the loch, rather rare.

ARTHROPODA.

CRUSTACEA.

The Crustacea were by far the most numerous and varied of the inverte-brate fauna of the loch.   Cladocera and Copepoda occurred in great pro-fusion all over and through the water.   Daphniæ were most abundant.
Cyclops, especially *C. strenuus*, were also plentiful ; Ostracoda were not so common in the loch itself as they were around its margin, particularly those parts that were more or less overgrown with vegetation, as round the north-east shore.   This part of the shore from near the mouth of the Powburn to old Leven mouth is very much overgrown with reeds and other aquatic plants ; in some places the vegetation is so tall and dense that it is with some difficulty one can force his way through it.   I saw here some fine trout hiding among the roots of the tall reeds as I waded through among them where they thinned outwards from the shore.
Seventeen species of Ostracoda were obtained along this part of the shore ; twelve species were obtained from the south shore, and only eight from the loch itself.   Among them were *Cytheridea lacustris*, a rather uncommon species in the eastern counties of Scotland, and one, with the somewhat curious name *Limnicythere sancti-patrici*, was also moderately common.
Among the Cladocera the rare and interesting *Leptodora hyalina* occurred in considerable numbers ; *Monospilus tenuirostris*—the 'rarest of all the Entomostraca'—was also frequent in the material collected at one or two places.   The following are the Crustacea obtained :—

AMPHIPODA.

Gammarus pulex (Linné).   Common all over the loch.

COPEPODA.

Diaptomus gracilis (G. O. Sars).   Common throughout the loch.
Cyclops crassicornis (Müller).   Frequent round the margin.
   ,,   *strenuus* (Fischer).   Frequent, especially round the margin.
   ,,   *serrulatus* (Fischer).   Also frequent with the last.
Canthocamptus minutus (Müller).   Common, especially round the margin.
Atheyella spinosa (Brady).   Frequent in the same localities as the last.

OSTRACODA.

Cypria serena (Koch).   Common, generally distributed.
   ,,   *ophthalmica* (Jurine).   Common with the last.
   ,,   *exsculpta* (S. Fischer).   West end, not very common
Erpetocypris tumefacta (Brady and Robertson).   North and south shores, not common.
   ,,   *reptans* (Baird).   Generally distributed, frequent.
   ,,   *strigata* (O. F. Müller).   Deep water, rare.
Cypridopsis vidua (Müller).   North and south shores, frequent.
   ,,   *villosa* (Jurine).   Not very common, west end.
Potamocypris fulva, Brady.   North and south shores, frequent.

Candona candida (Müller). Generally distributed, frequent.
,, *lactea*, Baird. Generally distributed, frequent.
,, *kingsleii*, Brady and Robertson. North shore, not common.
,, *pubescens* (Koch). Frequent and generally distributed.
Ilyocypris gibba (Müller). North side frequent ; south side rare.
Cytheridea lacustris (G. O. Sars). Frequent all over the bottom of the loch.
Limnicythere sancti-patricii, Brady and Robertson. Generally distributed, common.
,, *inopinata*, Baird. Generally distributed, not common.

CLADOCERA.

Daphnella brachyura (Lievin). Generally distributed, not common.
Daphnia pulex (Müller). Generally distributed, frequent.
,, *longispina*, Müller. Generally distributed, abundant.
Bosmina longirostris, Muller. In surface tow-nettings, rare.
Ilyocryptus sordidus (Lievin). South shore, frequent.
Eurycercus lamellatus (Müller). Generally distributed, not common.
Acroperus harpæ, Baird. North and south shores, frequent.
Alonopsis elongata, G. O. Sars. South shore, not common.
Alona quadrangularis, Müller. Common, north and south shores.
Alynella exigua, Lilljeborg. Frequent, north and south shores.
,, *nana*, Baird. South shore, scarce.
Pleuroxus trigonellus, Müller. Common in the loch.
,, *uncinatus*, Baird. In the middle of the loch, scarce.
Chydorus sphæricus, Müller. Generally distributed, common.
Monospilus tenuirostris, Fischer. South shore, scarce.
Polyphemus pediculus (Linné). Middle of the loch, scarce.
Bythotrephes longimanus, Leydig. In the middle of the loch, frequent.
Leptodora hyalina, Lilljeborg. In the middle of the loch, frequent.

INSECTA.

The larvæ of insects were abundant in the loch, especially the larvæ of the Iphemeridæ. The Libellulidæ and Phryganidæ were also represented in the larval stage more or less frequently. Some idea may be formed of the myriads of these organisms present in the loch when it is stated that a conspicuous ridge composed of cast-off skins of insect larvæ which had been washed ashore during the preceding stormy weather extended along the margin of the loch for a considerable distance. The curious so-called 'water-bears' (Tardigrada), now included in the class *Arachnida*, were common among the decaying vegetable matter at the bottom. Species of Notonectidæ or 'water-bugs,' and of aquatic Coleoptera were also more or less common, though their distribution seemed to be more localised.

VERMES.

This division of invertebrates was represented by several species—parasitic and non-parasitic. Among the former were *Schistocephalus solidus*, Crepl., obtained by my son from the body-cavity of a Stickleback (*Gasterosteus aculeatus*), and a species of tape-worm (*Bothriocephalus latus?*) several of which were found in the alimentary canal of the trout, six specimens being taken from one fish. The heads of the parasites were fixed at the extreme end of the cœca or blind tubes of the stomach, and their bodies were so elongated as to extend well down into the intestine. Usually one parasite occupied a cœcum. *Tubifex rivulorum* was very common in the loch. The following species of Annelidæ were also obtained :—

Clepsine 6—oculata.
Clepsine sp., a very small form.
Nephelis octoculata (*?N. reticulata*, Malm.).

s

PROTOZOA.

Rhizopoda were common all over the loch. Several forms were obtained in the dredged and hand-netted material ; the following are the more typical varieties observed :—

Difflugia pyriformis. This form varies from pear to balloon shape, but the pear shape is usually the more common.

Difflugia globularis.—The test is more or less globular, $\frac{1}{9}$ to $\frac{1}{3}$ being truncate and forming the aperture ; this form was much less frequent than the last.

Difflugia corona.—Test crown shaped, furnished posteriorly with few or more *cornua*. This appeared to be a scarce form.

Difflugia marsupiformis.—Test from the side view varies a good deal, but the most common form was that of a slightly eccentric and depressed hemisphere. The aperture is not central, but is situated more to one side, so that its outer margin is usually close to the edge of the test, while the inner margin is some distance from the opposite edge. This form of *Difflugia* was of frequent occurrence.

DIATOMACEA.

Diatomacea were abundant, especially in the deeper parts of the loch, and included a considerable number of species. J. Rae, M.D., and Mr Robert Kidston, F.R.S.E., F.G.S., have kindly furnished me with the following list of species :—

Amphora ovalis, Kütz.
Asterionella formosa, Hassal.
Cymbella cymbiformis, Ehr.
  ,,   *anglica,* Lagenstedt.
  ,,   *cistula,* Hempr.
Cyclotella operculata, var. *mesoleia,* Grun.
Ceratoneis arcus, Kütz.
Campylodiscus costatus, W. Sm.
Cocconeis scutellum, Ehr.
  ,,   *pediculus,* Ehr.
Cymatopleura solea (Breb.), W. Sm.
Diatoma vulgare, Borq.
  ,,   *tenue,* Kütz.
Epithemia turgida (Ehr.), Kütz.
  ,,   *zorex,* Kütz.
Encyonema cæspitosum, Kütz.
Fragilaria virescens, var. (?) *exiqua,* Grun.
  ,,   *construens,* var. *binodis,* Grun.
  ,,   *mutabilis,* (W. Sm.), Grun.
Gomphonema (*acuminata* var.), *Brebissonii,* Kütz.
Gomphonema intricatum, Kütz.
  ,,   *olivaceum,* Ehr.
Meridion circulare, Ag.

Melosira orichalcea, W. Sm.
  ,,   *varians,* Ag.
Navicula Smithii, Breb.
  ,,   *radiosa,* Kütz.
  ,,   ,, *acuta,* Kütz.
  ,,   *gastrum* (Ehr.), Donkin.
  ,,   *viridula,* Kütz.
  ,,   *mesolepta,* Ehr.
  ,,   *limosa,* Kütz.
  ,,   *gibba,* Ehr.
  ,,   *pusilla,* Sm.
  ,,   *Reinhardti,* Grun.
  ,,   sp.
Nitzschia sigmoidea (Ehr.), W. Sm.
  ,,   *thermalis* (Kütz), Grun.
Pleurosigma acuminatum (Kütz), Grun.
  ,,   *Spenserii,* W. Sm.
Stauroneis anceps, Ehr.
Surirella linearis, W. Sm.
  ,,   *biseriata,* Breb.
  ,,   *ovalis,* Breb.
Tabellaria flocculosa, (Roth.), Kütz.
Tryblionella angustata, W. Sm.
  ,,   ,, var. *curta,* Grun.

2. RAITH LAKE, CAMILLA LOCH, AND LOCHGELLY LOCH IN THE COUNTY OF FIFE.

These lochs were visited during the month of August, and were examined by hand-net from the shore.

RAITH LAKE.

Raith Lake, in the parish of Abbotshall, is situated within the pleasure grounds of the Raith estate, the property of Munro-Ferguson, Esq. M.P. and is therefore private. Permission to examine the loch was readily granted to me by Mr Prentiss, factor on the estate, who also kindly offered to give me the use of a boat, but unfortunately I was unable to take advantage of it for want of time. The site of the loch was originally an extensive hollow, as if scooped out in some measure for the purpose by the hand of Nature, and by filling the hollow up at some places, and deepening and extending it at others, the formation of the lake was in a short time completed. The extent of ground covered by it is not less than 21 acres, and the water is in some places 25 feet deep.[*] Trout, Perch, Pike, and Eels are common in the loch. It is beautifully situated, and is to some extent surrounded by trees and shrubbery ; in some places it is partly overgrown with vegetation, and the Bull-rush (*Typha latifolia*), a rare plant, grows luxuriantly at its lower end. At the time of my visit the water was teeming with Entomostraca, some of the more common species being *Sida crystallina, Daphnella brachyura, Scapholeberis cornuta, Pleuroxus uncinatus.* A great variety of Cladocera were obtained here. Copepoda were also common, and included *Cyclops affinis* and what appears to be an undescribed species. Ostracoda were scarcely so numerous as might have been expected. Mollusca were moderately common, but the species obtained belonged to those that are of more or less general occurrence throughout Scotland. The curious Diatom, *Gomphonema capitatum,* Ehr., was obtained here.

CAMILLA LOCH.

Camilla Loch, in the parish of Auchtertool, derives its name from the old house of Camilla, anciently called Hallyards, but which was changed to Camilla from its being the residence of one of the Countesses of Moray whose name was Campbell. The loch is situated a little to the north of the village of Auchtertool. The extent of its surface is about 18 acres, and its greatest depth is said to be 22 feet, the deepest part being towards the east end. The west end is much overgrown with vegetation. It contains perch, pike, and eels. The small stream that runs from the loch is the principal feeder of Raith Lake.

The invertebrate fauna of the loch appeared to be fairly abundant, but there was time for only a partial examination being made. Among the organisms collected, Entomostraca were largely represented. In some of the small sheltered bays, where the water was shallow, the common *Cypria serena* were observed swimming about in myriads. The Horse Leech, (*Hæmopsis corax,* Mog.), was common in the soft mud at the west end, where the water was shallow. Among the Cladocera observed were *Alona guttata* and the somewhat rare *Alona tenuicauda.* The beautiful *Volvox globator* was also observed in this loch.

LOCHGELLY LOCH.

This loch, also in the parish of Auchtertool and a short distance north-west of Camilla Loch, is about three miles in circumference, and is much overgrown with vegetation along the south side. Part of this side only

[*] *Old Stat. Account of Scot.,* vol. vi., 1793.

was examined. Invertebrate organisms were abundant, the larger part consisting, as in the case of the other lochs described, of Entomostraca. These included several rare forms not observed elsewhere, such as *Leydigia quadrangularis* and *Peracantha truncata* among the Cladocera, and *Cypris obliqua*, *Cypris pubera*, and a *Candona* that appears to be undescribed, among the Ostracoda. The following are the lists of species observed in the three lochs :—

MOLLUSCA.

Sphærium lacustre, Müller. Raith Lake, not common.
Pisidium pusillum, Gmelin. Raith Lake, Camilla Loch, Lochgelly Loch.
　,,　　*fontinale*, Drap. Raith Lake and Lochgelly Loch.
　,,　　*nitidum*, Jenyns. Raith Lake.
Valvata piscinalis, Müller. Raith Lake and Lochgelly Loch.
Planorbis albus, Müller. Raith Lake and Lochgelly Loch.
　,,　　*nautileus*, Linné. Raith Lake, Camilla Loch, Lochgelly Loch.
　,,　　*contortus*, Linné. Raith Lake, Camilla Loch, Lochgelly Loch.
Physa fontinalis, Linné. Raith Lake, not common.
Limnæa peregra, Müller. Raith Lake and Lochgelly Loch.
　,,　　*truncatula*, Müller. Raith Lake and Lochgelly Loch.

CRUSTACEA.

AMPHIPODA.

Gammarus pulex (Linné). Common in the three lochs.

COPEPODA.

Diaptomus gracilis, G. O. Sars. Raith Lake frequent, Lochgelly Loch.
Cyclops tenuicornis, Claus. Raith Lake, not very common ; Lochgelly Loch.
　,,　　*pholcratus*, Koch. Raith Lake (with ova), Lochgelly Loch.
　,,　　*affinis*, Sars. Raith Lake (with ova).
　,,　　*serrulatus*, Fischer. Common in the three lochs.
　,,　　*thomasi*, Forbes. Lochgelly and Camilla Lochs.
　,,　　*viridis*, Jurine. Frequent in Raith Lake and Lochgelly Loch.
　,,　　*crassicornis*, Müller. Frequent in Raith Lake and Lochgelly Loch.
Canthocamptus minutus. More or less frequent in the three lochs.
Atheyella spinosa, Brady. Lochgelly Loch.

OSTRACODA.

Cypria exsculpta, S. Fischer. Lochgelly Loch, frequent.
　,,　　*ophthalmica* (Jurine). Common in the three lochs.
　,,　　*serena* (Koch). Common in the three lochs, and especially Camilla.
　,,　　*lævis* (O. F. Müller). Lochgelly Loch, frequent.
Cypris pubera, O. F. Müller. Lochgelly Loch, not common.
　,,　　*obliqua*, Brady. Lochgelly Loch, frequent.
Erpetocypris reptans (Baird). Frequent in the three lochs.
Cypridopsis vidua (Müller). Common in the three lochs.
　,,　　*villosa* (Jurine). Frequent in Lochgelly and Raith Lakes.
Potamocypris fulva, Brady. Frequent in the three lochs.
Notodromas monacha (O. F. Müller). Frequent in Lochgelly and Camilla Lochs. Raith Lake, rare.
Candona candida (O. F. Müller). Common in the three lochs.
　,,　　*lactea*, Baird. Frequent in Lochgelly and Camilla Lochs.
　,,　　*pubescens* (Koch). Frequent in Lochgelly and Camilla Lochs.
　,,　　*rostrata*, Brady and Norman. Raith Lake, not very common.

Candona kingsleii, Brady and Robertson. Frequent in Lochgelly and Camilla Lochs.

   ,,   *fabæformis* (Fischer). Lochgelly and Raith Lake.

   ,,   *hyalina*, Brady and Robertson. Lochgelly Loch, rare.

   ,,   *ambigua*, n.s. Lochgelly Loch, not common.

Ilyocypris gibba (Ramdohr). Lochgelly and Raith Lake.

Limnicythere inopinata, Baird. Occurred in the three lochs.

Candona ambigua, provisional name (Pl. IV. figs. 7a–c). *Shell.*—Lateral view somewhat similar to *C. pubescens*, but the straight part of the dorsal margin is more central, and the slope from each end of it towards the extremities of the shell more equal than in that species. The flexure of the ventral margin is also more distinct. Height rather more than half the length. The outline as seen from above is very compressed, much more so than in *C. pubescens*. It is acuminate in front and narrowly rounded behind. *Width* a third of the length. *Length*, $\frac{1}{30}$ in. = ·83 mm.; height $\frac{1}{54}$ in. *Antennules.*—The length of the first, second, and fourth joints counting from the extremity is nearly equal, the third is about two-thirds the length of the second, the fifth is about equal to the third joint. *Antennæ.*—The ultimate scarcely half the length of the penultimate joint, which is about two-thirds the length of the preceding. I could not clearly make out the other joints of the antennules and antennæ in my dissections. The figures show approximately the proportional lengths of the joints.

CLADOCERA.

Sida crystallina (Müller). Raith Lake, common.

Daphnella brachyura (Lievin). Raith Lake, common.

Ceriodaphnia reticulata (Jurine). Raith Lake, frequent.

Scapholeberis mucronata (Müller). Raith Lake, frequent.

Simocephalus vetulus (Müller). Raith Lake, Lochgelly Loch.

Daphnia pulex, Linné. Raith Lake, frequent.

Eurycercus lamellatus (Müller). Raith Lake, Lochgelly Loch, frequent.

Acroperus harpæ, Baird. Frequent in the three lochs.

Leydigia quadrangularis (Leydig). Lochgelly Loch, not very common.

Lynceus tenuicaudis (G. O. Sars). Camilla Loch, not common.

   ,,   *quadrangularis*, Müller. Frequent in the three lochs.

   ,,   *guttatus* (G. O. Sars). Camilla Loch.

Alonella nana (Baird). Raith Lake, scarce.

Peracantha truncata (Müller). Lochgelly Loch, not common.

Pleuroxus trigonellus (Müller). Raith Lake, frequent.

   ,,   *uncinatus*, Baird. Raith Lake, not very common.

   ,,   *lævis* (G. O. Sars). Camilla Loch, not common.

Chydorus sphæricus (Müller). Frequent in the three lochs.

Polyphemus pediculus (Linné). Frequent in Raith Lake.

3. LOCH NESS, LOCH OICH, AND LOCH LOCHY.

These lochs occupy a considerable portion of the great valley that stretches across Scotland from the Moray Firth to Loch Linnhe, and they constitute about two-thirds of the entire length of the Caledonian Canal, the whole length of which is about 60 miles, while the combined length of the three lochs is little, if anything, under 40 miles.

For various reasons it was considered desirable that an effort should be made to examine the invertebrate fauna of these lochs while the 'Garland' was proceeding through the canal on its way to and from the west coast, but the stormy weather encountered by us, both going and returning, formed a serious hindrance to anything like a satisfactory investigation being made.

LOCH NESS.

The first of these lochs to be examined was Loch Ness, which occupies the eastern part of the valley. It is, with the exception of Loch Morar, the deepest fresh-water loch in Britain. The deepest part of the loch is in the vicinity of Foyers, and is, according to the chart of the Caledonian Canal, 129 fathoms, or 774 feet deep. The surface of the loch is 50 feet above the sea and the bottom is therefore fully 120 fathoms below sea level, or deeper than the deepest part of the North Sea anywhere within the British area except in the extreme north. It is 26 miles in length by about 1 mile in breadth.

From some apparently unknown cause angling for salmon in Loch Ness is usually not very successful, not because salmon do not visit the loch, for great numbers of them are known to pass through it on their way to the river Oich, and thence to the Garray, that falls into Loch Oich, both of which are excellent salmon rivers. The uniformly great depth of the loch may be partly accountable for this want of success, for, with the exception of about four miles at the east end and about two miles at the west end, and a very limited portion along each side, no part of the loch is less than 100 fathoms in depth.

The dredge was first let down a little east of Castle Urquhart, where the depth is given on the chart as 128 fathoms. A lining of fine gauze was fixed to the inside of the net for the purpose of preventing the smaller objects collected at the bottom from being washed out while the dredge was being hauled up. The dredge reached the bottom easily, no obstacle of any kind, such as under-currents, being encountered, and was hauled up choke-full of fine mud ; mixed up with the mud were numerous fragments of firm peaty matter containing pieces of partially decayed wood. The contents of the dredge were emptied into a tub, and a careful examination made of it. First, a quantity of the fine mud was put into a bottle with water and allowed to stand till the mud subsided ; a distinct stratum of flocculent matter was then observed on the surface of the mud, this was shown by the microscope to consist wholly of the exuviæ of Entomostraca, chiefly of Cladocera. Second, a large quantity of the mud was carefully washed through a fine muslin sieve, but though fragments of Entomostraca were obtained in considerable abundance no trace of any living organism could be detected. Third, some of the mud was dried and then redissolved in water, and the floating matter carefully collected and examined, but still no trace of any living thing—that is, no organism that had been alive when the mud was brought up—could be observed. The deep part (129 fathoms) in the vicinity of Foyers was also dredged with the same results.

It would seem as if the fragments of firm peaty matter referred to were part of an older and more compressed portion, and the fine mud the newer and continually increasing layer, of a deposit of peat that is being formed at the bottom of the loch. If this supposition be correct, it explains to some extent why living organisms, especially Entomostraca, were so conspicuous by their absence in the dredged material from the vicinity of Castle Urquhart and from the other parts of the loch examined. It has been proved that peaty mud is not a favourable habitat for non-pelagic Entomostraca, more especially Ostracoda.

Some Diatomacea were observed among the peaty ooze, but they appeared to consist only of dead tests.

Besides using the dredge, a tow-net was worked near the surface and another at a depth of about 60 to 70 fathoms, and both nets captured a considerable number of Entomostraca, including *Bosmina, Cyclops, Daphnia,*

Diaptomus, Holopedium gibberum, Polyphemus, and *Polytrephes.* Both tow-nettings were similar as regards variety of species, but in that from the deep water *Bosmina* and *Diaptomus* were more plentiful than in the surface gatherings.

Loch Oich.

This loch is about four miles long by about half a mile broad. Its depth varies considerably, but the deepest part is towards the east end, where 23 fathoms occurs. The loch is studded with islets and shoals, so that though the track for vessels is carefully indicated by buoys and beacons, it is rather intricate, and requires more than usual care and watchfulness on the part of those navigating it. The storm we experienced was so violent when passing through this loch that it would have been at considerable risk, if not actual danger, to have slowed the steamer for the purpose of dredging ; the best we could do under the circumstances was to use our tow-nets as frequently as possible. The tow nettings obtained, both as regards number and variety of species, were similar to those from Loch Ness. The level of Loch Oich is 50 feet higher than Loch Ness or 100 feet above sea-level, and the River Oich flows out of it into Loch Ness.

Loch Lochy.

Loch Lochy—the westmost of the three lochs under consideration—is much larger than Loch Oich, being about ten miles long by about one mile in breadth. It is of considerable depth, reaching to 76 fathoms at one place. We dredged in the deepest part and also in about 40 fathoms. The material brought up in the first dredge from the deep part resembled that dredged in Loch Ness, and was also similarly devoid of living organisms. The second dredge was a failure—the net being empty when hauled up. The tow-nettings from this loch were similar to those from the other two.

On our return journey a further effort was made to examine the lochs, but no additional information was obtained.

Though we were unable to find any trace of animal life at the bottom of Loch Ness and Loch Lochy, our experiments were too few to make it safe to found any theory upon them ; besides, they were not carried out under very favourable conditions. It is not improbable that living organisms may exist at the bottom of these lochs, but by being more or less localised in their distribution be easily missed ; still, if it be the case that the peaty deposit which is being formed in the deeper parts of Loch Ness and Loch Lochy is the same all over the bottom, there is little likelihood of animal life being very abundant, if present at all.

The following are the lists of species obtained in the tow-nettings from the three lochs :—

CRUSTACEA.

Copepoda.

Diaptomus gracilis, G. O. Sars. In the three lochs, more or less common.
Cyclops strenuus, Fischer. In the three lochs.
   ,,   *thomasi,* ? Forbes. Loch Oich, not common.

CLADOCERA.

Holopedium gibberum, Zaddach. Frequent in the three lochs.
Daphnia jardini, Baird. Loch Ness, frequent also in Loch Oich.
  ,,   *longispina*, ? Müller. Loch Ness.
Bosmina longirostris (Müller). Loch Ness, not very common.
Polyphemus pediculus (Linné). Frequent in the three lochs.
Bythotrephes longmanus, Leydig. Loch Ness and Loch Oich, common.

Diatomacea were also frequent in the lochs, especially in Loch Lochy—*Asterionella*, with the frustules arranged in their characteristic star-like clusters, being common, especially in Loch Lochy.

4. LISMORE ISLAND LOCHS, ARGYLLSHIRE.

There are three small lochs on Lismore Island, viz., Loch Bail nan Gobhann (or, as it is called in the *Sportsman's Guide*, Loch Balnagowan), Loch Kilcheran, and Loch Fiart. The first two were examined about the middle of July, the other later on.

LOCH BALNAGOWAN.

This loch, which is of small size, is moderately deep towards the south end, but the north end is shallow and much overgrown with vegetation, comprising water-lilies, *Polygonum*, *Scirpus*, grasses, &c. At the time of our visit the water was of a whitish colour and obscure, as if fine, light-coloured mud were held in suspension. On examining a portion of the water with the microscope, a pale-coloured or bleached-like conferva (?) in the form of small jointed rods, about 1 mm. long, was seen to be present in enormous numbers, and was no doubt the chief, if not the only, cause of the peculiar coloration and obscurity of the water.

The loch is leased by the Lorne Angling Association, who keep a boat on it, and we had the use of this boat for the examination of the loch. I am informed that many of the trout in this loch, which are said to be of fine quality, are more or less deformed by having the back-bone abnormally bent, usually in a lateral direction. The bottom of the loch consisted of a greyish marl, largely composed of dead molluscan shells. The entomostracan fauna was abundant. The curious and somewhat rare Infusorian, *Ceratium longicorne*, Perty, and also *Peridinium tabulatum*, Ehr., were common, as were also the Rotifera, *Anuraea cochlearis*, Gosse, and *Anuraea aculeata*, Ehr. *Ceratium longicorne* was discovered by Schrank and Perty in the Bernese Alps in 1848 ; in Calcutta by Major, now Colonel, Stewart-Wortley in 1859, and in 1879 in Olton Reservoir, near Birmingham, by Levick. I am unable to find any Scotch record for it.[*] Ostracoda were fairly common, and included the rather rare and pretty *Cypris obliqua*, Brady. Among the Cladocera observed was *Chydorus globosus*, a rare species.

LOCH KILCHERAN.

Loch Kilcheran, to the south of Loch Balnagowan, and nearly intermediate between it and Loch Fiart, is also comparatively small, being only about a mile in circumference. Its margin nearly all round, but especially at the north and south ends, is overgrown with tall reeds and

[*] Mr J. Hood, Dundee, informs me that he has observed this *Ceratium* in Black Loch, near Blairgowrie, but has not recorded it.

other aquatic plants, so that it is difficult to get near the loch in some places. The water of this loch at the time of my visit was clear and pellucid, and thus formed a marked contrast to Loch Balnagowan. Loch Kilcheran was examined only by hand-net from the shore, and some rather rare Ostracoda were obtained, including *Erpetocypris robertsoni*, Norman and Brady, *Darwinella stevensoni*, Brady and Robertson, and *Scottia browniana* (Jones).

LOCH FIART.

Loch Fiart, which is situated near the south end of the island, is about the same size as the last, and like it has a reedy margin, and can only be fished properly from a boat. The trout in this loch are said to resemble those of Loch Leven, and to attain a similar size. The water of Loch Fiart, like that of Loch Balnagowan, is of a whitish colour, but only during summer. In this case the whitish colour appears to be caused by fine calcareous mud being held in suspension by the water. In winter the water is said to be colourless and transparent. This alternate obscurity and transparency of the water is rather remarkable, but it may be partly or wholly accounted for in this way. The basin of the loch being formed in rock consisting more or less of limestone, a more than usual quantity of calcareous matter in the form of an acid carbonate will be held in solution by the water. During spring-time and summer, when plant life is vigorous, the plants will decompose the carbonic acid, and part of the lime may remain as a fine precipitate suspended in the water. During winter the decaying vegetation may give off carbonic acid, free or combined, in sufficient quantity to redissolve the suspended lime, thus causing the water again to become clear.[*] Whether this be the true explanation of the phenomona observed or not, the effects described are interesting enough to deserve careful study.

The following are the lists of organisms obtained and identified in the three lochs referred to :—

MOLLUSCA.

Pisidium pusillum, Gmelin. In each loch.
Valvata piscinalis, Müller. Loch Balnagowan.
Planorbis albus, Müller. Loch Balnagowan.
    ,,   *contortus*, Linné. Loch Balnagowan.
Limnæa peregra, Müller. In each loch.

[*] The water of inland lochs usually contains a certain amount of carbonic acid (H_2CO_3) derived from various sources. When the loch is formed in rocks consisting more or less of limestone ($CaCO_3$), the carbonic acid acts on the limestone, forming an acid carbonate of lime, which is soluble and at the same time unstable ; so long as the water contains excess of carbonic acid it will remain clear and pellucid. The soluble acid carbonate of lime being unstable, very little change in the temperature or otherwise is sufficient to decompose it. For instance, should aquatic vegetation be abundant in the loch, this vegetation, when in vigorous growth during spring and summer, will in the course of its development decompose, in the presence of sunlight, a considerable amount of the carbonic acid which is required to keep the carbonate of lime in solution. (During the summer months, there being very little darkness, this action will be more or less continuous and of course the warmer the weather is the greater the reaction will be.) On the decomposition taking place, the carbonate of lime is precipitated in the form of a fine white powder, which may remain suspended in the water, imparting to it a milky appearance. The following chemical equations represent the reactions which probably take place—

$$H_2CO_3 + CaCO_3 = CaH_2(CO_3)_2 \qquad CaH_2(CO_3)_2 = CaCO_3 + H_2O + C + O_2$$

On the other hand, during autumn and winter the vegetation in the course of its decomposition gives off a certain amount of carbon dioxide, either free or combined, which may form with the water sufficient carbonic acid (H_2CO_3) to redissolve the precipitated carbonate of lime, so that the water will again become transparent.—ANDREW SCOTT.

CRUSTACEA.

AMPHIPODA.

Gammarus pulex (Linné).   Frequent in and about the three lochs.

COPEPODA.

Diaptomus gracilis.   Loch Balnagowan.
Cyclops tenuicornis.   Loch Kilcheran.
  „   *serrulatus.*   Frequent in the three lochs.
  „   *crassicornis.*   Lochs Balnagowan and Kilcheran.
Canthocamptus minutus.   Frequent in the three lochs.

OSTRACODA.

Cypria ophthalmica (Jurine).   Frequent in the three lochs.
  „   *serena* (Koch).   Frequent with the last.
Cyclocypris globosa (G. O. Sars).   Balnagowan and Kilcheran Lochs.
Scottia browniana (Jones).   Loch Kilcheran, not common.
Cypris obliqua, Brady.   Loch Balnagowan, frequent.
  „   *incongruens* (O. F. Müller).   Loch Balnagowan, rare.
Erpetocypris reptans (Baird).   In the three lochs.
  „   *olivacea*, Brady and Norman.   Lochs Balnagowan and Kilcheran.
  „   *robertsoni*, Brady and Norman.   Kilcheran Loch, scarce.
  „   *tumefacta* (Brady and Robertson).   Lochs Balnagowan and Kilcheran.
Cypridopsis vidua (Müller).   In the three lochs.
  „   *villosa* (Jurine).   Lochs Balnagowan and Kilcheran.
Potamocypris fulva, Brady.   In the three lochs, frequent.
Notodromus monacha (Müller).   Loch Balnagowan, frequent.
Candona candida (Müller).   Frequent in the three lochs.
  „   *lactea*, Baird.   Lochs Balnagowan and Kilcheran.
  „   *pubescens* (Koch).   In the three lochs, not uncommon.
  „   *kingsleii* (Brady and Robertson).   Distribution as the last.
Ilyocypris gibba (Ramdohr).   Lochs Balnagowan and Kilcheran.
Darwinella stevensoni, Brady and Robertson.   Loch Kilcheran, rare.
Limnicythere inopinata (Baird).   In the three lochs, frequent.

CLADOCERA.

Simocephalus vetulus (Müller).   Loch Kilcheran.
Ceriodaphnia punctata (P. E. Müller).   Loch Balnagowan, not common.
Bosmina longirostris (Müller).   Loch Balnagowan.
  „   *longispina*, Leydig.   With the last.
Lynceus quadrangularis (Müller).   Lochs Balnagowan and Kilcheran.
Chydorus globosus, Baird.   Loch Balnagowan, not common.
  „   *sphæricus* (Müller).   In the three lochs.

As already stated, Infusoria and also Diatomacea were present in
Loch Balnagowan in considerable abundance.   The more conspicuous
of the Infusoria have already been referred to ; the more common of the
Diatomacea were *Asterionella formosa*, with its star-like arrangement of the
frustules, *Pediastrum selenæa*, various species of *Navicula*, &c.   Desmids
were also of frequent occurrence.

5. LOCH STRATHBEG.

This loch, which is situated one and a half miles west from Rattray
Head, Aberdeenshire, is about two miles in length by about three-quarters
of a mile broad.   It is separated from the sea by a ridge of bent-covered
sandhills, nearly a mile in width.   In the Old Statistical Account of
Scotland (1793), it is said "there is a tradition that in the beginning of

last century Dutch busses frequented the loch of Strathbeg." At present the sea is said to flow into it only during high-water of spring tides through a sort of canal-like outlet, and the water is thus rendered somewhat brackish. The loch is moderately deep in some parts, but at the time of our visit it was so shallow in some places at the north-east end that it was necessary to propel the boat we were using by punting it. It is also stated in the Old Statistical Account that an effort was once made to drain the loch, but from some cause the attempt failed. It contains perch and sea-trout but is little fished. The loch was partially examined early in September last year (1890), while the 'Garland' was windbound in Fraserburgh Harbour. Entomostraca were very abundant —*Ceriodaphnia reticulata* and *Cypridopsis aculeata* being very common species. The bottom of the part examined consisted very much of blackish mud, having its surface covered in some places by a rust-coloured deposit. Considerable portions of the bottom were overgrown by a luxuriant crop of *Chara*. I have no doubt that a careful examination of the invertebrate fauna of this loch would yield interesting results. The following are the organisms obtained during our visit in September :—

MOLLUSCA.

Pisidium pusillum (Gmelin). Frequent.
Planorbis spirorbis (Müller). Rather rare.
Limnæa peregra (Müller). Common.
 ,, *truncatula* (Müller). Rather rare.

CRUSTACEA.

COPEPODA.

Cyclops strenuus, Fischer. Frequent.
 ,, *tenuicornis*, Claus. Not common.
 ,, *viridis*, Jurine. Frequent.
 ,, *serrulatus*, Fischer. Frequent, especially round the margin.
 ,, *crassicornis*, O. F. Müller. Frequent round the margin.
Canthocamptus, sp. Rather rare round the margin.

OSTRACODA.

Cypria serena (Koch). Common.
 ,, *lævis* (O. F. Müller).
 ,, *ophthalmica* (Jurine). Frequent.
Erpetocypris reptans (Baird). Frequent.
Cypridopsis vidua (Müller). Frequent.
 ,, *villosa* (Jurine). Frequent.
 ,, *aculeata* (Lilljeborg). Very common.
Candona candida (Müller). Frequent.
Ilyocypris gibba (Ramdohr). Not common.
Limnicythere inopinata (Baird). Frequent.

CLADOCERA.

Ceriodaphnia reticulata (Jurine). Very common.
Simocephalus vetulus (Müller). Frequent.
Daphnia pulex (Linné).
 ,, *longispina*, Müller. Frequent.
Eurycercus lamellatus (Müller). Not very common.
Pleuroxus trigonellus (Müller). Not very common.

6. LOCH ACHNACLOICH.

Loch Achnacloich, four or five miles north of Invergordon, Ross-shire, is of small size, and easy of access. It was visited one day towards the end of May last year, when the 'Garland' was at Invergordon wind-bound. Owing to the very limited time at my disposal the examination of the loch was necessarily incomplete; nevertheless, the results were fairly satisfactory. Some of the Entomostraca have not previously been recorded from so far north. Dr Sutherland of Invergordon informs me that there are a good many trout in the loch. The loch was examined by hand-net from the shore. The following species were obtained and identified :—

MOLLUSCA.

Pisidium nitidum, Jenyns.   Not very common.
„   *pusillum* (Gmelin).   Frequent.
„   *fontinale* (Drap.).   Not very common.
Valvata cristata, Müller.   Frequent.
Planorbis contortus (Linné).   Common.
„   *spirorbis*, Müller.   Rather rare.
„   *nitidus*, Müller.   Rather rare.
Limnæa truncatula (Müller).   Rather rare.
Vertigo antivertigo (Drap.).   Rather rare.

CRUSTACEA.

COPEPODA.

Cyclops serrulatus, Fischer.   Frequent.
„   *crassicornis*, O. F. Müller.   Frequent.
„   *thomasi*, Forbes.   Rather rare.
Canthocamptus minutus (O. F. Müller).
„   sp.   A small form, with ova.

OSTRACODA.

Cypria serena (Koch).   Frequent.
„   *ophthalmica* (Jurine).   Frequent.
„   *exsculpta* (S. Fischer).   Common.
Cypris reticulata, Zaddach.   Frequent.
Erpetocypris reptans (Baird).   Rather rare.
„   *tumefacta* (Brady and Robertson).   Common.
„   *strigata* (O. F. Müller).   Rather rare.
Cypridopsis villosa (Jurine).   Frequent.
Potamocypris fulva, Brady.   Frequent.
Candona candida (Müller).   Frequent.
„   *rostrata*, Brady and Norman.   Common.
„   *kingsleii*, Brady and Robertson.   Rather rare.
„   *hyalina*, Brady and Robertson.   Rare.
„   *fabæformis* (Fischer).   Frequent.

CLADOCERA.

Daphnia pulex (Linné).   Not common.
Chydorus sphæricus (Müller).   Frequent.

7. LOCHS IN THE VICINITY OF WICK, CAITHNESS-SHIRE— LOCH HEMPRIGGS.

This loch is three miles or so south of Wick, and within a short distance of the highway to Lybster; it is somewhat circular in outline and covers a considerable area, and the surrounding country is flat and bare. Towards the south the loch is bordered by black peaty and heath-covered moorland, but round the north and east are cultivated fields. The principal inlet, which is at the south-west end, is a sluggish stream, two or three yards wide, with soft boggy ground, thickly covered with grass and reeds on each side. The outlet is by Hempriggs burn, which, flowing east and north, joins Wick River near the town of Wick. This loch was examined about the middle of September, and by hand-net from the shore.

Loch Wester.

Loch Wester is about five miles north of Wick and a short distance off the road from Wick to Kiess. It is about one and a half mile in circumference, and is so little above the sea that during high spring tides the sea is said to flow into it. The River Wester flows through it, and falls into the sea at Sinclair Bay, scarcely a mile from the loch. It contains both sea and loch trout. At the time of my visit, in September last year, the water of the loch appeared to be perfectly fresh, its density being 1000·4, and its temperature 13·6 C. The density of the water for domestic use in Wick, tested at the same time and with the same hydrometer, was 1000·6, and the temperature of the water when tested 11·1 C.* It was therefore interesting to find in the fresh water of the loch that *Mysis vulgaris* was quite common. This species is sometimes common in the Forth above Queensferry. I have also taken it in Granton Harbour, along the shore at Portobello, where it was plentiful, and in an old brickfield near Dunbar to which the tide has free access, and at other places. It would thus appear that *Mysis vulgaris* is capable of surviving under very varied conditions. The loch was examined by hand-net from the shore, but with the exception of the *Mysis* referred to, comparatively few organisms were obtained.

The shores of Wick River, in the vicinity of the town of Wick, were also examined about the same time as the two lochs here described. The invertebrata obtained are recorded in the following lists, *Cypris prasina*, Fischer, was a very common species in the Wick River.

MOLLUSCA.

Sphærium corneum (Linné). Frequent.
Pisidium fontinale (Drap.). Loch Hempriggs.
Valvata piscinalis (Müller). Loch Wester and Loch Hempriggs, rather rare.
Planorbis albus (Müller). Loch Wester, common
  ,, *contortus* (Linné). Loch Hempriggs, common.
  ,, *vortex* (Linné). Rather rare.
Limnæa peregra (Müller). Common in both lochs.
  ,, *truncatula* (Müller). Loch Wester, common also in Wick River.
Succinea putris (Linné). Loch Wester, frequent.

CRUSTACEA.

Schizopoda.

Mysis vulgaris, Thompson. Loch Wester, frequent.

Amphipoda.

Gammarus pulex (Linné). Loch Hempriggs, common.

* It is important to note, in connection with the density observations referred to in this paper, that the temperature of the water was tested and recorded at the *time* the density observations were taken.

COPEPODA.

Cyclops viridis, Jurine. Loch Hempriggs, frequent.
,,   *serrulatus*, Fischer. Loch Hempriggs, frequent.
,,   *crassicornis*, O. F. Müller. Loch Hempriggs, not common.
Canthocamptus, sp. Loch Hempriggs.

OSTRACODA.

Cypria serena (Koch). Loch Hempriggs, Loch Wester, side of Wick River.
,,   *lœvis* (O. F. Müller). The same localities as the last.
,,   *ophthalmica* (Jurine). With the previous two species frequent.
,,   *exsculpta* (S. Fischer). Loch Hempriggs, scarce.
Cypris fuscata (Jurine). Loch Hempriggs, rare.
Cypris virens (Jurine). Side of Wick River, frequent.
,,   *prasina*, Fischer. With the last common.
Erpetocypris reptans (Baird). Loch Hempriggs and Loch Wester, frequent.
,,   *tumefacta* (B. and R.). Loch Hempriggs, frequent.
Cyprutopsis vidua (Müller). Loch Wester, frequent.
,,   *aculeata* (Lilljeborg). Loch Hempriggs, Loch Wester, side of Wick
        River.
,,   *villosa* (Jurine). Loch Hempriggs, frequent.
Potamocypris fulva, Brady. Loch Hempriggs, Loch Wester.
Candona candida (Müller). Loch Hempriggs, Loch Wester, side of Wick River.
,,   *lactea*, Baird. With the last, but not so common.
,,   *pubescens* (Koch). Also with the last.
,,   *kingsleii*, B. and R. Loch Hempriggs, Wick River.
,,   *acuminata* (Fischer). Loch Wester, rare.
Ilyocypris gibba (Ramdohr). Loch Hempriggs, Loch Wester.
Limnicythere inopinata (Baird). Loch Hempriggs, Loch Wester, side of Wick
        River.
Cythere pellucida (Baird). Wick River, not common.

CLADOCERA.

Lynceus elongatus. Loch Hempriggs.
,,   *excisa.* Loch Hempriggs.

LOCH OF STENNESS, ORKNEY.

" Loch Stenness is a great sheet of water about 15 miles in circumference, including its upper and lower divisions. The name is sometimes applied to designate both the divisions of the loch, and sometimes it is applied only to the lower loch which communicates with the sea ; while the upper loch, which is entirely fresh, is termed the Loch of Harray. The banks of these lakes, like all the Orcadian lakes, are bare and treeless ; and the upper loch is divided from the lower by two long narrow promotories that jut out from opposite sides, and so nearly meet in the middle as to be connected by a low bridge, called the Bridge of Bogar, over which the roadway passes. The area of the Loch of Stenness is 1792 acres, and that of the Loch of Harray 2432 acres ; or, together, 4224 acres. . . . . The former is nearly 4 miles long, with a maximum breadth of $1\frac{1}{2}$ mile ; while the latter is $4\frac{2}{3}$ miles long, and varying in breadth from 3 furlongs to $1\frac{1}{2}$ mile."[*]

LOCH STENNESS (proper).

I made an examination of this loch by means of a small sailing-boat— (the weather was too stormy to attempt rowing)—on the 27th September 1890. The loch was tow-netted from above the Bridge of Wraith to near

[*] Mr A. Young, Inspector of the Scottish Salmon Fisheries, in Fifth Annual Report of the Fishery Board for Scotland, pp. 366-371 (1887).

its upper end, the dredge being also used at several places. Copepoda—
chiefly *Dias longiremus*—were moderately common in the tow-net material,
both surface and bottom, and *Cyclops æquorus* was frequent in the dredged
material. The Lamellibranch mollusc, *Mya arenaria*, was abundant in all
stages of growth, from an almost microscopic size up to about 1½ inch,
in breadth, but no adult specimens were obtained, the reason no doubt
being that the dredge could not penetrate deep enough into the mud to
capture them, for where the conditions are favourable they burrow to a
considerable depth. The bottom appeared to have a dense growth of algæ
covering it in many places, for the dredge, on being hauled up on several
occasions, was filled with little else than growing weed, which was crowded
with young *Mya*. *Hydrobia ulvæ* was also present in considerable numbers.
The weed referred to evidently shelters a numerous micro-fauna, and therefore forms a rich feeding-ground for the various fishes inhabiting the loch.
It is reported that at least 12 kinds of food-fishes occur in Loch Stenness.
It would be interesting to have this loch examined at certain intervals, as,
for instance, about or shortly after the time of spring-tides, and about the
time of low neap-tides, because from its peculiar position the salinity of
the water must vary considerably, and the varying condition of the water
will no doubt react to some extent on the fauna, both vertebrate and
invertebrate. The loch is of no great depth, the deepest part dredged
by us being little over 5 fathoms. The following physical observations
were taken : (*a*) near the upper end of the loch, density 1012·7, temperature of water when density was taken 11·2 C.; (*b*) near lower end, density
1012·6, temperature 11·5 C., which shows that the water was decidedly
brackish at the time of our visit.

Loch Harray.

This loch was visited on September the 30th, and was examined by
means of a rowing-boat, kindly placed at my service by Mr John John
ston, of Vetquoy, near the head of the loch. I tow-netted and dredged
from the upper end of the loch down as far as Tenstone Ness. I had
the previous year partially examined the lower end in the vicinity of the
Bridge of Brogar. The microfauna of this loch showed a marked contrast
to that of Loch Stenness; I did not obtain a single Cladoceran in Loch
Stenness, but here they were in myriads, almost to the exclusion of everything else, the prevailing form being *Daphnia longispina* (?). Copepoda were
also present, and in the vicinity of Tenstone Ness *Cypria serena* likewise
occurred in the tow-net collections; but neither were very plentiful.
Loch Harray is very shallow, the deepest part, I was informed, being
little over 2 fathoms. I found the bottom to be very rough and stony,
with occasional intervening patches of clean ground which consisted of
fine blackish mud or muddy sand; the bottom is said to be very uneven,
and this was partly borne out by what was observed by us during our
examination. Pond-weed, water-milfoil, and other plants were of frequent
occurrence in the loch, some of the plants having stems 6 to 7 feet in
length. At the time of our visit the water was nearly though not quite
fresh, as the following density observations show. Thus (*a*) near the head
of the loch, density 1001·3, temperature of water at the time the density
was taken 8·5 C.; (*b*) off Tenstone Ness (near the middle of the loch), density 1001·6, temperature 8·6 C. For comparison it may be stated that the
water from a pump-well at Stromness had a density of 1000·4, temperature
14·7 C.

The following are the lists of invertebrata obtained in the two
lochs :—

MOLLUSCA.

Mya arenaria, Linné.   Loch Stenness, abundant.
Pisidium roseum, Scholtz.   Head of Loch Harray.
Planorbis nautileus (Linné).   Head of Loch Harray.
Valvata cristata, Müller.   Head of Loch Harray.
Neritina fluviatilis (Linné).   Loch Harray at Bridge of Brogar, common.
Hydrobia ulvæ, Pennant.   Loch Stenness, common.

CRUSTACEA.

COPEPODA.

Dias longiremus, Lillj.   Common in Loch Stenness.
Diaptomus gracilis, G. O. Sars.   Loch Harray, frequent.
Cyclops viridis, Jurine.   Frequent in Loch Harray.
   ,,   *æquorus*, Fischer.   Frequent in Loch Stenness.

OSTRACODA.

Cypria serena (Koch).   Loch Harray, generally distributed.
   ,,   *lævis* (O. F. Müller).   Loch Harray, not so common as the last.
   ,,   *ophthalmica* (Jurine).   Loch Harray, frequent.
Cypris pennsina, Fischer.   Loch Harray, near Bridge of Brogar.
Erpetocypris reptans (Baird).   Loch Harray, upper end.
Cypridopsis villosa (Jurine).   Loch Harray, generally distributed.
Potamocypris fulva, Brady.   Loch Harray, lower end.
Candona candida (Müller).   Loch Harray, generally distributed.
Candona lactea, Baird.   Loch Harray, upper end.
   ,,   *pubescens* (Koch).   Loch Harray, generally distributed.
   ,,   *hyalina*, B. and R.   Loch Harray, rare, but generally distributed.
   ,,   *kingsleii*, B. and R.   Loch Harray, rare, upper end.
Ilyocypris gibba (Ramdohr).   Loch Harray, frequent, generally distributed.
Limnicythere inopinata (Baird).   Loch Harray, generally distributed, but not
   very common.
Cytheridea torosa (Jones).   Loch Stenness, common ; Loch Harray, lower end.
Cythere pellucida, Baird.   Loch Stenness, frequent.
   ,,   *confusa*, B. and N.   Loch Stenness, frequent.
   ,,   *gibbosa*, Brady and Robertson.   Loch Stenness, rare.
Cytherura nigrescens (Baird).   Loch Stenness, frequent.
   ,,   *gibba* (Müller).   Loch Stenness, rare.

CLADOCERA.

Daphnia pulex.   Loch Harray, very abundant.

PROTOZOA.

FORAMINIFERA.

Miliolina agglutinans.   Loch Stenness, not common.
   ,,   *subrotunda*.   Loch Stenness, not common.
Truncatulina lobatula.   Loch Stenness, not common.
Polystomella umbilicatula.   Loch Stenness, not common.
Nonionina depressula.   Loch Stenness, not common.

CATALOGUE OF CLADOCERA REFERRED TO IN THE PRE-
CEDING REPORT, WITH SYNONYMS AND DESCRIPTIVE
NOTES.

SIDIDÆ.

Sida, Straus.

Sida crystallina (Müller).
    1776. *Daphne crystallina*, Müller, Zool. Dan. Prod., No. 2405.
    1821. *Sida crystallina*, Straus, Mem. Mus. Hist., v.
    1850. *Sida crystallina*, Baird, Brit. Entom., p. 107, pl. xii. figs. 3, 4 ; pl. xiii.
        fig. 1 *a-h*.

A large and easily observed species, readily distinguished from the next
by the external branch of the antenna having three and the other two articu-
lations. It is somewhat local, but is sometimes found in abundance.

Daphnella, Baird.

Daphnella brachyura (Lievin).
    1848. *Sida brachyura*, Lievin, Branch. d. Danziger Geg.
    1850. *Daphnella wingii*, Baird, Brit. Entom., p. 109, pl. xiv. figs. 1-4.
    1884. *Daphnella brachyura*, Herrick, Crust. of Minnesota, p. 21.

This is much smaller than the last, and at times equally numerous ;
both branches of antennæ have only two joints. The shorter branch
has the appearance of a third very short joint at the proximal end, but
the articulation is not complete.

HOLOPEDIDÆ.

Holopedium, Zaddach.

Holopedium gibberum, Zaddach.
    1855. *Holopedium gibberum*, Zaddach, Wiegmann. Archiv. für Naturges., Bd.
        21, p. 159, pl. viii. fig. 9.
    1865. *Holopedium gibberum*, G. O. Sars, Norg. Ferskvand. Clad. Cten., p. 57,
        pl. iv.
    1883. *Holopedium gibberum*, C. Beck, Jour. R. Mic. Soc., p. 778, pl. xi.
    1884. *Holopedium gibberum*, Herrick, *loc. cit.*, p. 22, pl. N, fig. 11.

This is a remarkable species, and differs from all other known Cladocera.
It "has the brood cavity greatly elevated, and the whole upper part of the
animal is covered by a jelly-like mass secreted as a protection or float." It
was added to the British fauna by C. Beck in 1881, who found it in
Grasmere, Cumberland, and afterwards in a few other places. Sars gives
a fine figure of the species in the work quoted above.

DAPHNIDÆ.

Ceriodaphnia, Dana.

Ceriodaphnia reticulata (Jurine).
    1820. *Monoculus reticulatus*, Jurine, Hist. Nat. Monoc., p. 139, pl. xiv. figs. 3, 4.
    1850. *Daphnia reticulata*, Baird, Hist. Entom., p. 97, pl. vii. fig. 5 ; pl. xii.
        fig. 1 (♂) and fig. 2 (var. quadrangula).
    1884. *Ceriodaphnia reticulata*, Herrick, *loc. cit.* ⁊. 38, pl. A, fig. 21.

The extremity of the post-abdomen is less rounded in this species than in C. (Daphnia) rotunda ; the test is also less strongly reticulated. "The members of this genus are danger signals from a hygienic point of view, for they frequent water containing decaying matter ; as many as 1,400 were counted in a single quart of such water" (Herrick).

? *Ceriodaphnia punctata* (P. E. Müller). Pl. i. figs. 3a, b.
    1886. *Ceriodaphnia punctata*, Herrick, *loc. cit.*, p. 39, pl. A, fig. 13.
    A form of *Ceriodaphnia*, with the carapace finely and closely punctate, occurred in
    Loch Balnagowan. It appears to differ from *C. reticulata* and *C. rotunda*. I have
    referred it, provisionally, to *C. punctata*, P. E. Müller, with which it seems to
    agree in the sculpture of the carapace and length of the anterior antennæ.

Scapholeberis (?), Schödler.

Scapholeberis mucronata (Müller). Pl. i. figs. 2, a–c.
    1776. *Daphne mucronata*, Müller, Zool. Dan. Prod., No. 2404.
    1850. *Daphnia mucronata*, Baird, Brit. Entom., p. 99, pl. x. figs. 2, 3.
    1863. *Scapholeberis mucronata*, Schödler, Neue Beitrage zur Naturges. der Clad.
    1884. *Scapholeberis mucronata*, Herrick, *loc. cit.*, p. 42, pl. iii. fig. 5., and
    1884. *Scapholeberis cornuta*, idem ibidem, p. 43, pl. T, fig. 6 (*forma*).

S. mucronata and *cornuta* are apparently forms of the same species. In Raith Lake I find specimens both with and without the spine on the front of the head, and also intermediate forms with the spine very little developed. The free margins of the shell, the head, and the antennæ are of a dark sooty colour.

Simocephalus, Schödler.

Simocephalus vetulus (Müller).
    1776. *Daphne vetula*, Müller, Zool. Dan. Prod., No. 2399.
    1785. *Daphnia sima*, Müller, Entomostraca, p. 91, t. 12, figs. 11, 12.
    1850. *Daphnia vetula*, Baird, Brit. Entom., p. 95, pl. x. figs. 1–1a.
    1863. *Simocephalus vetulus*, Schödler, Neue Beit. zur Naturges. der Clad.
    1884. *Simocephalus vetulus*, Herrick, *loc. cit.*, p. 46.

The shell of this species is finely and distinctly striate ; the posterior end is usually obliquely truncate and without a spine. It seems to be a widely distributed species.

Daphnia, Müller.

Daphnia pulex (Linné).
    1758. *Monoculus pulex*, Linné, Syst. Nat., 10th ed., vol. i. p. 635, No. 4.
    1776. *Daphne pulex*, Müller, Zool. Dan. Prod., p. 199, No. 2400.
    1850. *Daphnia pulex*, Baird, Brit. Entom., p. 89, pl. vi. figs. 1–3.
    1884. *Daphnia pulex*, Herrick, *loc. cit.*, p. 56.

A very common and variable species.

Daphnia longispina, Müller.
    1785. *Daphnia longispina*, Müller, Entomostraca, p. 88, t. 12, figs. 8–10.
    1850. *Daphnia pulex*, var. *longispina*, Baird, Brit. Entom., p. 89, pl. vii.
    figs. 3, 4.
    1860. *Daphnia longispina*, Leydig, Naturges. der Daphniden.
    1884. *Daphnia longispina*, Herrick, *loc. cit.*, p. 58.

This comes very near *D. pulex*, and may only be a form of that species ; at any rate, I find specimens that appear to be intermediate between these two forms.

Daphnia jardinii, Baird. Pl. i. figs. 4, 4a.
    1857. *Daphnia jardinii*, Baird, Edin. New Philos. Journ., vol. vi.

1862. *Daphnia cucullata*, G. O. Sars, Om Crust. Clad. iagttagne i Omegnen af
   Christi.
1866. *Hyalo-daphnia bavolinensis*, Schödler, Archiv. für Naturges., p. 24, pl. ii.
   fig. 8 ; pl. iii. fig. 15.
1884. *Daphnia cucullata*, Herrick, *loc. cit.*, p. 63.

This is distinguished from other British Daphnids by the head being
produced upward to a sharp point.

BOSMINIDÆ.

Bosmina, Baird.

Bosmina longirostris (Müller).
   1776. *Lynceus longirostris*, Müller, Zool. Dan. Prod., No. 2394.
   1861. *Bosmina longirostris*, G. O. Sars, Om de i Omegnen af Christi. forekom.
      Clad., p. 11.
   1867. *Bosmina longirostris*, Nor. and Brady, Nat. Hist. Trans. of Northumb.
      and Durham, vol. i. p. 357, pl. xxii. fig. 4.

This is a small species. The anterior antennæ are comparatively long and
curved. It occurs sometimes in considerable abundance.

Bosmina longispina, Leydig.
   1850. (?) *Bosmina longirostris*, Baird, Brit. Entom., p. 105, pl. xv. fig. 3.
   1860. *Bosmina longispina*, Leydig, Naturges. der Daphn., p. 207, pl. viii.
      fig. 62.
   1867. *Bosmina longispina*, Nor. and Brady, *loc. cit.*, p. 358, pl. xxii. figs. 1, 2.

This a much larger species than the last, being about double the size.
The anterior antennæ are not so curved and proportionally stouter ; the
spines of the postero-ventral angle are longer and stouter ; the outer or
convex margin is frequently furnished with two indistinct notches ; the
surface of the carapace is faintly striate longitudinally and sometimes
reticulate. This appears to be the form described by Baird. The two
species come very near each other, and may be only forms of one. I
found both forms in Loch Balnagowan, Lismore Island.

LYNCODAPHNIA.

Ilyocryptus, G. O. Sars.

Ilyocryptus sordidus (Lievin).
   1858. *Acanthocercus sordidus*, Lievin.
   1863. *Acantholeberis sordidus*, Norman, Ann. Nat. Hist., ser. iii. vol. xi. pl. xi
      figs. 6–9.
   1861. *Ilyocryptus sordidus*, G. O. Sars, *op. cit.*, p. 12; Andet Bidrag., p. 34
      (1862).
   1867. *Ilyocryptus sordidus*, Nor. and Brady, *loc. cit.*, p. 368.

Ilyocryptus is apparently not a swimmer. I did not observe any of those
captured making even an attempt to swim though kept alive for several
days; but if clumsy in their movements, which consisted of a kind of half-
crawling, half-tumbling process, they were by no means inactive. Adult
specimens were usually more or less coated with mud.

LYNCEIDÆ.

Eurycercus, Baird.

Eurycercus lamellatus (Müller).
   1776. *Lynceus lamellatus*, Müller, Zool. Dan. Prod., No. 3396.
   1850. *Eurycercus lamellatus*, Baird, Brit. Entom., p. 124, pl. xv. fig. 1.
   1884. *Eurycercus lamellatus*, Herrick, lo cit., p. 80, pl. H, figs. 5, 6.

A large and easily identified species. The post-abdomen is "a broad flattened plate with a very closely serrated margin."

Acroperus, Baird.

Acroperus harpæ, Baird.
    1835. Lynceus harpæ, Baird, Trans. Berw. Nat. Club, vol. i. p. 100, pl. ii.
        fig. 17.
    1841. Lynceus leucocephalus, Koch, Deutsch. Crust., Myriap. u Arach., p. 36,
        pl. x.
    1850. Acroperus harpæ, Baird, Brit. Entom., p. 129, pl. xvi. fig. 5.
    1884. Acroperus leucocephalus, Herrick, loc. cit., p. 81, pl. E, fig. 5, pl. i.
        fig. 9.

This species is generally distributed and moderately common. For full description, see Baird, loc. cit., p. 129.

Alonopsis (G. O. Sars).

Alonopsis elongata, G. O. Sars.
    1848. Lynceus macrurus, Lievin, Die Branch. der Dan. Geg., p. 41, pl. x. fig. 1.
    1862. Alonopsis elongata, G. O. Sars, Om. de i Omeg. af Christi. forekom. Clad.
        Andet Bidrag., p. 41.
    1867. Lynceus elongatus, Nor. and Brady, loc. cit., p. 376, pl. xviii. fig. 1,
        pl. xxi. fig. 2.
    1884. Alonopsis elongata, Herrick, loc. cit., p. 85.

This species is larger than Lynceus quadrangularis, for which it has probably sometimes been mistaken. It is easily distinguished "by the presence of the three spines, which spring from the terminal claws of the abdomen, and which are very conspicuous under the microscope." It does not seem to be a rare species.

Leydigia (Kurz).

Leydigia quadrangularis (Leydig). Pl. ii. figs. 5, a-b.
    1860. Lynceus quadrangularis, Leydig, Naturges. der Daphn., p. 221, pl. viii.
        fig. 59.
    1863. Alona leydigii, Schœdler, Neue Beit. zur Naturges. der Clad., p. 27.
    ?   Leydigia quadrangularis, Kurz, Dodekas Neuer Cladoceren.
    1884. Leydigia quadrangularis, Herrick, loc. cit., p. 88, pl. H, fig. 4.

This species is easily identified by the remarkably broad and almost semicircular post-abdomen. Norman and Brady include Alona leydigii, Schœdler, with Lynceus acanthocercoides, Fischer, and reproduce Fischer's figures of the species. These figures do not agree with Herrick's figure of post-abdomen of Leydigia quadrangularis, whereas the Lochgelly specimens agree perfectly with Herrick's description and figure. I have followed Herrick in adopting Kurz's name for this Lynceid.

Lynceus, Müller.

Lynceus tenuicaudis (G. O. Sars).
    1862. Alona tenuicaudis, G. O. Sars, op. cit., Andet Bidrag., p. 37.
    1867. Lynceus tenuicaudis, Nor. and Brady, loc. cit., p. 376, pl. xix. fig. 3.
    1884. Alona tenuicaudis, Herrick, loc. cit., p. 95, pl. i. fig. 2.

This species is easily recognised by the form and armature of the post-abdomen, which is long, with the sides nearly parallel, "incised below, lower angle armed with about six strong teeth, the remainder of the series small."

Lynceus quadrangularis, Müller.
    1776. *Lynceus quadrangularis*, Müller, Zool. Dan. Prod., p. 199, No. 2393.
    1850. *Alona quadrangularis*, Baird, Brit. Entom., p. 131, pl. xvi. fig. 4.
    1860. *Lynceus affinis*, Leydig, Naturges. der Daph., p. 223, pl. xi. figs. 68, 69.
    1867. *Lynceus quadrangularis*, Nor. and Brady, *loc. cit.*, p. 377, pl. xxi. fig. 5.
    1884. *Alona quadrangularis*, Herrick, *loc. cit.*, p. 97, pl. E, figs. 1, 2.

A widely distributed and common species, with which *Alonopsis elongatus* has probably been included, though this latter species is easily distinguished by its greater size and by the form of the post-abdomen, and the armature of the terminal claws of the same.

Lynceus costatus (G. O. Sars).
    1862. *Alona costata*, G. O. Sars, *op. cit.*, Audet Bidrag., p. 38.
    1867. *Lynceus costatus*, Nor. and Brady, *loc. cit.*, p. 379, pl. xviii. fig. 2 ; pl. xxi. fig. 7.
    1884. *Alona costata*, Herrick, *loc. cit.*, p. 97.

This species is considered by Herrick as a not very satisfactory one, being " founded practically upon the absence of the eighth seta of the antennæ." The post-abdomen is short, moderately broad, and of nearly equal breadth as far as the superior marginal spines extend, and the termination of the superior margin is produced to form a distinct angle.

Lynceus guttatus (G. O. Sars).
    1862. *Alona guttata*, G. O. Sars, *op. cit.*, Audet Bidrag., p. 38.
    1867. *Lynceus guttatus*, Nor. and Brady, *loc. cit.*, p. 380, pl. xviii. fig 6 ; pl. xxi. fig. 10.
    1884. *Alona guttata*, Herrick, *loc. cit.*, p. 94.

This comes near to the last, but the post-abdomen is shorter and more hollowed out on the superior margin immediately in front of the marginal spines. The surface of the carapace is frequently ornamented by cellular or pit-like impressions it is also rather smaller than *L. costatus*.

Alonella, Sars.

Alonella exigua (Lilljeborg).
    1853. *Lynceus exiguus*, Lillj., De Crust. in Scania Occurr., p. 79, pl. vii. figs. 9, 10.
    1854. *Lynceus excisus*, Fischer, Bull. de Soc. Imp. des Nat. de Moscou, p. 428, pl. iii. figs. 11-14.
    1862. *Alonella excisa*, G. O. Sars, *op. cit.*, Audet Bidrag., p. 52.
    1867. *Lynceus exiguus*, Nor. and Brady, *loc. cit.*, p. 384, pl. xviii. fig. 3 ; pl. xxi. fig. 3.
    1884. *Alonella excisa* and *exigua*, Herrick, *loc. cit.*, pp. 103-105, pl. E, fig. 6 ; pl. G, figs. 10, 11.

The two forms here referred to are by some authors considered as distinct species, but so far as I can make out from the descriptions and figures of the authors cited, the chief difference between *A. exigua* and *A. excisa* is, that in the first, the carapace is smooth or nearly so, whereas in the second it is distinctly reticulated and marked between the reticulations by distinct and close-set striæ. This difference does not seem to be of sufficient importance to be specific. The post-abdomen, so far as I can make out from the specimens examined, is similar in form in both, the antennæ also of both are similar. I have therefore included both in Lilljeborg's species.

Alonella nana (Baird). Pl. ii. fig. 6.
    1850. *Acroperus nanus*, Baird, Brit. Entom., p. 130, pl. xvi. fig. 6.
    1860. *Lynceus nanus*, Leydig, Naturges. der. Daphn., p. 228.
    1862. *Alonella pygmæa*, G. O. Sars, *op. cit.*, Audet Bidrag., p. 48.
    1867. *Lynceus nanus*, Nor. and Brady, *loc. cit.*, p. 396, pl. xviii. fig. 8 ; pl. xxi. fig. 3.
    1884. *Alonella pygmæa*, Herrick, *loc. cit.*, p. 105, pl. H, fig. 7.

The sculpture of this exceedingly small species is like that of no other *Lynceid*. It consists of bold and obliquely curved striæ, and not merely impressed lines, which impart a beautifully fluted appearance to the shell when the light strikes across the striæ. My specimens differ somewhat from that figured by Norman and Brady in the work cited, in the rostrum being less curved—being in fact nearly straight—whereas the figure shows the rostrum curving inwards considerably. My specimens agree better with the figure of *A. pygmœa* in Herrick's report. It seems to be a somewhat rare species.

Peracantha, Baird.

Peracantha truncata (Müller).
  1781. *Lynceus truncatus*, Müller, Entomostraca, p. 75, pl. ii. figs. 4–6.
  1850. *Peracantha truncata*, Baird, Brit. Entom., p. 136, pl. xvi. fig. 1.
  1867. *Lynceus truncatus*, Nor. and Brady, *loc. cit.*, p. 387, pl. xxi. fig. 9.
  1884. *Pleuroxus truncata*, Herrick, *loc. cit.*, p. 112.

This is one of the more distinctly characterised species, the posterior extremity of the carapace is subtruncate and "armed all along the edge with a series of large tooth-like processes, the hindermost of which are directed backwards; surface of shell divergently striated." I have followed Norman in retaining Baird's generic name, Peracantha.*

Pleuroxus, Baird.

Pleuroxus lævis, G. O. Sars.
  1844. *Lynceus trigonellus*, Zaddach, Syn. Crust. Pruss. Prod., p. 28.
  1861. *Pleuroxus lævis*, G. O. Sars, Om. de i Omeg. af Christi., forekom. Clad., p. 22.
  1867. *Lynceus lævis*, Nor. and Brady, *loc. cit.*, p. 389, pl. xviii. fig. 5 ; pl. xxi. fig. 14.
  1884. *Pleuroxus hastatus*, Herrick, *loc. cit.*, p. 108, pl. i. fig. xvi.

This species may be distinguished from the others by the form of the post-abdomen, which narrows gradually towards the claws. The claws are furnished with two basal spines. With a low power of the microscope the carapace appears faintly costate, but under a moderately high power ($\frac{1}{4}$ inch) the surface is seen to be reticulate, besides being finely striate longitudinally.

Pleuroxus trigonellus (Müller).
  1776. *Lynceus trigonellus*, Müller, Zool. Dan. Prod. No. 2395.
  1850. *Pleuroxus trigonellus*, Baird, Brit. Entom., p. 134, pl. xvii. fig. 3 (fem.).
  1850. *Pleuroxus hamatus*, idem ibidem, p. 136, pl. xvii. fig. 5 (mas.).
  1867. *Lynceus trigonellus*, Nor. and Brady, *loc. cit.*, p. 391, pl. xxi. fig. 11.
  1884. *Pleuroxus trigonellus*, Herrick, *loc. cit.*, p. 108.

A widely distributed species, but seldom very abundant. The post-abdominal claws are furnished with one basal spine.

Pleuroxus uncinatus, Baird.
  1850. *Pleuroxus uncinatus*, Baird, Brit. Entom., p. 135, pl. xvii. fig.
  1867. *Lynceus uncinatus*, Nor. and Brady, *loc. cit.*, p. 393, pl. xviii. fig. 9 ; pl. xxi. fig. 13.
  1884. *Pleuroxus uncinatus*, Herrick, *loc. cit.*, p. 114.

This seems to be widely distributed though not very common. It varies a good deal in the shell sculpture. There are also slight modifications in the form of the carapace ; the greater number of our specimens have the rostrum more closely appressed to the shell than in those figured by Norman and Brady, and in this respect they agree more closely with Baird's figure. The post-abdomen is similar to Norman and Brady's figure.

* See *Museum Normanianum*, pt. iii. (1886).

Chydorus, Baird.

Chydorus globosus, Baird.
  1850. *Chydorus globosus*, Baird, Brit. Entom., p. 127, pl. xvi. fig. 7.
  1867. *Lynceus globosus*, Nor. and Brady, *loc. cit.*, p. 398, pl. xx. fig. 5.
  1884. *Chydorus globosus*, Herrick, *loc. cit.*, p. 116, pl. F, figs. 1, 2, 3, and 9.

This fine species does not appear to be very common in Scotland. It was obtained by Mr D. Robertson, F.L.S., many years ago in the Paisley Canal (where I also found it later on), and in the Hebrides by Rev. A. M. Norman. It is recorded from several places both in England and Ireland.

Chydorus sphæricus (Müller).
  1776. *Lynceus sphæricus*, Müller, Zool. Dan. Prod., No. 2392.
  1850. *Chydorus sphæricus*, Baird, Brit. Entom., p. 126, pl. xvi. fig. 8.
  1884. *Chydorus sphæricus*, Herrick, *loc. cit.*, p. 116, pl. F. figs. 4, 7, 8, 10.

Chydorus sphæricus is one of the commonest species of the Lynceidæ.

Monospilus, G. O. Sars.

Monospilus tenuirostris (Fischer). Pl. i. fig. 1.
  1854. *Lynceus tenuirostris*, Fischer, Bull. de Soc. Imp. des Nat. de Moscou, p. 427, pl. M, figs. 7–10.
  1861. *Monospilus dispar*, G. O. Sars, *op. cit.*, p. 23.
  1867. *Monospilus tenuirostris*, Nor. and Brady, *loc. cit.*, p. 403, pl. xix. fig. 2 ; pl. xx. fig. 9.
  1884. *Monospilus dispar*, Herrick, *loc. cit.*, p. 119, pl. i. fig. 21.

This Lynceid differs from all the other species in having only the larval eye present during all stages of growth, and in the ecdysis not being complete, so that the old shell, instead of being cast off, remains attached to the new one, which projects more or less beyond the old. The result is that there are in adult specimens the appearance of more or less regular growth lines, similar to that observed in some Lamellibranch molluscan shells. The head is articulated to the carapace so slightly as to easily become detached from it. Herrick speaks of it as the "rarest of all Entomostraca." He regards it "as a degraded offshoot of the more typical stem of the Lynceidæ."

POLYPHEMIDÆ.

Polyphemus, Müller.

Polyphemus pediculus (Linné).
  1746. *Monoculus pediculus*, Linné, Faun. Suec., No. 2048.
  1776. *Polyphemus occulus*, Müller, Zool. Dan. Prod., No. 2417.
  1850. *Polyphemus pediculus*, Baird, Brit. Entom., p. 111, pl. xvii. fig. 1.
  1884. *Polyphemus pediculus*, Herrick, *loc. cit.*, p. 121, pl. B, figs. 4–6.

This is a moderately common species, especially in large sheets of water.

Bythotrephes, Leydig.

Bythotrephes longimanus, Leydig.
Bythotrephes longimanus, Leydig.
Bythotrephes cederstromii, Schœdler.

Bythotrephes often occurs in considerable abundance ; it is easily distinguished by its having an abdominal spine of enormous length.

LEPTODORIDÆ.

Leptodora, Lilljeborg (1861).

Leptodora hyalina, Lilljeborg.
  1874. *Leptodora hyalina*, Weismann, Bau und Lebenser, von Leptodora hyalina.
  1884. *Leptodora hyalina*, Herrick, *loc. cit.*, p. 123, pl. N, figs. 6, 7.

This is a large species ; some of my specimens from Loch Leven measure nearly half an inch in length. Both branches of the antennæ are four-jointed ; the body is elongated, not curved as in *Bythotrephes*. It is considered to be a somewhat rare species, but it was not very rare in Loch Leven.

EXPLANATION TO PLATES.

PLATE V.

Fig. 1.	*Monospilus tenuirostris*,		× 57
Fig. 2.	*Scapholebris mucronata*,	head with spine,	× 40
Fig. 2A.	,, ,,	head without spine,	× 40
Fig. 2B.	,, ,,	intermediate form of spine,	× 40
Fig. 2C.	,, ,,	post-abdomen,	× 110
Fig. 3.	(?) *Ceriodaphnia punctata*,		× 56
Fig. 3A.	,, ,,	post-abdomen,	× 132
Fig. 3B	,, ,,	part of carapace, highly magnified.	
Fig. 4.	*Daphnia jardini*,		× 40
Fig. 4A.	,, ,,	post-abdomen,	× 50

PLATE VI.

Fig. 5.	*Leydigia quadrangularis*,		× 52
Fig. 5A.	,, ,,	antenna,	× 150
Fig. 5B.	,, ,,	post-abdomen,	× 87
Fig. 6.	*Alonella nana*,		× 165
Fig. 7.	*Candona ambigua*, n. sp.,	seen from left side,	× 65
Fig. 7A.	,, ,,	seen from above.	× 65
Fig. 7B.	,, ,,	antenna,	× 165
Fig. 7C.	,, ,,	antennule,	× 145
Fig. 8.	,, *pubescens*,	seen from right side (for comparison),	× 60
Fig. 8A.	,, ,,	seen from above (for comparison),	× 60

PLATE V.

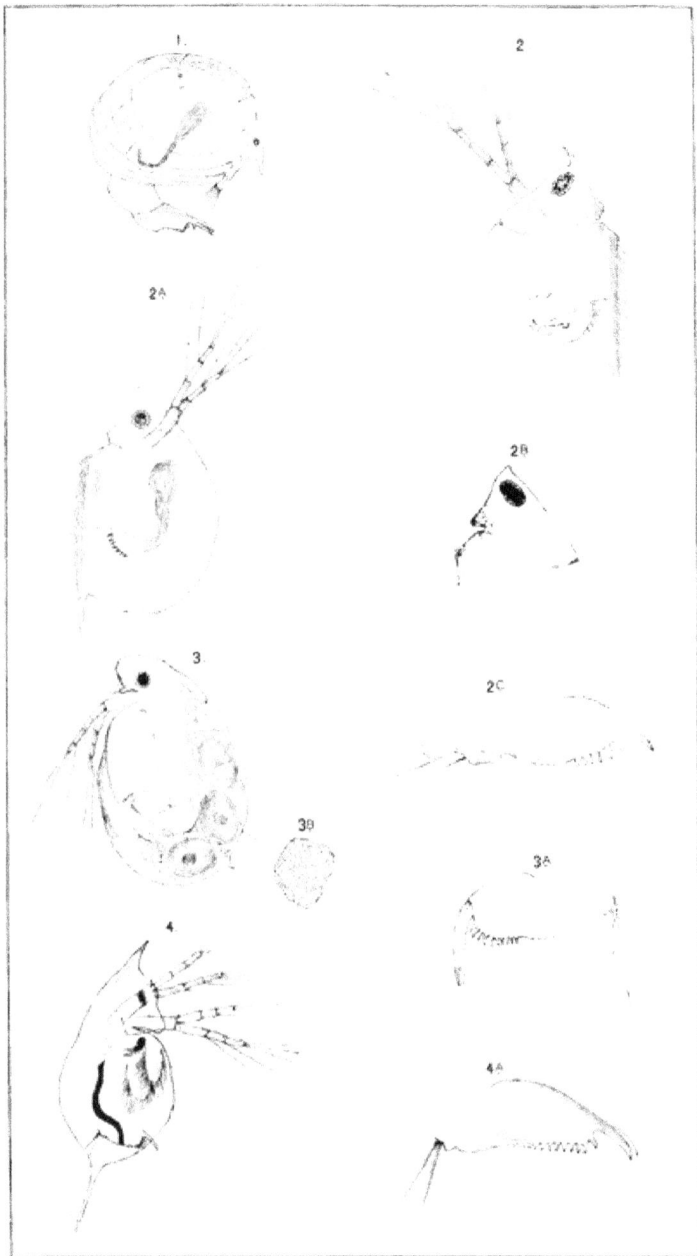

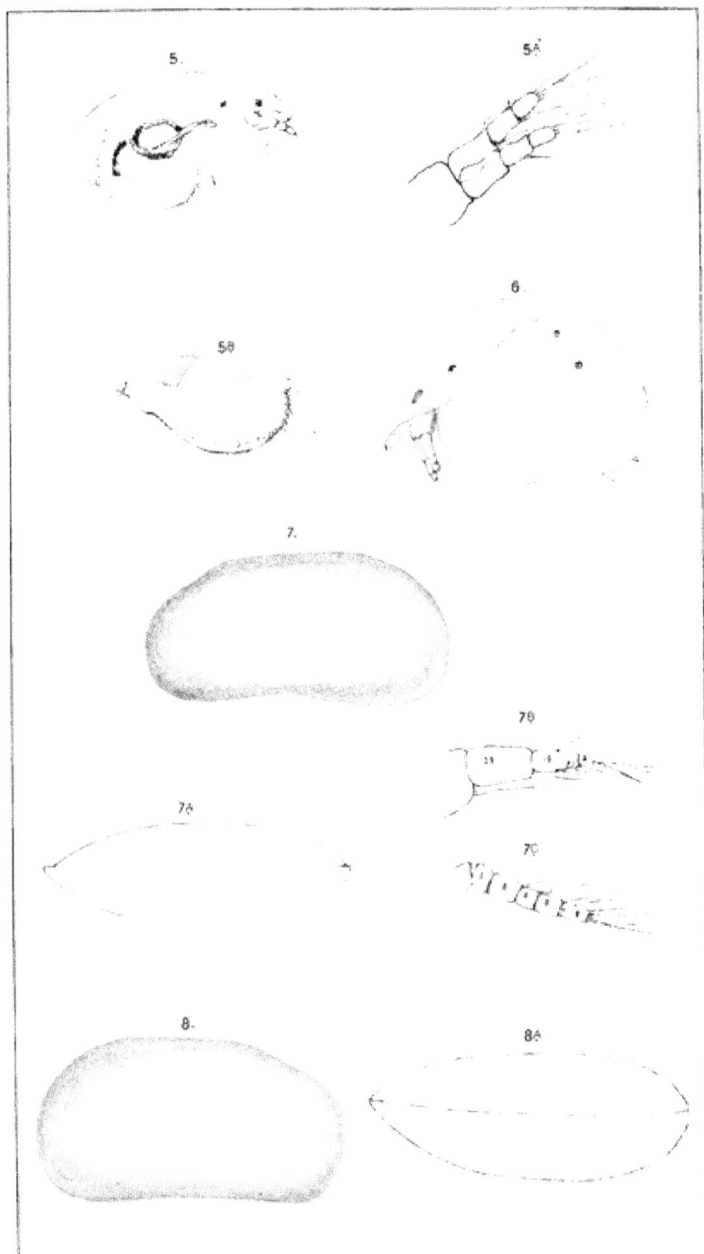

PLATE VI.

5.

5A

5B

6

7.

7B

7A

7C

8.

8A

.

V.—ADDITIONS TO THE FAUNA OF THE FIRTH OF FORTH.

Part III.   By Thomas Scott, F.L.S.

Forty-two species of Crustacea and one species of Mollusca are in this paper recorded for the Firth of Forth. Though several groups of the Crustacea are represented among those now recorded, by far the greater part belong to the *Copepoda*, of which there are 31 species, while the *Ostracoda, Cladocera, Amphipoda, Isopoda,* and *Schizopoda* are each represented by 1 species, and the *Decapoda* by 3.

It will be observed from this and previous papers that the Firth of Forth possesses a rich crustacean fauna. Previous to 1887 about 120 species had been recorded for the Forth ; whereas now there are, including those mentioned in the present paper, about 370 species, besides several which are still undetermined. But though the number of species has within recent years been considerably increased, I do not think it can yet be considered as approximately representative of the crustacean fauna of our estuary ; and I venture to predict that when the Firth of Forth becomes more thoroughly and systematically worked up the number of Crustacea will be little, if at all, short of 500 species.

COPEPODA.

GNATHOSTOMA.

Calanidæ.

Metridia armata, Boeck.

> 1865. *Metridia armata*, Boeck, 'Oversigt over de ved Norges Kyster iagttagne Copepoder,' p. 14.
>
> 1873. *Paracalanus hibernicus*, Brady and Robertson, 'Ann. and Mag. Nat. Hist.,' ser. 4, vol. xii. p. 126, pl. viii. figs. 1–3.
>
> 1878. *Metridia armata*, Brady, 'Mon. Brit. Copep.,' vol. i. p. 42, pl. ii. figs. 1–12 ; pl. lvi. figs. 19, 20.

Habitat.—In various parts of the Forth, between Inchkeith and May Island, I have obtained this species several times during the past year with both the surface and bottom tow-nets, but not very common. It is readily distinguished, especially the female form, by the elongated joints of the abdomen, and several of the first joints of the anterior antenna being strongly toothed on the upper edge.

Acartia discaudatus, Giesbrecht.

> 1881. *Dias discaudatus*, Giesbrecht,[*] 'Die Freilebenden Cope-poden der Kieler Foehrde,' p. 148, pl. iii. figs. 4, 22, 23 ; pl. v. fig. 18 ; and other plates.

Habitat.—Between Portobello and Cockenzie, well inshore. I obtained several specimens of this species ; the peculiar form of the caudal furca first attracted my attention. In the female the last abdominal segment is much broader posteriorly ; the furca are broad flattened plates, which together are in breadth about equal to the length of the furca and last abdominal segment. This appears to be the first British record of *Acartia discaudatus.*

---

[*] *Vierter Bericht der Commission zur wissenschaftlichen Untersuchung der deutschen Meere, in Kiel,* 1887–1881.

CYCLOPIDÆ.

Oithona setiger (Dana).

1852. *Acartia setiger*, Dana, 'U.S. Expl. Exped.'

Habitat.—East of Inchkeith, among bottom tow-net material, several specimens of this species were obtained in March of this year (1891). They were readily distinguished from the more common *O. spinirostris*, Claus, by the extremely long and slender setæ of the anterior antennæ and of the fifth feet. This appears to be the first British record of *Oithona setiger*.

NOTODELPHYIDÆ.

Doropygus normani, Brady.

1878. *Doropygus normani*, Brady, 'Mon. Brit. Copep.,' vol. i. p. 136, pl. xxxii. figs. 1–14.

Habitat.—In the branchial cavities of Ascidians, obtained off Mussel-burgh by the trawl-net and dredge. This *Doropygus* is comparatively common, and is the only species observed, but I do not often find it in other parts of the Forth. I have examined many Ascidians from deep water in mid channel east of Inchkeith, and find them much freer of entomostracan parasites than those from the inshore waters of the south side of the Firth.

BUPRORIDÆ.

Enterocola eruca, Norman.

1868. *Enterocola eruca*, Norman. Last Shetland Dredging Report, 'Brit. Assoc. Report,' p. 300

1878. *Enterocola eruca*, Brady, *loc. cit.*, p. 147.

Habitat.—In the intestine of Ascidians taken by trawl-net off Mussel-burgh in March of this year (1891). Four specimens were obtained from as many Ascidians—one in each Ascidian—and from *inside* the intestine, thus differing in their habitat very markedly from the other entomostracan parasites found in Ascidians. Dr G. S. Brady, to whom I sent a specimen, identifies it as the *Enterocola eruca*, Norman. Dr Norman obtained a single specimen adhering to the intestine of *Ascidia intestinalis* while dredging among the Shetland Islands, which appears to be the only previous British record. Dr Norman, in his 'Report to the British Association in 1868,' says that it is 'allied to *Enterocola fulgens*, Van Beneden ("Recherches sur la Faune littorale de Belgique Crustacés," 1861, p. 149, pl. xxvi.), but is apparently distinct.' *

HARPACTICIDÆ.

Jonesiella spinulosa (Brady and Robertson).

1875. *Zosime spinulosa*, B. and R. 'Brit. Assoc. Report,' p. 196.

1880. *Jonesiella spinulosa*, Brady, *loc. cit.*, vol. ii. p. 41, pl. xlviii. figs. 14–18 ; pl. xlix. figs. 14, 15.

Habitat.—Largo Bay, dredged, several specimens, ♂ ♀. This is quite distinct from *Jonesiella fusiformis* (B. and R.).

Ectinosoma atlanticum (Brady and Robertson).

1873. *Microstella atlantica*, Brady and Robertson, 'Ann. and Mag. Nat. Hist.,' ser. 4, vol xii. p. 130, pl. ix. figs. 11–16.

* *Brit. Assoc. Report*, 1868, p. 300.

1880. *Ectinosoma atlanticum,* Brady, *loc. cit.,* vol. ii. p. 13, pl. xxxviii. figs. 11-19.

Habitat.—East of Inchkeith, in surface tow-net, November 1890; frequent. This is a very small species, and consequently may be easily overlooked; but though frequent in a tow-net collection in November, I have obtained it very rarely since. I am inclined to believe that some of the recent additions to the Forth Crustacea, at least in the case of pelagic species, are only occasional visitants, through having wandered out of their way, or been carried by currents within the confines of the estuary. I may state in proof of this that *Thalestris serrulatus,* Brady, of which several specimens were obtained in a surface tow-netting in 1889, has not since been observed by me anywhere within the Firth of Forth. Though this appears to be the first recorded observance of *Ectinosoma atlanticum* on the east of Scotland, it has been obtained from several places on the West Coast, especially the Loch Fyne district. Dr T. Wemyss Fulton has obtained it in Upper Loch Fyne, and Mr I. C. Thompson, F.L.S., records it from Lower Loch Fyne; also from Loch Striven, off Kirn, and south of Arran, Firth of Clyde. It was first described by Brady and Robertson from specimens collected by Mr E. C. Davison in the open sea, to the west and south-west of Ireland.

Tachidius discipes, Giesbrecht.

1853. *Tachidius brevicornis,* Lilljeborg, 'Decr.,' p. 196 (non-Müller).
1880. *Tachidius brevicornis,* Brady, *loc. cit.,* vol. ii. p. 20, pl. xxxvii.
1881. *Tachidius discipes,* Giesbrecht, *op. cit.,* p. 108, pl. ii. fig. 4; pl. iv. figs. 25, 28, &c.

Habitat.—Brackish-water pools at the mouth of Cocklemill Burn, Largo Bay, July 1890.

Canthocamptus palustris, Brady.

1880. *Canthocamptus palustris,* Brady, *loc. cit.,* vol. ii. p. 53, pl. xxxix. figs. 13-23.

Habitat.—Brackish-water pools, May Island, 1889. This, though like the last, a brackish-water species, and inhabiting similar localities, need not be mistaken for it. *Canthocamptus palustris* is more slender, and the anterior antennæ are longer and eight-jointed; the fifth feet are also different.

Dactylopus tenuiremus, Brady and Robertson.

1875. *Dactylopus tenuiremus,* B. & R., 'Brit. Assoc. Report,' p. 197.
1880. *Dactylopus tenuiremus,* Brady, *loc. cit.,* vol. ii. p. 115, pl. lvi. figs. 12-18.

Habitat.—Kirkcaldy Bay, dredged 27th March 1891, readily distinguished from other species of the genus by the form of the anterior antennæ, which are slender and comparatively long.

Dactylopus flavus, Claus.

1866. *Dactylopus flavus,* Claus, 'Die Copepoden-Fauna von Nizza, p. 28, t. iii. figs. 13-16.
1880. *Dactylopus flavus,* Brady, *loc. cit.,* vol. ii. p. 116, pl. lvi. figs. 1-11.

Habitat.—With the last, and also Largo Bay. This *Dactylopus* is easily distinguished from the others by its broad depressed form; it is a small species, being little over half a millimetre in length.

Dactylopus brevicornis, Claus.

    1866. *Dactylopus brevicornis,* Claus, *op. cit.,* p. 29, t. iii. figs. 20–25.

    1880. *Dactylopus brevicornis,* Brady, *loc. cit.,* vol. ii. p. 118, pl. lvii. figs. 10–12 ; pl. lviii. fig. 14.

Habitat.—Largo Bay, dredged April 1891. The dredge brought up a quantity of broken weeds, zoophytes, &c., which yielded a large number of Entomostraca and other organisms, including several not previously observed in the Forth. *D. brevicornis* has very short and stout anterior antennæ, by which it is readily distinguished.

Dactylopus minutus, Claus.

    1863. *Dactylopus minutus,* Claus, 'Die frei lebenden Copepoden,' p. 126, t. xvi. figs. 14, 15.

    1880. *Dactylopus minutus,* Brady, *loc. cit.,* vol. ii. p. 119, pl. lxvii. figs. 12–14.

Habitat.—Largo Bay, with the last, a small species with two ova-sacs ; only a few specimens were obtained. A somewhat large but slender form of Dactylopus was of frequent occurrence in the dredged material from Largo Bay, which, as far as I could make out, appears to be a variety of *D. tisboides,* Claus. The more typical form of the species having the pellucid markings on the outer branch of the fifth feet also occurred.

Thalestris helgolandica, Claus.

    1863. *Thalestris helgolandica,* Claus, 'Die frei lebenden Copepoden,' p. 131, t. xvii. figs. 12–21.

    1880. *Thalestris helgolandica,* Brady, *loc. cit.,* vol. ii. p. 123, pl. lxi. figs. 9–14.

Habitat.—Largo Bay, with the others. One or two specimens that appeared to belong to this species were obtained. The form of the second foot-jaw, the long slender spines of the first feet, and the peculiar form of the fifth feet, are sufficiently characteristic to allow of this species being readily distinguished from other *Thalestris* ; it seems to be a rare species.

Thalestris clausii, Norman.

    1868. *Thalestris clausii,* Norman, 'Brit. Assoc. Report,' p. 297.

    1873. *Parathalestris clausii,* Brady and Robertson, 'Ann. and Mag. Nat. Hist.,' vol. xii. p. 136.

    1880. *Thalestris clausii,* Brady, *loc. cit.,* vol. ii. p. 128, pl. lxii. figs. 1–12.

Habitat.—Largo Bay, frequent among weed dredged in 7 to 8 fathoms. This is a robust species, the first feet are stout, with strong and comparatively short terminal claws.

Thalestris rufo-violascens, Claus.

    1866. *Thalestris rufo-violascens,* Claus, 'Die Copepoden-Fauna von Nizza,' p. 33, t. iv figs. 18–22.

    1880. *Thalestris rufo-violascens,* Brady, *loc. cit.,* vol. ii. p. 131, pl. lxi. figs. 1–8.

Habitat.—Largo Bay, obtained in the same dredging with the last. This is apparently a rare species ; as yet I have only observed one or two specimens. The only other British localities where it has been obtained, so far as I know, are those mentioned in Dr Brady's monograph, viz., 'Shetland and Firth of Clyde.' Dr Brady has kindly examined one of the Forth specimens, and corroborated my diagnosis.

Porcellidium fimbriatum, Claus.

> 1863. *Porcellidium fimbriatum*, Claus., 'Die frei lebenden Cope-
> poden,' p. 140, t. xxii. fig. 1.
> 1880. *Porcellidum fimbriatum*, Brady, *loc. cit.*, vol. ii. p. 167, pl.
> lxx. figs. 1–4.

Habitat.—Largo Bay, among weed brought up by the dredged ; a few
specimens only were obtained. The caudal segments of the Forth specimens
have four small setæ at the outer and one at the inner angles ; the outer
margins of the last thoracic segments have a deep notch nearly midway
between their extremities and the extremities of the preceding segment,
and a stout ciliated tooth projects from the upper edge of the notch ;
the last thoracic segment has its outer margins also much more strongly
ciliated than the margins of any of the preceding segments.

Lichomolgus furcillatus, Thorell.

> 1859. *Lichomolgus furcillatus*, Thorell, 'Om Krustaceer i Ascedier,'
> p. 74, t. xiii. fig. 20.
> 1880. *Lichomolgus furcillatus*, Brady, *loc. cit.*, vol. iii. p. 49, pl.
> lxxxviii. figs. 10–14.

Habitat.—In branchial sac of an Ascidian, brought up in the trawl-net
west of May Island, 31st April 1890 ; one specimen.* The anterior
antennæ of this species are six-jointed, which thus differs from *L. fucicolus*
Moreover, this last appears to be 'an entirely free-living' species.
Another species, *L. forficula*, that I have frequently obtained in large
Ascidians in East Loch Tarbert, Loch Fyne, and in Scapa Flow, Orkney,
has not yet been observed in the Forth.

CORYCÆIDÆ.

Monstrilla rigida (J. C. Thompson), ♀.

> 1887. *Cymbasoma rigida*, J. C. Thompson, 'Jour. Linn. Soc.
> (Zool.),' vol. xx.. p. 154, pl. xiii. figs. 1–4.
> 1890. *Monstrilla rigida*, Bourne, 'Quart. Jour. Microscopical
> Science,' vol. xxx. (new ser.), p. 565, pl. xxxvii. figs. 8,
> 11–12.

Habitat.—Off Musselburgh, in bottom tow-net. A specimen, apparently
belonging to this species, was obtained 25th March this year (1891).† It
has three setæ to each of the caudal furca, the penultimate abdominal
segment is considerably longer and stouter than the last one, and furnished
with a genital appendage. The genital appendage (setæ) is very little
longer than the furcal setæ. Several other specimens belonging to this
interesting genus have been obtained in the Firth of Forth and in St
Andrews Bay during the past year, but I have not yet had time to
examine them.

PARASITA.

CALIGIDÆ.

Caligus diaphanus, Nordmann.

> 1832. *Caligus diaphanus*, Nordmann, 'Microgr. Beitr.,' ii. p. 26.
> 1850. *Caligus diaphanus*, Baird, ' Brit. Entom.,' p. 269, pl. xxxii.
> fig. 1.

---

* I have recently obtained a few more specimens east of Inchkeith in Ascidians.
† J. C. Thompson, F.L.S., F.R.M.S., Liverpool, to whom I sent the specimen,
confirms my diagnosis. Little or nothing is yet known of the internal structure of
Monstrilla ; though the mouth is quite distinct, there is little, if any, trace of an
alimentary track, but there surely must be such a track, however indistinct.

Habitat.—On cod, pollack, and other fishes. This appears to be the largest of the British *Caligi ;* the carapace of some of my specimens measures 8 mm. in breadth. It is not so common as the next species.

Caligus rapax, M. Edwards.

> 1840. *Caligus rapax,* M. Edwards, 'Hist. Nat. Crust.,' iii. p. 453, No. 6, t. 38, fig. 9.
> 1850. *Caligus rapax,* Baird, 'Brit. Entom.,' p. 270, pl. xxxii. figs. 2, 3.

Habitat.—On various fishes, as the cod, saith, pollack, lumpsucker, taken also with surface and bottom tow-nets. A common species. The young (*Chalimus scombri,* Burmeister) frequent on the lumpsucker.

Caligus mülleri, Leach.

> 1816. *Caligus mülleri,* Leach, 'Enc. Brit. Supp.,' i. t. 20, fig. 68.
> 1850. *Caligus mülleri,* Baird, *loc. cit.,* p. 271, pl. xxxii. figs. 4, 5.

Habitat.—The abdomen of this species is decidedly shorter than that of *Caligus rapax,* the shape of the carapace is also more oblong. It does not appear to be very common.

Lepeoptheirus pectoralis, Müller.

> 1776. *Lernæa pectoralis,* Müller, 'Zool. Dan.,' i. p. 41, t. 33, fig. 7.
> 1832. *Lepeoptheirus pectoralis,* Nordmann, 'Mikr. Beitr.,' ii, p. 30.
> 1850. *Lepeoptheirus pectoralis,* Baird, *loc. cit.,* p. 275, pl. xxxii. fig. 10.

Habitat.—Adhering to the pectoral fins of the plaice (*Pleuronectes platessa*), frequent. *Lepeoptheirus* differs from *Caligus* by not having sucking disks on the frontal plates.

Lepeoptheirus thompsoni, Baird.

> 1850. *Lepeoptheirus thompsoni,* Baird, *loc. cit.,* p. 278, pl. xxxiii. fig. 2.

Habitat.—On the gills of the turbot; frequent. In this species the thorax is larger and the abdomen longer than in the previous one.

CECROPIDÆ.

Cecrops latreillii, Leach.

> 1816. *Cecrops latreillii,* Leach, 'Enc. Brit. Supp.,' vol. i. t. 20, figs. 1–5.
> 1850. *Cecrops latreillii,* Baird, *loc. cit.,* p. 293, pl. xxxiv. figs. 1, 2.

Habitat.—Attached to the gills of the short sunfish (*Orthagoriscus molæ*) caught in the Forth, October 1890 (Andrew Scott). This species has been taken by Edward on the gills of both the *short* and *oblong* sunfish in the Moray Firth. Specimens of *C. latreillii,* from the Forth are also in the Edinburgh Museum (E. Clark).

Læmargus muricatus, Kroyer.

> 1838. *Læmargus muricatus,* Kroyer, 'Tidsskrift,' vol. i. p. 487, t. 5, figs. A, B, C, D.
> 1850. *Læmargus muricatus,* Baird, *loc. cit.,* p. 295, pl. xxxiv. figs. 3, 4.

Habitat.—Attached to cavities hollowed out of the flesh of a short sunfish caught in the Forth, October 1890 (Andrew Scott). It has also been recorded by Edward from the Moray Firth, from the same species of sunfish. *Læmargus* appears to be more restricted in its habitat than *Cecrops.*

CHONDRACANTHIDÆ.

Lernentoma cornuta (Müller).
1776. *Lernæa cornuta*, Müller, ' Zool. Dan.,' vol. i. t. 33, fig. 6.
1817. *Chondracanthus cornutus*, Cuvier, ' Regne. An.,' vol. iv.
    p. 258.
1850. *Lernentoma cornuta*, Baird, *loc. cit.*, p. 328, pl. xxxv. fig. 2.

Habitat.—Attached to the gills of the long rough dab.

Lernentoma lophii (Johnston).
1836. *Chondracanthus lophii*, Johnston, Loudon's ' Mag. Nat.
    Hist.,' vol. ix. pp. 81, 82, figs. 16, *a–c.*
1850. *Lernentoma lophii*, Baird, *loc. cit.*, p. 330, pl. xxxv. fig. 3.

Habitat.—On *Lophius piscatorius*, caught in the Forth; frequent. It
has long twisted ovigerous appendages. It is commonly found inside
the gill-pouches of the *Lophius.*

Lernentoma asellina (Linné).
1761. *Lernæa asellina*, Linn., ' Fauna Suecica,' 2101.
1822. *Lernentoma asellina*, Blainville, ' Journ. Phys.,' xcv. 441.
1850. *Lernentoma asellina*, Baird, *loc. cit.*, p. 329, pl. xxxv. fig. 4.

Habitat.—On the gills of the common gurnard (*Trigla gurnardus*)
caught in the Firth of Forth, not common. Also recorded for the Moray
Firth, by T. Edward.

Trebius caudatus, Kröyer.
1837. *Trebius caudatus*, Kröyer, ' Tidsskrift,' vol. ii. p. 30, t. 1.,
    fig. 4.
1850. *Trebius caudatus*, Baird, *loc. cit.*, p. 280, pl. xxxiii.
    figs. 3, 4.

Habitat.—On the back of a large grey skate (*Raia batis*) caught
near May Island.

ANCHORELLIDÆ.

Anchorella uncinata (Müller).
1776. *Lernæa uncinata*, Müller, ' Zool. Dan.,' vol. i. t. 33, fig. 2,
1832. *Anchorella uncinata*, Nordmann, ' Mikr. Beitr.,' vol. ii. p.
    102, t. 8, figs. 8, 9 ; t. 10, figs. 1–5.
1850. *Anchorella uncinata*, Baird, *loc. cit.*, p. 337, pl. xxxv. fig. 9,

Habitat.—Adhering to the inside of the mouth, on the gills, gill-
covers, and other parts of the cod ; also found on the haddock. This
species is of frequent occurrence, especially on large cod taken in the
Firth of Forth.

Anchorella rugosa, Kröyer.
1837. *Anchorella rugosa*, Kröyer, ' Tidsskrift,' vol. i. p. 284, t. 2,
    fig. 7.
1850. *Anchorella rugosa*, Baird, *loc. cit.*, p. 338, pl. xxxv. fig. 8.

Habitat.—Attached to the gills of the cat-fish (*Anarrhichas lupus*) ;
rather rare.

PENELLADÆ.

Lerneonema spratta (Sowerby).
1806. *Lernæa spratta*, Sowerby, ' British Miscellany,' t. 68.
1840. *Lerneonema monillaris*, M. Edwards, ' Hist. Nat. Crust.,'
    vol. iii. p. 525, t. 41, fig. 5.
1850. *Lerneonema spratta*, Baird, *loc. cit.*, p. 341, pl. xxxv. fig. 10.

Habitat. — Attached to the eye of a sprat (?) in one of the Leith dry docks. One day in August last year, my son (John Scott) observed in one of the Leith dry docks (one of the two belonging to the Leith Dock Commission), from which the water was being pumped out, a parasite adhering to the eye of a young herring or sprat— one of a small shoal that had entered the dock while the gates were open — which, from his description, appears to belong to this species. 'The parasite,' he says, 'had a long slender body, and long ovisacs.' The eye to which the parasite was attached appeared to be blind, for on one occasion he reached out his hand so close to the fish as almost to touch it. As he was about to lay hold of the fish it happened to turn so as to see his hand, and instantly darted off. A considerable number of the fish were left dead after the water was pumped out of the dock, but he failed to find among them the one with the parasite —it had probably been carried out with the water (as many of the other fish were) that had been pumped from the dock. *Lerneonema spratta* seems to be very rare in the Forth, although I have examined hundreds of sprats, I have failed to observe a single specimen of the parasite. So far as my son can remember, the body of the parasite he saw was of a greenish colour. He and others saw the fish with the parasite attached swimming about for a considerable time before the water was all pumped out of the dock, but ultimately lost sight of it.

LERNEOCERADÆ.

Lernæa branchialis, Linné.
    1767. *Lernæa branchialis,* 'Systema Naturæ,' edit. 12th.
    1850. *Lernæa branchialis,* Baird, *loc. cit.,* p. 344, pl. xxxv. fig. 12.

Habitat.—Attached to the gills of the cod, haddock, and whiting. Frequent on such fish taken by us in the Forth, mostly on fish in poor condition, sometimes so many are adhering to the gills, that the gill-covers are kept from closing because of them.

The parasitic Copepoda are an interesting group, and doubtless play an important part in the economy of nature. Dr Baird's work, which I have mainly followed in the preceding notes, is the only monograph we have on the British species, though several records have since been published through scientific societies and otherwise.

OSTRACODA.

CYTHERIDÆ.

Cythere gibbosa, Brady and Robertson.
    1869. *Cythere gibbosa,* Brady and Robertson, 'Ann. and Mag. Nat. Hist.,' ser. iv., vol. iii. p. 368, pl. xxi. figs. 1–3.
    1889. *Cythere gibbosa,* Brady and Norman, 'Mon. M. and F.-W. Ostrac. of the N. Atlantic and N.-W. Europe,' p. 136, pl. xiv. figs. 30, 31.

Habitat.—Brackish pools at the mouth of the Cocklemill Burn, Largo Bay ; obtained July 12, 1890 ; frequent. This is very much a brackish-water species. It has been recorded from localities more or less purely marine. It would be interesting to know if specimens from such localities were living ; those now recorded were so.

CLADOCERA.

POLYPHEMIDÆ.

Podon polyphemoides, Leuckart.
 1859. *Podon polyphemoides*, Leuckart, 'Arch. f. Naturg. I.,' p.
  263, t. vii. fig. 5.
 1859. *Podon polyphemoides*, P. E. Müller, *idem*, p. 220, t. vi.
  figs. 5, 6.
 Habitat.—Firth of Forth, in surface tow-net east of Inchkeith; not
very common; December 1890. This species has also been recorded for
St Andrews Bay by Professor M'Intosh.

AMPHIPODA.

DULICHIIDÆ.

Dulichia porrecta (Spence Bate).
 1857. *Dyopedos porrecta*, Spence Bate, 'Ann. Nat. Hist.,' 2nd ser.,
  vol. xix. p. 151.
 1868. *Dulichia porrecta*, Bate and Westwood, 'Brit. Sess.-Eyed
  Crust.,' vol. ii. p. 21.
 Habitat.—Off Musselburgh, Firth of Forth, among trawled material
(weeds, zoophytes, &c.); a few specimens only; April 1891. The propodos of the second Gnathopoda are armed with the two prominent forward-directed teeth characteristic of the species.

ISOPODA.

ASELLIDÆ.

Leptaspidia brevepes, Spence Bate and Westwood.
 1868. *Leptaspidia brevepes*, Spence Bate and Westwood, 'Brit.
  Sess.-Eyed Crust.,' vol. ii. p. 333.
 1888. *Leptaspidia brevepes*, Robertson, 'Amphipoda and Isopoda
  of the Firth of Clyde,' p. 76.
 Habitat.—Off Musselburgh, Firth of Forth; rare; April 1891. This
species, seen from above, is broadly ovate; it is also considerably
depressed. The species is a very small one, being little over the one-
twentieth of an inch in length. It was first obtained by Mr Robertson
at Cumbrae, and described in 'British Sessile-Eyed Crustacea.' I have
observed it in East Loch Tarbert, Loch Fyne; but it does not appear to
be a common species anywhere. There is no danger of mistaking it for
Munna, though it seems to be nearly allied to that genus.

SCHIZOPODA.

MYSIDÆ.

Mysidopsis angusta, G. O. Sars.
 1863. *Mysidopsis angusta*, G. O. Sars, 'Zool.-Reise,' p. 30.
 1872. *Mysidopsis angusta*, G. O. Sars, 'Mon. over Norges My-
  sider,' pt. ii. p. 27, pl. viii. figs. 14–24.
 1876. *Mysidopsis angusta*, G. O. Sars, 'Middelhavet's Invert.
  Fauna,' p. 101.

Habitat.—East of Inchkeith. Rare. Specimens have been obtained from other parts of the estuary, but seldom more than one or two at a time. It is easily distinguished from the other two British species by the slightly forked telson. I have also obtained *Mysidopsis angusta* in Loch Fyne and in the Moray Firth. There are now twenty well-authenticated species of Schizopoda included in the Forth fauna.

DECAPODA.

CRANGONIDÆ.

Crangon fasciatus, Risso.*

  1816. *Crangon fasciatus*, Risso, 'Crust. de Nice,' p. 82, t. iii. fig. 5.
  1853. *Crangon fasciatus*, Bell, ' Brit. Stalk-Eyed Crust.,' p. 259.

Habitat.—Off Musselburgh, Firth of Forth. Two specimens of this pretty little species were captured with the bottom tow-net in 4 to 5 fathoms water. The contrast of colour was very marked in these specimens, the cephalo-thoracic and fourth abdominal segments being dark chocolate brown, and a band of the same colour across the tail appendages, while the other parts were nearly white. The squarely-truncate rostrum is a well-marked characteristic of the species.

Crangon neglectus, G. O. Sars.*

  1882. *Crangon neglectus*, G. O. Sars.

Habitat.—Largo Bay ; several specimens among weeds dredged in about 8 to 9 fathoms water. This is quite distinct from the previous species, and also from *C. nanus*. The rostrum is broadly rounded in this species. It does not appear to have been previously recorded for the East Coast of Scotland.

Nika edulis, Risso.

  1816. *Nika edulis*, Risso, ' Crust. de Nice,' p. 85, pl. lxxxiii. fig. 3.
  1853. *Nika edulis*, Bell, ' Brit. Stalk-Eyed Crust.,' p. 275.

Habitat.—Seven to eight miles east of May Island ; rare. It has been obtained by Edward in the Moray Firth, where I also observed it. It is recorded from Loch Fyne, and from Brodick Bay, Firth of Clyde, by Professor J. R. Henderson,† but nowhere does it appear to be very common.

MOLLUSCA.

BULLIDÆ.

Cylichna nitidula, Lovèn.

  1846. *Cylichna nitidula*, Lovèn, ' Ind. Moll. Scand.,' p. 10.
  1853. *Cylichna nitidula*, F. and H., ' Brit. Moll.,' vol. iii. p. 515, pl. cxiv. fig. 6.
  1867. *Cylichna nitidula*, Jeffreys, ' B   Conch.,' vol. iv. p. 412 ; vol. v. pl. xciii. fig. 2.

Habitat.—West of May Island ; rare. A few specimens were obtained from the stomachs of haddocks—a source from which a few other rarities have been obtained.

---

  * See also Rev. A. M. Norman's paper in *Fishery Board's Fifth Annual Report* (1886), p. 156, for further references.
  † ' Decapod and Schizopod Crust. of the Firth of Clyde,' p. 33 (1886).

ADDITIONAL NOTES.—Since the preceding notes were in the hands of the printer I have been enabled to add the following species, the Rev. A. M. Norman, F.R.S., having kindly named them for me.

Caligus isonyx, Steenstrup and Lütken.

Habitat.—Inside the gill-covers of the common gurnard (*Trigla gurnardus*). I find this *Caligus* of frequent occurence on specimens of the common gurnard taken by us in the Forth, but on no other kind of fish.

Hæmobaphes cyclopterina, Fabr

Habitat.—Attached to the gills of the pogge (*Agonus cataphractus*). This interesting species was obtained by Mr Peter Jamieson, Assistant Naturalist to the Fishery Board, adhering to the gills of a pogge taken from the stomach of a cod at Dunbar, during April of this year (1891). Dr A. M. Norman informs me that this is the first time *Hæmobaphes* has been observed in the British seas. The arrangement of the ovaries differs from that of all other species with which I am acquainted. Each ovary resembles a coiled-up rope, the coils being of equal diameter throughout, and resting the one on the other, thus forming a miniature cylinder. The length of the parasite, including the ovaries, is fully half an inch.

Charopinus Dalmanni, Retzius.

Habitat.—Attached to the inside of one of the spiracles of a large grey skate (*Raia batis*) caught by the 'Garland's' trawl S.E. of May Island. This is a large species. The specimen obtained measures about 2 inches in length, including the ova-sacs.

Tauria medusarum (O. Fab.), [*Hyperia tauriformis*, Bate and Westwood, Brit. sess.-eyed Crust., ii. p. 519 (1868)]. Taken with the surface tow-net in the vicinity of the Bass Rock, November 1890. Several specimens were obtained. Both gnathopods of this species are chelate, the lower part of the carpus being produced forwards so as to reach the extremity of the propodos. The inner margins of the produced part of the carpus and of the propodos is serrated. This species was first obtained in British waters by Thomas Edward of Banff.

PART IV. By Thomas Scott, F.L.S. (Plates VII.-XIII.).

This, the fourth contribution towards a better knowledge of the fauna of the Firth of Forth, especially the invertebrate fauna, includes among other interesting forms several species of *Copepoda* now described for the first time, as well as a few not previously recorded for the east of Scotland ; also a few species of *Amphipoda*, rare, or not previously recorded for the East Coast.

The species here recorded or described for the first time for the Firth of Forth comprise 25 species of *Copepoda*, 9 species of *Amphipoda*, and a rare species of *Actiniadæ*.

A description (with figures) is also given of a species of *Copepod* previously recorded in Part III. of the Eighth Annual Report, p. 320, in order to indicate more satisfactorily its position in the classification.

In the preparation of this paper I have again the pleasure of gratefully acknowledging the kindness of Professor G. S. Brady, F.R.S., also of the Rev. A. M. Norman, F.R.S., Rev. T. R. R. Stebbing, M.A., and A. O. Walker, F.L.S. I am also much indebted to Dr T. Wemyss Fulton, whose active interest in and sympathy with my work is a source of much encouragement. I also desire to say that not a little of my success in the study of the organisms recorded in this paper is due to the hearty co-operation of Captain R. E. Simpson, and to the intelligent interest shown by the mate in the investigations carried out on board the 'Garland.' My son, Mr A. Scott, has prepared the drawings which accompany this paper. He has also largely assisted me with the preparation of the dissections (a troublesome work) from which the drawings were made. Without the drawings it would have been difficult to realise the important and striking characters of the species mentioned, even though these characters have been, where necessary, fully described.

CRUSTACEA.

I. COPEPODA.

GNATHOSTOMA.

Family CALANIDÆ.

Acartia bifilosus (Giesbrecht). (Pl. VII. fig. 14).

1881. *Dias bifilosus*, Giesbrecht, 'Die Freilebenden Copepoden der Kieler Foehrde,' p. 147, pl. iii. figs. 4, 22, 23, &c. *

Habitat.—In the vicinity of Culross, near the head of the Forth estuary, a number of specimens were obtained among material collected with a small beam-trawl-like tow-net, designed by Professor M'Intosh,† and worked from a rowing or small sailing boat. *Acartia bifilosus* closely resembles *Acartia longiremis*, and requires to be very carefully diagnosed to distinguish it from that species. The inner spines of the fifth pair of

---

* *Vierter Bericht der Commission zur wissenchaftlichen Untersuchung der deutschen Meere, in Kiel*, 1887–1881.

† We find this net a most effective apparatus for capturing micro-organisms and young fish should any be present to capture.

feet in the female of *A. longiremis* are usually long and bent, or geniculate, near the middle ; in *A. bifilosus*, on the other hand, the inner spines are much shorter and are not geniculate (fig. 14). The male fifth feet do not differ much in the two species, except that in *A. bifilosus* they are rather stouter than those of *A. longiremis*. The caudal stylets are usually shorter in *A. bifilosus*, and the last thoracic segment appears to be destitute of setæ. After examining a large number of specimens of both forms, I find the difference between them to be comparatively unimportant, and coincide with Dr Brady in considering the differences as of varietal value only. The characters which distinguish *Acartia discaulata* (Giesbrecht) —a form which I have already recorded from the Forth—are more marked, and show a greater divergence from *A. longiremis*.

Eurytemora affinis (Poppe).

> 1881. *Temora affinis*, S.A. Poppe, Ueber Eine neue Art der Calanaden-Gattung *Temora*, Baird, p. 55, pl. iii. figs. 1–14. ‡
>
> 1881. *Eurytemora hirundo*, Giesbrecht, *loc. cit.*, p. 152,§ pl. ii. figs. 7, 12, 19, &c.
>
> 1891. *Eurytemora affinis*, Brady, Brit. F.-W. Cyclop. and Calan., p. 42, pl. xiii. figs. 6–9. ¶

Habitat.—In the upper reaches of the Forth, about Culross and between Kincardine-on-Forth and Alloa. It was moderately common in some tow-nettings collected in July 1891, and again in February this year (1892). ♂ and ♀ were nearly equally common, and many of the latter were carrying ova-sacs. *Eurytemora affinis* is readily distinguished from other British species of *Calanidæ* by the elongate abdomen (which is thickly clothed with very small stout setæ) and caudal stylets. The terminal spines of the swimming feet are very faintly serrate on the outer margin.

It is strange that the occurrence of *Eurytemora affinis*, which is such an easily distinguished species, should have been so long overlooked, especially as it is at times comparatively common in the upper parts of the Forth estuary.

Stephos, nov. gen. (provisional name).**

Like *Pseudocalanus*, except in the following particulars :—

The anterior antennæ are twenty-four-jointed. The female possesses a fifth pair of feet, which are simple, one-branched, and two-jointed, and the same on both sides. The fifth pair in the male form powerful grasping organs ; they are one-branched and dissimilar on the two sides.

The posterior antennæ and mouth organs are similar to those of *Calanus*. The outer branches of the first four pairs of swimming feet are three-jointed, the inner branches of the first pair are one-jointed, of the second pair two-jointed, of the third and fourth pairs three-jointed as in *Pseudocalanus*.

Stephos minor (nov. gen. et sp. provisional name). (Pl. VII. figs. 1–13.)

Length ·74 mm. (₁/₃₀ of an inch). Cephalothorax robust, the body segment about half as long again as the combined length of the next three. Forehead rounded. Anterior antennæ about as long as the cephalothorax,

---

‡ *Abhandl. des Naturw. Ver.*, Bremen, vii.

§ See also *loc. cit.*, p. 167.

¶ *Nat. Hist. Trans., Northumb., Durham, and Newcastle-upon-Tyne*, vol. xi. Part I.

** Στέφος garland. After the name of our little steamer—the Garland—by means of which we have, with more or less success, investigated the fauna of the Forth.

twenty-four-jointed, the proportional length of the joints as in the formula

$$\frac{18 \cdot 20 \cdot 6 \cdot 6 \cdot 5 \cdot 5 \cdot 5 \cdot 7 \cdot 4 \cdot 4 \cdot 4 \cdot 7 \cdot 6 \cdot 7 \cdot 6 \cdot 6 \cdot 5 \cdot 5 \cdot 6 \cdot 6 \cdot 7 \cdot 9 \cdot 10 \cdot 8}{1 \cdot 2 \cdot 3 \cdot 4 \cdot 5 \cdot 6 \cdot 7 \cdot 8 \cdot 9 \cdot 10 \cdot 11 \cdot 12 \cdot 13 \cdot 14 \cdot 15 \cdot 16 \cdot 17 \cdot 18 \cdot 19 \cdot 20 \cdot 21 \cdot 22 \cdot 23 \cdot 24}$$

Sparingly setiferous; there appears to be a depressed lobe-like process upon the distal end of the first or proximal end of the second joint (fig. 2). Antennæ the same in both sexes; posterior antennæ nearly as in *Calanus finmarchicus*, but the primary branch is somewhat shorter proportionally; mouth organs also as in that species. First four pairs of swimming feet as in *Pseudocalanus elongatus*, fifth pair in the female simple, one-branched, two-jointed, small; first joint about one and a half time longer than broad; the second joint about twice as long as the first, diminishing in breadth from the base to the apex, and bearing two small marginal spines —one opposite the other—on the distal half. The female fifth feet resemble somewhat those of *Candace pectinata*. Fifth pair of feet in the male long and forming a powerful grasping organ; both feet are one-branched and four-jointed; the two last joints of the right foot are elongate and slender, the ultimate joint being strongly curved outward in its upper half and forming a long powerful claw. The left foot is rather shorter than the other, and terminates in two digitiform processes between which the claw-like terminal joint of the right foot interlocks. Abdomen short; in the female four-, in the male five-jointed, the last segment shorter than either of the others. Caudal stylets short, length about equal to the breadth, and furnished with four long subequal setæ, and a few small hairs.

Habitat.—Off St Monans, Firth of Forth. Several specimens were obtained.

This comes very near *Pseudocalanus*, and but for the presence of a fifth pair of feet in the female, and the powerfully developed fifth feet of the male, would have become a member of that genus; as it is, the affinities of *Stephos minor* seem to be with *Pseudocalanus* on the one hand, and *Candace* or *Acartia* on the other.

Family MISOPHRIADÆ, Brady (1878).

Pseudocyclopia, nov. gen. (provisional name).

Body robust, and resembling *Pseudocyclops* in general appearance. Head anchylosed with thorax. Basal joint of the anterior antennæ very large and nearly half the entire length of the antenna. The primary branch of the posterior antennæ three-jointed, the middle joint long; secondary branch large but scarcely so long as the primary branch, five-jointed, the third and fourth joints small. Mouth organs nearly as in *Calanus*. The outer branches of the first four pairs of swimming feet three-jointed, and longer than the inner branches; the inner branch of the first pair one-jointed, of the second pair two-jointed, of the third and fourth pairs three-jointed; the first basal joint of the third pair bears a long stout spine on the inner distal angle, longer than the inner branch. The fifth pair of feet in the female are small, one-branched, two-jointed, the first joint short, subrotund; the fifth feet in the male, elongate, one- or two-branched, unequal on the two sides, and forming powerful grasping organs. Abdomen in the female four-, in the male five-jointed.

Pseudocyclopia crassicornis, n. sp. (provisional name). (Pl. VII. figs. 15-20).

Length, exclusive of caudal setæ, ·66 mm. Cephalo-thorax robust, four-jointed, the first segment more than twice the combined length of the other three. Abdomen small, five-jointed in the male, four-jointed in the female; rostrum short, directed downwards. Anterior antennæ short, sixteen-jointed ; basal joint large and furnished with three elongate, stout, marginal sensory filaments and several small setæ ; the second, sixth, tenth, and last joints are each also provided with a sensory filament, but smaller than those of the basal joint. The proportional length of the joints are very nearly as shown by the annexed formula

$$\frac{60 \cdot 6 \cdot 3 \cdot 3 \cdot 3 \cdot 3 \cdot 4 \cdot 4 \cdot 4 \cdot 4 \cdot 4 \cdot 4 \cdot 6 \cdot 8 \cdot 6 \cdot 6}{1 \cdot 2 \cdot 3 \cdot 4 \cdot 5 \cdot 6 \cdot 7 \cdot 8 \cdot 9 \cdot 10 \cdot 11 \cdot 12 \cdot 13 \cdot 14 \cdot 15 \cdot 16}$$

Posterior antennæ three-jointed, the middle joint elongate with two small setæ on the exterior margin, and the last joint with a number of apical setæ. Secondary branch large, five-jointed, but shorter than the primary branch, the third and fourth joints very small. Mandibles small, consisting of a broad biting part, and a two-branched palp—one of the branches being two-, the other three-jointed. Anterior foot-jaw small, four-jointed, with several marginal setiferous processes. The basal joint of the posterior foot-jaw elongate, the lower distal angle produced, with a blunt tooth-like process ; second joint also elongate, slender ; the last four joints small and setiferous. The outer branch of the first pair of swimming feet three-jointed, each joint armed with a stout spine at the outer distal angle, the inner branch one-jointed and rather longer than the first joint of the outer branch. The outer branch of the second pair is also three-jointed. Each of the first and second joints bear one, and the last joint four, stout spines of variable length, that of the second joint and the terminal spine of the last joint being larger than the others ; the inner branch is two-jointed and shorter than the outer one, and the first joint is rather smaller than the second. The third and fourth pairs have both branches three-jointed. A stout and nearly straight spine—longer than the inner branch—springs from the inner distal angle of the first basal joint of the third pair, otherwise the third and fourth pairs are similar. The fifth pair in the female is one-branched, two-jointed, the first joint short and somewhat dilated ; the extremity of the second is produced into two elongate spiniform processes (these are not spines articulated to the end of the joint but are prolongations of it), the inner one much longer than the other ; there is also a subapical spine exterior to the two processes and shorter than either. Fifth pair in the male also one-branched, four-jointed, and elongate ; that of the left (?) very slender. The first joint of the right (?) foot is short and dilated, the second and third long, the last very small and furnished with a marginal hooklet and a subapical digitiform process. Caudal stylets short, each bearing four long, plumose, terminal setæ, the two middle ones being stout and spiniform. *Spermatophore* elongate, narrow, curved, and showing under the microscope a beautifully reticulated structure (fig. 29).

Habitat.—Off St Monans, Firth of Forth. Several specimens were obtained.

Pseudocyclopia minor, n. sp. (provisional name). (Pl. VIII. figs. 1–10).

Length, exclusive of caudal setæ, ·43 mm. Cephalothorax robust, four-jointed, first segment large, more than twice the combined lengths of the other three. Anterior antennæ short, setiferous, seventeen-jointed, the basal joint large, provided with a hook-like spine on the outer margin and near the middle of the joint, and with a sensory filament at the outer distal angle ; the fourth, seventh, ninth, and thirteenth joints are also each furnished with a small sense-organ. The proportional length of the joints are very nearly as shown in the annexed formula

$$\frac{30 \cdot 2 \cdot 2 \cdot 2 \cdot 2 \cdot 2 \cdot 3 \cdot 3 \cdot 3 \cdot 3 \cdot 3 \cdot 2 \cdot 2 \cdot 3 \cdot 4 \cdot 3 \cdot 3}{1 \cdot 2 \cdot 3 \cdot 4 \cdot 5 \cdot 6 \cdot 7 \cdot 8 \cdot 9 \cdot 10 \cdot 11 \cdot 12 \cdot 13 \cdot 14 \cdot 15 \cdot 16 \cdot 17}$$

Posterior antennæ three-jointed, middle joint long, secondary branch five-jointed, shorter than the primary branch. Mouth organs as in *Pseudocyclopia crassicornis*. In the first pair of swimming feet the first joint of the outer branch is about as long as the other two together, while the one-jointed inner branch is longer than the first joint of the outer one. Each of the three joints of the outer branch is armed with a large spine at the outer distal angle; both branches are furnished with several plumose setæ. The second pair is similar to those of *Pseudocyclopia crassicornis*. The third and fourth pairs are also similar to those of that species, but the spine which springs from the inner distal angle of the first basal joint of the third pair is curved, and is longer and more powerful, and extends beyond the extremity of the outer branch. The fifth pair of feet in the female are very small and somewhat resemble those of *Pseudocyclopia crassicornis*, but the extremity is bluntly rounded and provided with three spinous setæ, the middle one of which is the longest. The fifth pair in the male form very powerful grasping organs; the left (?) foot consists of two very long branches, one of which is four-jointed, and one five-jointed; the basal point of the first (the four-jointed branch) is moderately short and dilated, the second joint is very small, the third elongate and geniculate, and bearing a curved spine at the inner distal angle; the last joint is long and slender, with a rounded extremity; the third and fourth joints of the other branch (which is rather longer than the first) are elongate and slender, while the last joint is very short and produced into a digitiform process. The right (?) foot consists of a single four-jointed branch, the breadth of the first two joints of which is rather greater than the length; the third joint is elongate, and bears exteriorly on its lower half a dense fringe of plain spinous hairs, and two stout spines interiorly. The last joint, which is very short, has three small subapical lobes. Abdomen in the male five-jointed, in the female four-jointed. The second and third joints of the female abdomen are produced posteriorly on each side of the median dorsal line into sharp angular processes as shown in the figures (fig. 9); the male abdomen wants the dorsal processes possessed by that of the female. Caudal stylets short, each furnished with four long, plumose, terminal hairs, the two middle ones being stout and spiniform.

Habitat.—Off St Monans, Firth of Forth. Several specimens of this species were obtained.

Family HARPACTICIDÆ.

Neobradya, nov. gen. (provisional name).

Near *Bradya*, Boeck, in form and structure. Anterior antennæ nine- or ten-jointed, scarcely if at all longer than the first body segment; those of the male hinged and adapted for grasping. Posterior antennæ large, three-jointed; secondary branch of posterior antennæ, four-jointed, the first joint as long as the entire length of the other three. Mandibles well developed, possessing a broad biting part, and a large two-branched palp, one of the branches of which is one- and the other four-jointed. Maxillæ somewhat as in *Longipedia*. Anterior foot-jaws stout, five-jointed, the first joint rather longer than the second, and furnished with three digitiform lobes, the three last joints small. Posterior foot-jaws not uncinate, resembling somewhat those of *Bradya*. Both branches of the first pair of swimming feet three-jointed and about equal in length. The outer branches of the second, third, and fourth pairs three-jointed; the inner branches two-jointed; the fifth pair small foliaceous.

Neobradya pectinifer*, nov. gen. et sp. (provisional name) (Pl. XIII. figs. 19-32).

Female.—Body elongate, cylindrical ; length, exclusive of caudal stylets, 1·2 mm. and composed of nine segments. The first cephalo-thoracic segment longer than the next two together. Rostrum short, obtusely rounded. Anterior antennæ nine-jointed, about as long as the first body segment, stout, and well furnished with setæ ; the proportional length of the joints are as shown by the formula

$$\frac{13 \cdot 22 \cdot 10 \cdot 5 \cdot 3 \cdot 4 \cdot 3 \cdot 2 \cdot 5}{1 \cdot 2 \cdot 3 \cdot 4 \cdot 5 \cdot 6 \cdot 7 \cdot 8 \cdot 9}.$$

One side of the fourth joint is produced to form the base of a long olfactory appendage. Posterior antennæ large, three-jointed, the extremity of the last joint furnished with one plain and five plumose hairs ; the secondary branch is four-jointed ; the first joint is as long as all the other three together ; the first joint bears two setæ, the second and third one setæ each, and the last two very small marginal and two long terminal setæ. The mandible is well developed, having a broad biting part and a large two-branched palp—one of the branches is four-, the other one-jointed ; both the basal part and the branches of the palp are furnished with setæ. Maxillæ nearly as in *Longipedia coronata*. Anterior foot-jaw stout, five-jointed, the first joint large and possessing three marginal digitiform lobes, each of the lobes with three strong, nearly equal terminal hairs, the second joint much smaller than the first, and produced to form a stout process similar to those on the first joint, and also, like them, provided with three stout, subequal, terminal hairs ; the three last joints are very small, and furnished with four moderately long hairs. Posterior foot-jaws very small, three-jointed, armed with several appressed and short, stout, blunt-pointed, marginal spines, each of which is furnished with a fringe of short hairs arranged in a pectinate manner along the upper margin (fig. 27). All the swimming feet two-branched and nearly alike in both sexes. Both branches of the first pair of nearly equal length and three-jointed, the second, third, and fourth pairs have the outer branch three-jointed ; the inner branch, which is rather shorter, is two-jointed, the first joint of both branches of the first four pairs longer than any of the other joints ; the second joint of the basal part of each of the four pairs is very short, that of the first pair armed with a spine on the inner distal angle ; that of the second, third, and fourth pairs provided with a small setæ instead of a spine ; the last joint of each branch of all the four pairs is furnished with one or two long plumose setæ and one or two smaller hairs. Fifth pair of feet small, foliaceous, the produced inner portion of the basal joint rather smaller than the outer semicircular joint, and provided with two elongate, stout, plumose setæ of unequal length. The exterior lobe of the same joint bears a very long, slender, curved hair at its apex. A long, stout, plumose hair springs from the inner distal angle of the outer semicircular joint, and three others from its outer margin. Abdomen four-jointed, the first and third segments longer than either of the other two. Caudal stylets short and furnished with a long slender terminal hair and several very small ones.

Male.—The male differs little from the female except in the form of the anterior antennæ which are distinctly geniculated and form powerful grasping organs (fig. 22).

Habitat.—Off St Monans, Firth of Forth. Obtained from dredged material from 14 fathoms water.

---

* Referring to the comb-like arrangement of the hairs on the marginal spines of the posterior foot-jaws.

Tachidius crassicornis, n. sp. (provisional name).   (Pl. VIII. figs. 14–27).

Length, exclusive of tail setæ, ·7 mm. Body moderately stout, first cephalo-thoracic segment longer than the next two together, the forehead produced into a short rostrum. Anterior antennæ shorter than the first body segment ; that of the female six-jointed, stout, and densely setiferous towards the extremity, a small sensory filament springs from fifth joint. The proportional length of the joints are nearly as in the formula

$$\frac{20 \cdot 10 \cdot 9 \cdot 5 \cdot 3 \cdot 9}{1 \cdot 2 \cdot 3 \cdot 4 \cdot 5 \cdot 6}$$

The anterior antennæ in the male form powerful grasping organs, closely resembling those of *Tachidius brevicornis* (fig. 17). Posterior antennæ short, three-jointed, the last joint nearly as long as the preceding two together ; a small one-jointed secondary branch springs from the end of the first joint. Mouth organs nearly as in *Tachidius brevicornis*. The first four pairs of swimming feet nearly alike, both branches three-jointed, the first joint of the inner branches of all the four pairs smaller than either the second or third joints. The fifth pair in the female moderately large and foliaceous, furnished with three equal and plumose terminal setæ ; a plumose seta springs from a rounded basal part on the anterior margin of the female fifth pair, which may represent a rudimentary second branch. The fifth pair in the male are very small, subquadrate, and furnished with one small and two moderately long setæ near the inner angle and one at the outer angle ; the first abdominal segment in the male is armed with prominent lateral appendages, which are easily observed without dissection, and which consist of a broad, but short, basal part bearing three unequal spiniform and plumose marginal setæ, the inner one being longer than either of the other two. Caudal stylets short, about as long as the last abdominal segment, and furnished with four setæ,—the inner and outer being plain and very small, the other two plumose and elongate ; the inner of the two principal setæ is much longer than the other ; and the basal part of the proximal half is broader than the remaining portion ; the broad part, which is of nearly equal breadth throughout, merges abruptly into the more slender portion as shown in the figure. Ovisac single, large, with a number of large ova.

Habitat.—Near Culross on the upper estuary of the Forth ; not very rare. Obtained February 1892.

This species comes near *Tachidius brevicornis* (Müller). but differs in the form of the anterior antennæ, which are rather stouter and shorter and six-jointed ; in the first joint of the inner branches of the first four pairs of swimming feet being smaller than the other two joints ; and in the form of the fifth feet in the female.

*Ameira longicaudata,** n. sp. (provisional name).   (Pl. IX. figs. 1–18).

Body slender ; length, exclusive of tail setæ, 1 mm. (25th of an inch). Anterior margin of first body segment squarely truncate ; forehead produced into a short blunt rostrum. Anterior antennæ longer than the first cephalo-thoracic segment, elongate, and sparingly setiferous ; that of the female eight-jointed, of the male nine-jointed ; the male antennæ are distinctly hinged between the sixth and seventh joints, and indistinctly between the third and fourth joints. A long sensory filament springs from the end of the fourth joint in both sexes ; the porportional length of the joints of the female and male antennæ are nearly as in the annexed formulæ

---

* Referring to the long caudal stylets.

Female	30 · 19 · 14 · 10 · 5 · 8 · 5 · 10
	1 · 2 · 3 · 4 · 5 · 6 · 7 · 8 · 9
Male	30 · 19 · 10 · 14 · 3 · 7 · 8 · 4 · 10

Posterior antennæ of moderate length, three-jointed; joints nearly equal, a small one- (? or two-) jointed secondary branch springs from the end of the first joint, and bears three subequal terminal hairs; two of these hairs arise from a common and somewhat dilated basal part which may possibly represent a rudimentary second joint, but this is doubtful. Mandibles moderately stout, the biting part broad with several strong tooth-like processes, and a divergent, marginal, setiferous spine; the palp with two small branches and one or two terminal hairs. Maxillæ small; the terminal part, which is comparatively broad, is furnished with several spiniform teeth on the inner distal margin, and exteriorly with three small marginal setiferous lobes. Anterior foot-jaw small, two-jointed; the first joint with two marginal setiferous lobes, the last joint small and produced into an elongate slender process, bearing at its apex a stout plumose hair, and exteriorly, near the base, a plain slender seta. Posterior foot-jaw strong, and armed with a powerful clawed spine. The first four pairs of swimming feet have both branches three-jointed and elongate; the first joint of the inner branch of the first pair longer than the entire outer branch, and furnished with an elongate seta on the lower half of the inner margin; the two last joints are short, the second being the shorter of the two. Inner branches of each of the other three pairs shorter than the outer,—especially in those of the fourth pair; all the four pairs furnished with moderately long plumose setæ. The inner part of the basal joint of the female fifth pair moderately broad, furnished with four elongate setæ on its inner margin; the outer part is laterally produced and attenuated, and forms the base of a single elongate seta. The second joint is long and slender (fig. 12), and furnished with five setæ,—three on the outer margin, one on the inner margin near the apex, and an apical seta. The fifth pair in the male are very small; the basal joint is scarcely produced posteriorly, and bears three subterminal setæ the lateral produced part bears a single hair, the second joint narrow, ciliate on the outer margin, and furnished with one terminal seta, and another on the inner margin, both being of moderate length. The first abdominal segment bears two small setiferous lateral appendages, as shown in fig. 16. Caudal stylets elongate, slender, longer than the last abdominal segment, each with one extremely long and a few short terminal setæ. The posterior margins of all the cephalothoracic and abdominal segments are more or less distinctly denticulate.

A variety occurs, somewhat smaller than that described (figs. 17, 18), which has the antero-lateral angles of the first body segment rounded instead of angular; the posterior margins of all the body segments spiniferous instead of denticulate, and also armed at the postero-lateral angles with two strong spines and several small setæ. To distinguish this variety I have named it var. *spinosa.*

Habitat.—Off St Monans, Firth of Forth. Frequent. I first obtained this species two or three years ago, but for want of time to study its structure and affinities, it was laid aside, along with some others, till a more convenient season. With the assistance of my son, I am now able to describe this and several other interesting members of the Forth fauna.

Paramesochra,[*] nov. gen. (provisional name).

Body subpyriform; anterior antennæ short, seven-jointed in the female;

[*] Near *Mesochra,* Boeck, which it resembles in several important points, especially in the structure of the first four pairs of swimming feet.

modified, and forming powerful grasping organs in the male ; posterior antennæ, with the primary branch three- or four-jointed, secondary branch very small, one-jointed. Mandibles well developed, and possessing a two-branched palp. Maxillæ small. Anterior foot-jaw with several marginal setiferous processes. Posterior foot-jaw small, feebly clawed. All the five pairs of swimming feet two-branched ; both branches of the first pair two-jointed, the inner branch longer than the outer, first joint of the inner branch elongate, the last very small and imperfectly hinged ; the outer branches of the second, third, and fourth pairs three-jointed, the inner branches two-jointed and shorter than the outer. Fifth pair foliaceous. The abdomen in the female four, in the male five jointed.

Paramesochra dubia, n. sp. (provisional name).   (Pl. XII. figs. 18–32.)

Female.—Body subpyriform ; length about ·65 mm. The postero-lateral angles of the cephalo-thoracic segment spiniform and produced backward beyond the next somite ; two lenses—one on each side near the postero-lateral angles, as shown in the figure—can be easily made out with a ⅛th or ⅒th inch objective. Anterior antennæ short ; seven-jointed basal joint very large and stout, the upper distal angle produced so as to form a stout prominent tooth, the remaining joints small, the proportional lengths of which are as shown in the formula

$$\frac{12 \cdot 4 \cdot 4 \cdot 9 \cdot 2 \cdot 3 \cdot 3}{1 \cdot 2 \cdot 3 \cdot 4 \cdot 5 \cdot 6 \cdot 7}$$

Posterior antennæ three- (or four?) jointed ; secondary branch small, slender, one-jointed. Mandible well developed, consisting of a biting part (the apex of which is armed with several long teeth) and a two-branched palp—one branch three- the other one-jointed. Maxillæ small, with a lateral two-jointed lobe, serrate at the apex, and an intermediate appendage furnished with two short terminal hairs. Anterior foot-jaw four-jointed, with several marginal setiferous processes ; posterior foot-jaw three-jointed ; last joint very small, and armed with three nearly equal setæ. All the five pairs of swimming feet two-branched ; both branches of the first pair two-jointed ; the outer branch of the second, third, and fourth pair three-jointed ; the inner branch two-jointed ; the first joint of the inner branch of the first pair elongate ; the last very small and imperfectly hinged ; inner branch longer than the outer one, the inner branch of the following three pairs shorter than the outer. The fifth pair foliaceous ; basal joint large, its inner lobe with one plain and one plumose terminal seta. The exterior lobe, which is small, is also furnished with two small setæ ; the second joint with four stout marginal hairs. Abdomen four-jointed ; first segment large, composed of two coalescent joints ; the last segment very small. Caudal stylets longer than the two last abdominal segments, and about six times longer than broad, furnished with one very long and three short unequal terminal hairs.

Male.—Rather smaller than the female ; length about ·6 mm. Anterior antennæ forming powerful prehensile organs. The basal joint of the fifth pair of swimming feet much smaller than in the female, and wanting the two setæ. Abdomen five-jointed. With these exceptions the description of the female is equally applicable to the male.

Habitat.—Firth of Forth, west of May Island, February 1892. Several specimens were obtained.

Tetragoniceps (?) *maleolata*, Brady.   (Plate VIII. figs. 11, 12.)

A Copepod answering to the description and figures of *Tetragoniceps maleolata*, except in the two following particulars, was obtained in material dredged off St Monans.

1st. The anterior antennæ are nine-jointed, four small joints precede the last one in the Forth specimen (fig. 11) instead of three as described for *Tetragoniceps maleolata.* 2d. The fifth pair of swimming feet are two-branched (fig. 12) in the Forth specimen, but in *T. maleolata* they are one-branched. This difference is a more important one than that between the anterior antennæ, because the one-branched fifth feet form one of the principal characters that distinguish *Tetragoniceps* from *Normanella.* Our specimen, even though possessing a three-jointed posterior antennæ, might have been ascribed to that genus, but the general contour of the animal is that of *Tetragoniceps,* and decidedly different from either *Normanella* or *Cletodes.* It is worth noting also that the general outline of the fifth foot of our specimen—leaving out of account its two-jointed structure—has a close resemblance to the fifth foot of *Tetragoniceps.*

Tetragoniceps macronyx,† n. sp. (Pl. X. fig. 19–28.)

Length, ·54 mm. (¹⁄₅₀th of an inch). Body slender. Rostrum small. Anterior antennæ slender, nine-jointed in the male, eight-jointed in the female, the proportional length of the joints as in the formula

$$
\text{Male,} \qquad \frac{15 \cdot 16 \cdot 11 \cdot 2 \cdot 6 \cdot 2 \cdot 5 \cdot 4 \cdot 8}{1 \cdot 2 \cdot 3 \cdot 4 \cdot 5 \cdot 6 \cdot 7 \cdot 8 \cdot 9}
$$
$$
\text{Female,} \qquad 15 \cdot 15 \cdot 3 \cdot 15 \cdot 5 \cdot 5 \cdot 5 \cdot 9
$$

The male antennæ are hinged between the second and third and sixth and seventh joints. Posterior antennæ are of moderate length and three-jointed; secondary branch very rudimentary (fig. 22). Mandible palp small, one- or (?) two-branched. Anterior foot-jaw small, furnished with two marginal bi-lobed setiferous processes, and bearing at the apex a long, slender, filamentous hair and a claw-like spine. Posterior foot-jaw elongate, armed with a long, slender, sinuous, terminal clawed spine, which has a long delicate seta springing from its base. The outer branches of the first four pairs of swimming feet three-jointed—that of the first pair being shorter than those of the other three pairs; three slender subequal setæ spring from the end of the second joint of the outer branch of the fourth pair; the inner branch of the first pair are elongate, two-jointed; first joint nearly as long as the outer branch, and bearing a single delicate seta near the middle of the outer margin; second joint fully half the length of the first, and furnished with two elongate terminal hairs. The inner branches of the following three pairs are short, two-jointed, and armed with a moderately long, stout terminal spine. Feet of fifth pair foliaceous, elongate, narrow-triangular. Caudal stylets rather longer than the last abdominal segment, and furnished with a moderately long and a few small setæ. Ovisac single, and containing a few large ova.

Habitat.—Off St Monans, Firth of Forth. A few specimens only were obtained among dredged material from about 14 fathoms water, bottom clean sand.

*Tetragoniceps Bradyi,** n. sp. (Pl. IX. fig. 19–32.)

Length, exclusive of tail seta, 1 mm. (²⁄₅th of an inch). In general form like *Tetragoniceps maleolata,* but the first cephalo-thoracic segment is scarcely so angular in front. Rostrum very short, anterior antennæ about as long as the first cephalo-thoracic segment, nine-jointed, the second joint produced into a strong claw on the under side (fig. 20); the proportional length of the joints are nearly as in the annexed formula

---

* The name is given in compliment to Professor G. S. Brady, who instituted the genus, and to whose untiring and disinterested kindness the author of these notes owes much of his success in the study of the Entomostraca.

† Μακρός, long, and ὄνυξ, claw, referring to the long claw of the posterior foot-jaw.

$$\frac{27 \cdot 10 \cdot 7 \cdot 5 \cdot 3 \cdot 3 \cdot 2 \cdot 2 \quad 11}{1 \cdot 2 \cdot 3 \cdot 4 \cdot 5 \cdot 6 \cdot 7 \cdot 8 \quad 9}$$

The fourth joint is produced so as to form the base of a long and stout sensory filament. All the joints except the first are more or less setiferous. Posterior antennæ three-jointed, the joints subequal; a small secondary one-jointed branch springs from the end of the first joint. Mandible palp distinctly two-branched—one of the branches much larger than the other (fig. 22). Maxillæ with a broad biting part and a four-lobed branchial appendage. Anterior foot-jaw five-jointed; the broad first and second joints bear five marginal, digitiform, setiferous lobes arranged in two groups—three lobes in the one and two in the other, with a clear space between. The last three joints, which are very small, are furnished with a number of small setæ. Posterior foot-jaw three-jointed, last joint forming a base for a moderately long terminal claw and a small seta; a plumose seta springs from the inner margin, and near the middle of the second joint, anterior to the plumose seta, are a number of fine marginal cilia. The first joint is furnished with two subterminal plumose hairs. The first four pairs of swimming feet are nearly as in *Tetragoniceps malcolata*. The fifth pair, which are one-branched, are in the form of large, foliaceous concave plates, the length of which is about one-third the length of the whole animal (fig. 30). Their breadth is about equal to half their length. The extremity and outer margin are provided with a few setæ, the inner terminal seta being plumose, the others plain. A strong muscle extends down the exterior side and across the extremity, and sends off branches to the marginal setæ. Inclosed within the feet were a number of ova, having apparently no other covering than that of the enclosing large foliaceous plates. Abdomen five-jointed; the posterior ventral margin of the third segment is produced so as to form a prominent fold which extends about half-way over the next segment. Caudal stylets about as long as the last abdominal segment, and having the outer margin nearly straight and the inner strongly sigmoid; each stylet bears a long terminal seta, the base of which is considerably dilated, and a few very small hairs, as shown in figure 32. No males were obtained.

Habitat.—Off St Monans. Rare. The nine-jointed anterior antennæ, with the strong claw-like process of the second joint, together with the remarkably large, foliaceous fifth feet, render this a well-marked species.

Tetragoniceps incertus. (Pl. XII. figs. 1–17).

Female.—Body elongate, cylindrical; length, exclusive of caudal setæ, 1 mm. First cephalo-thoracic segment about as long as the next two together, forehead produced into a sharp-pointed rostrum. Anterior antennæ about as long as the first body segment, seven-jointed, the proportional length of the joints as shown in the formula

$$\frac{20 \cdot 18 \cdot 12 \cdot 7 \cdot 4 \cdot 5 \cdot 8}{1 \cdot 2 \cdot 3 \cdot 4 \cdot 5 \cdot 6 \cdot 7}.$$

All the joints except the first sparingly setiferous; a moderately long olfactory filament springs from the end of the fourth joint. Posterior antennæ short, two- (or three- ?) jointed, and possessing a very small one-jointed secondary branch which bears two terminal setæ. The apex of the last joint of the primary branch is furnished with five setæ, the three longest of which are bent near the middle, the outer one of the three having a small forward-directed spine at the bend. Mandible dilated at the base, the apex truncate, and armed with several blunt-pointed teeth; mandible palp one-branched, long, and slender. Maxillæ small, simple, with two small lateral appendages. Anterior foot-jaw small, armed with a stout curved terminal spine and two marginal setiferous lobes. Posterior foot-jaw uncinate, forming a moderately strong prehensile organ, the terminal claw slender and strongly curved. The inner branch of the first pair of swimming feet elongate, two-jointed, the last joint small, the first nearly twice the length of the three-jointed outer branch. A small seta springs from the inner margin of the second basal joint, and another from the inner margin and near the proximal end of the elongate first joint of the inner branch. Two slender hairs, one of which is setiferous, spring from the extremity of the last joint. Each of the three joints of the outer branch is armed near the exterior distal angle with a short spinous seta ; three hairs —two of which are long and setiferous and bent near the middle—spring from the extremity of the last joint. The inner branches of the second, third, and fourth pairs are one-jointed, that of the fourth rudimentary; the outer branches are three-jointed, the joints subequal and more strongly setiferous than those of the first pair. Fifth pair foliaceous,—the same on the both sides,—one-branched, and furnished with three hairs on the outer margin and four on the inner—the upper of the four being densely setiferous. The extremity of each branch terminates in a stout blunt-pointed spine nearly as long as the branch to which it appears to be articulated. Abdomen four-jointed, the first segment composed of two coalescent joints, and about twice the length of the next two together, the second, third and fourth segments subequal. Caudal stylets fully half as long as the last abdominal segment, slightly divergent, each stylet furnished with a long geniculated terminal seta and several small hairs.

Male closely resembling the female but smaller (·87 mm). Anterior antennæ eight-jointed, the two first joints long, as in the female, the fifth shorter than any of the other joints, and furnished with an olfactory appendage. The antennæ are distinctly hinged between the sixth and seventh joints, and indistinctly hinged between the third and fourth. The posterior antennæ, mouth-organs, and first pair of swimming feet as in the female. The last joint of the outer branch of the second pair of swimming feet like that of the female, but furnished with an additional and moderately stout plumose hair, the normal position of which appears to be that shown in the figure (fig. 12). A long spiniform appendage springs from the basal joint of the third pair, and close to, but inside of, the one-jointed inner branch (fig. 14). This appendage is more than twice the length of the inner branch, and as long as the two first joints of the outer branch. The fifth pair of feet is furnished with fewer marginal hairs than those of the female, and the terminal spine seems to be continuous with, and not articulated to, the basal part of the foot. Abdomen five-jointed, caudal stylets and setæ as in the female.

Laophonte horrida (Norman).

        1869–70. *Cleta minuticornis*, Buchholz, 'Die zweite deutsche Nord-
            polar-fahrt,' p. 393, pl. xv. fig. 3.
        1876. *Cleta horrida*, Norman, 'Report of the Valorus Expedition,'
            p. 206 (Proc. Roy. Soc.).
        1880. *Laophonte horrida*, Brady, *loc. cit.*, ii. p. 74, pl. xxiv.
            figs. 1–11.

Habitat.—Washed from a large root of sea-weed brought up in the trawl-net near the middle of the estuary between Fidra and St Monans during February last (1892). This remarkable species is readily distinguished by the strong dorsal armature of the body segments. The first pair of feet have the basal part long and rather slender. The rostrum is prominent and has the apex somewhat tri-lobed; the middle lobe projects forward considerably beyond the lateral ones.

It has been obtained from various parts of Great Britain. The following are some of the localities—Off the Island of Cumbrae; at Portincross, Ayrshire; Mulroy Loch, Donegal (G. S. Brady); Oban (A. M. Norman); East Loch Tarbert, Loch Fyne (Mibi). *Laophonte horrida,* so far as I have been able to know its habits, is no swimmer, but appears to frequent the muddy roots of weed and zoophytes, among which it crawls and finds food and shelter; it is usually more or less coated with mud.

Laophonte inopinata,* n. sp. (provisional name). (Pl. XI. figs. 1–12.)

Female.—Length, exclusive of caudal setæ, ·5 mm. Viewed dorsally, the body is elongate and becoming gradually narrower posteriorly, composed of ten segments, the first segment about as long as the next three together, and furnished with a few small spinous setæ at the antero-lateral angles. Rostrum short, obtuse. Anterior antennæ short and stout, six-jointed, the first three joints large, subequal, the fourth and fifth small. The proportional length of the joints are as in the annexed formula

$$\frac{7 \cdot 8 \cdot 7 \cdot 2 \cdot 2 \cdot 6}{1 \cdot 2 \cdot 3 \cdot 4 \cdot 5 \cdot 6}$$

The fourth joint produced on one side to form the base of an elongate olfactory filament. Posterior antennæ stout, three-jointed, with four long geniculated terminal setæ and one short curved terminal spine. The margin of the last joint is also fringed with short hairs and provided with a spine near the distal end. The secondary branch, which springs from near the middle of the second segment of the primary branch, is small, one-jointed, furnished with one marginal and three short, plumose terminal setæ. Anterior foot-jaw small, two-jointed, armed with a terminal clawed spine and two elongate marginal lobes. Posterior foot-jaw two-jointed, and bearing a long terminal claw. The first pair of swimming feet nearly as in *L. similis.* The second, third, and fourth pairs nearly alike, moderately stout; fifth pair small. The basal joint is furnished with several small marginal hairs, a moderately long plumose terminal hair, and three subterminal, spinous setæ toothed near the extremity; the second joint small and provided with one long and four short terminal hairs. Caudal stylets short, each with a long curved, spreading terminal seta, beset for two-thirds of its length with numerous wooly-like curled filaments; a short terminal seta plumose on one side; and a few very short hairs. The integument is thickly covered with minute hairs, and the posterior margins of the body segments are, besides being fringed with cilia, furnished with a number of small hairs placed at regular intervals along the margin of each segment as shown in the enlarged figure.

Male.—The chief difference between the female and male is in the form of the anterior antennæ, which in the latter are distinctly hinged, and constitute powerful grasping organs.

Habitat.—Washed from a large seaweed root brought up in the trawl-net a few miles west of May Island. Several ♂ and ♀ specimens were obtained; some of the latter carried ovisacs. The long, spreading, and neatly curved caudal setæ serve to distinguish this species at a glance, and especially so when examined under the microscope; the wooly-like curled filaments with which they are covered give them a very striking character.

---

* *Inopinata,* unexpected.

Cletodes lata,[*] n. sp. (provisional name). (Pl. X. figs. 10–18).

Length ·7 mm., body depressed, moderately broad, the last thoracic and first abdominal segments rather narrower than those that precede or follow ; all the segments, but especially the three first abdominal segments, have the postero-lateral angles more or less sharply angular ; the last abdominal segment nearly as long as the second and third together ; the first body segment broadly triangular, the breadth being rather greater than the length. Anterior antennæ shorter than the first body segment, stout, six-jointed, the second and fourth joints smaller than any of the others, the proportional length of the joints as in the formula

$$\frac{10 \cdot 3 \cdot 8 \cdot 2 \cdot 4 \cdot 10}{1 \cdot 2 \cdot 3 \cdot 4 \cdot 5 \cdot 6}.$$

All the joints, with the exception of the first, are armed with stout spiniform setæ, and a stout elongate sensory filament springs from the third joint. Posterior antennæ two-jointed, secondary branch obsolete, and represented by a small hair arising from a slightly produced part of the margin, and near the middle of the first joint of the primary branch. Mandible with three strong teeth ; mandible palp small, cylindrical, one-jointed (fig. 13). Posterior foot-paw furnished with a long slender curved terminal claw. The first joint of the outer branch of the first pair of swimming feet half as long again as either the second or the third joint ; the inner branch, which consists of two short equal joints, is about as long as the first joint of the outer branch. The middle joint of the outer branches of the second, third, and fourth pairs is shorter than either the first or last joints ; the first joint of the inner branches is not half the length of the second. Fifth pair foliaceous, the inner lobe of the basal joint broad, bearing two elongate, stout, subterminal setæ ; the outer lobe is in the form of an elongate cylindrical process, bearing a moderately long terminal setæ ; second joint elongate, ovate, the outer margin with three small hairs widely apart, a moderately long apical seta, and a very small hair on the inner margin. A variety (?) occurs having the second joint very narrow, with the apical and three marginal hairs very long. The caudal stylets short, widely apart, and bearing one moderately long and a few small setæ.

Habitat.—Off St Monans, Firth of Forth. Several specimens were obtained among dredged material.

Thalestris harpactoides, Claus. (Pl. XI. figs. 13–16).

    1863. *Thalestris harpactoides,* Claus, 'Die frei lebenden Copepoden,' p. 133, pl. xix. figs. 2–12.

    1880. *Thalestris harpactoides,* Brady, ' Brit. Copep.,' vol. ii. p. 127, pl. l. figs. 9–16 ; pl. lix. fig. 1.

Habitat.—Off St Monans, Firth of Forth. A few species were obtained among dredged material. It somewhat resembles *Th. rufocincta,* but is more slender. The colour of the Forth specimens was bluish. There is also a narrow but distinct belt, due to difference of colour or structure, along the margins of the body segments. Its slender form, the form of the posterior foot-jaws (fig. 13), of the first pair of feet, and of the fourth and fifth pairs, serve to distinguish this from other British species of *Thalestris.* The marginal spines of the outer branches of the second, third, and fourth swimming feet of *Th. rufocincta* are strongly setose, of *Th. harpactoides* finely ciliated. In the posterior foot-jaw in *Th. rufocincta* the terminal claw has three prominent though slender setæ spring-

---

[*] *Lata,* broad, referring to its comparatively broad outline when viewed dorsally.

R

ing from the base, but the terminal claw of the same appendage in *Th. harpactoides* is provided with only a single delicate seta. From *Th. mysis* this species is at once distinguished by the form of the fifth feet in both ♀ and ♂ specimens.

Scutilidium fasciatum (Boeck).

> 1864. *Porcilidium fasciatum*, Boeck, 'Oversigt Norges Copepoder,' p. 56.
> 1868. *Aspidiscus fasciatus*, Norman, 'Brit. Assoc. Report,' p. 298.
> 1880. *Scutilidium fasciatum*, Brady, 'Monog. Brit. Copep.,' ii. p. 178, pl. lxviii. fig. 11 ; pl. lxix. figs. 1–9.

Habitat.—In a shore-gathering from Dunbar, collected by Mr Peter Jamieson, assistant naturalist. This and *S. tisboides* may be distinguished from most other British Copepoda by the peculiar form of the first pair of feet.

Cylindropsyllus lævis, Brady (Pl. XIII. figs. 1–18.)

> 1880. *Cylindropsyllus lævis*, Brady, 'Monog. Brit. Copep.,' vol. iii.

Female.—Length 1·4 mm. Animal elongate, cylindrical, cephalo-thorax five-jointed, not distinctly separated from the abdomen, which is four-jointed ; the first body segment about as long as the next two together ; forehead produced into a sharp rostrum. Anterior antennæ short, scarcely longer than the first body segment, seven-jointed, the proportional length of the joints as in the formula

$$\frac{10 \cdot 22 \cdot 9 \cdot 6 \cdot 5 \cdot 4 \cdot 8}{1 \cdot 2 \cdot 3 \cdot 4 \cdot 5 \cdot 6 \cdot 7}$$

Sparingly setiferous, a long slender olfactory appendage springs from the end of the fifth joint. Posterior antennæ two-jointed, basal joint long and moderately stout, bearing near the proximal end a very small one-jointed secondary branch which is furnished with one long terminal setæ ; the last joint of the primary branch is armed with several spiniform hairs. Three of those which spring from the extremity of the joint are long and bent near the middle, where, on the outer one of the three, is a produced spine-like process, which looks to be a continuation of the straight proximal half of the hair. Mandibles well developed, consisting of a stout biting part furnished with several small teeth, and a small one-branched palp bearing three moderately long terminal hairs. There is anterior to the mandibles a peculiar organ possessing at its anterior edge two subtriangular appendages which are crenate on the outer margin and measure in breadth very nearly ·01 mm. These subtriangular appendages resemble somewhat the sucking disks on the forehead of *Caligus* and may function as such, but this is very doubtful. Our dissection shows a slender muscle extending upwards and, terminating between the two appendages where it becomes dilated, and seems to connect the two.

The maxillæ (fig. 7) consist of flattened plates, ciliate on the inner margin, the cilia being bounded externally by a small spine. Anterior foot-jaws one-jointed, small, and furnished with one or two marginal processes ; posterior foot-jaws stout, bearing a prominent, somewhat clawed terminal spine, and two marginal setiferous processes. Outer branches of first four pairs of swimming feet three-, inner branches two-jointed ; fifth pair one-branched, foliaceous. (For description of swimming feet, caudal stylets, and setæ see 'British Copepoda,' iii. p. 30.) Ovisacs two, each containing three large ova arranged as shown in the figure. The integument of cephalo-thorax and abdomen closely beset with minute papillæ, of which fig. 18 is an enlarged representation.

Male.—Body similar to that of the female but smaller (1·3 mm.). Anterior antennæ eight-jointed ; the proportional length of the joints are as in the formula

$$\frac{10 \cdot 22 \cdot 5 \cdot 8 \cdot 3 \cdot 8 \cdot 5 \cdot 5}{1 \cdot 2 \cdot 3 \cdot 4 \cdot 5 \cdot 6 \cdot 7 \cdot 8} ;$$

distinctly hinged between the sixth and seventh joints, and indistinctly between the third and fourth. The fifth joint, which is very short, bears a long olfactory filament. The posterior antennæ, mouth organs, and first pair of swimming feet as in the female. The last joint of the outer branches of the second pair of swimming feet bears at nearly right angles a long curved appendage closely resembling the blade of a reaping-hook, and setose on the inner margin (fig. 13). The basal joint of the inner branches of the third pair is furnished internally with a long slender process, which extends beyond the extremity of the branch, and is armed on the inner margin near the distal end with two barb-like teeth. The fourth and fifth pairs as in the female, except that the fifth is rather smaller and furnished with fewer setæ. The abdomen five-jointed ; the posterior margin of the first abdominal segment bears a foliaceous appendage armed with one long and two short, stout setæ. Caudal stylets and setæ as in the female.

Habitat.—Off St Monans, Firth of Forth, in 14 to 15 fathoms water ; bottom clean coarse sand. Not uncommon.

This interesting Copepod, which was described by Professor Brady in his monograph of the British Copepoda in 1880, is apparently local in its distribution, and is probably rare as well as local. The generic and specific descriptions given in the monograph were prepared from the examination of one specimen only—a female—and, as pointed out by Professor Brady, both descriptions were necessarily somewhat incomplete. Having some time ago in a single haul with the dredge secured a considerable number of specimens including both males and females, the opportunity was taken advantage of to make a careful examination of both sexes so as to gain some knowledge of the affinities of the species. In the monograph alluded to *Cylindropsyllus* was provisionally placed among the *Pœcilostoma* because of its apparently close relationship to that group, but as the structure of some of the mouth organs had not been satisfactorily made out no distinct place in the classification was assigned to it.

By the careful dissection of a number of specimens I have been able, with the assistance of my son, to prepare a fairly complete description, with a set of drawings, of the more important and characteristic appendages that distinguish *Cylindropsyllus lævis* from other Copepoda. It will be observed by referring to the description and drawings that there are one or two characters which render the position of *Cylindropsyllus* among the *Pœcilostoma* untenable. These are the distinctly hinged male anterior antennæ, the presence of a secondary branch on the posterior antennæ, and the form of the mandibles,—characters which indicate a closer affinity with the *Harpacticidæ* than with either the *Pœcilostoma* or the *Siphonostoma*. If, on the other hand, the appendages of the peculiar organ described as situated anterior to the mandibles be sucking disks, the position of *Cylindropsyllus* in the classification would be somewhat anomalous, as these appendages would indicate a tendency towards parasitism in this Copepod,—a tendency suggested by Dr Brady. No indication of parasitical habits has, however, been observed hitherto in any of the specimens obtained.

Cylindropsyllus minor. ♀ (Pl. XI. figs. 17–24).

Female.—In the female the body is cylindrical and elongate. Length, exclusive of tail setæ, 1 mm., the first thoracic segment nearly as long as the next three together, rostrum short with a rounded apex. Anterior antennæ about as long as the first thoracic segment, seven-jointed; the comparative length of the joints are as shown in the formula

$$\frac{7 \cdot 18 \cdot 9 \cdot 4 \cdot 5 \cdot 4 \cdot 8}{1 \cdot 2 \cdot 3 \cdot 4 \cdot 5 \cdot 6 \cdot 7}.$$

Sparingly setiferous. An olfactory filament springs from the end of the fourth joint, which is produced to form a base for the filament. The posterior antennæ, mandibles, maxillæ, and anterior foot-jaws as in *Tetragoniceps incertus*. Posterior foot-jaw two-jointed and armed with a long, stout, terminal claw which is ciliate on the inner margin of the distal half. A spine springs from the end of the first joint and projects forward so as to be opposed to the extremity of the terminal claw. The outer branch of the first four pairs of swimming feet three-jointed, the inner branch of the first and fourth pairs two-jointed, of the second and third one-jointed, the inner and outer branches of first pair nearly equal, but the inner rather longer, sparingly setiferous. The one-jointed inner branch of the second and third pairs shorter than the first joint of the outer branch, and terminating in a short stout spine; a moderately long slender hair springs from near the middle of the one-jointed inner branch of the third pair. The outer branch of the fourth pair is nearly twice as long as that of any of the preceding pairs. The two first joints are about equal in length; the last is rather shorter, and furnished with one very short and three long setæ. The inner branch, which is two-jointed, is scarcely longer than the first joint of the outer branch, and provided with a short terminal plumose spine or stout setæ. Fifth pair foliaceous, small, one-branched, the posterior margin armed with six plain setæ, and, exteriorly, with a stout spine. Abdomen four-jointed, first segment rather longer than any of the other three. Caudal stylets about as long as the last abdominal segment, slightly divergent, and bearing a lanceolate spiniform terminal process and a few very small hairs. Ovisacs two, each with four ova placed end to end as shown in the figure.

(?) *Lichomolgus littoralis,*[*] n. sp. (Pl. X. figs. 1–9).

Male.—Length about 1·12 mm. (exclusive of tail setæ). In form somewhat like *Lichomolgus arenicolus.* Anterior antennæ short, seven-jointed, the proportional length of the joints as shown by the formula [†]

$$\frac{16 \cdot 24 \cdot 12 \cdot 23 \cdot 15 \cdot 18 \cdot 16}{1 \cdot 2 \cdot 3 \cdot 4 \cdot 5 \cdot 6 \cdot 7}.$$

All the joints more or less setiferous; the second joint is furnished with a number of moderately short, and two long setæ. Posterior antennæ stout, four-jointed, the length of the joints gradually decreasing, the last about as broad as long, quadrangular, and bearing about six long, unequal, and plain terminal hairs and one plumose seta, the exterior margin of the second and third joints ciliated. Two spines, one of which is stout and strongly curved, and two setæ, spring from the exterior distal angle of the third joint as shown in the figure. There is no secondary appendage to the posterior antennæ. Mandible stout, consisting of a broad

---

[*] Of or belonging to the shore.

[†] When first examined the three last joints of the anterior antennæ were observed to be nearly equal in length, but the last one became detached before the joints were measured. The length stated, though closely approximate, may therefore not be quite correct.

basal part (from the end of which spring two submarginal plumose setæ) and a strong claw-like tooth armed with a few setæ on its outer aspect, as shown in the figure. Maxillæ well developed, the biting part with three apical processes, finely serrate on the margin, the palp with several terminal plumose setæ. Second foot-jaw strong, two-jointed, last joint broadly triangular, the inner margin armed with a double row of small teeth, terminal claw stout, curved, as long as the joint from which it springs, and forming with it a powerful grasping organ ; the first joint is provided with a stout plumose seta on the inner margin. Both branches of the first four pairs of swimming feet three-jointed and nearly alike ; the outer branch rather shorter than the inner. The fifth pair broadly foliaceous, truncate at the end, and furnished with four stout terminal hairs ; both the margins are ciliated. Abdomen five-jointed, the first segment considerably larger than the next, and armed with two short unequal spines at the postero-distal angles ; the last four segments gradually decrease in length and breadth. Caudal stylets short and broad, about as long as the last abdominal segment, and furnished with one long and two short terminal plumose setæ and three very short hairs.

Habitat.—Vicinity of Culross, on the north side of the Forth. One specimen only was obtained. The remarkable form of the posterior foot-jaws, so closely resembling the Gnathopods of some of the Amphipoda, and the broad fifth pair of swimming feet, enable the species to be readily distinguished. A full-sized drawing of the animal was to have been prepared, but unfortunately the cover-glass of the slide on which the Copepod was mounted preparatory to being figured was accidentally pressed down so that the thorax became abnormally flattened ; for this reason a correct full-sized drawing could not be prepared.

Lichomolgus concinnus,[*] n. sp. (provisional name).   (Pl. XI. figs. 25-23).

Female.—Length, exclusive of caudal setæ, ·9 mm. Cephalo-thorax broadly ovate. Abdomen short, narrow, four-jointed, first abdominal segment large, longer than the following three together, and as broad as long, the postero-lateral angles not produced nor furnished with spines. Caudal stylets stout, about as long as the last abdominal segment, and provided each with one marginal and four terminal setæ. Forehead rounded. Anterior antennæ shorter than the first cephalo-thoracic segment, seven-jointed, sparingly setiferous, the proportional length of the joints as in the formula

$$\frac{12 \cdot 28 \cdot 7 \cdot 18 \cdot 15 \cdot 12 \cdot 7}{1 \cdot 2 \cdot 3 \cdot 4 \cdot 5 \cdot 6 \cdot 7}.$$

A short olfactory filament springs from near the middle of the fifth joint. Posterior antennæ stout, four-jointed, and armed with a short and strong terminal claw ; the second joint is longer than the next two together, the third is small. There is no secondary branch. The anterior foot-jaw is in the form of an elongate curved spine, having a dilated base and a long slender extremity ; it resembles in form and marginal pectination the anterior foot-jaw of *Lichomolgus liber.* Posterior foot-jaw three-jointed, similar to that of *L. thorellii.* Second joint dilated and bearing two slender spines ; the last joint smaller, with two stout terminal spines, one of which is setose. The first three pairs of swimming feet as in *L. liber,* the inner branch of the fourth pair two-jointed, the second joint twice as long as the first. The foot of the fifth pair consists of a stout cylindrical joint bearing two elongate terminal hairs, which are articulated near the base ; ovisacs two. No male has been observed.

Habitat.—Off St Monans, Firth of Forth.   Rare.

[*] *Concinnus,* neat.

Lichomolgus arenicolus, Brady.

> 1872. *Boeckia arenicola,* Brady, ' Nat. Hist. Trans. Northumb
>   ' and Durham,' vol. iv. p. 430.
> 1880. *Lichomolgus arenicolus, idem.,* 'Brit. Copep.,' vol. iii. p. 46,
>   pl. lxxxvii. fig. 1–7.

Habitat.—Off St Monans, Firth of Forth. One specimen only of this fine and distinct species was obtained. It occurred among material dredged in about 14 fathoms water ; bottom, clean sand.

Cyclopicera gracilicauda, Brady.

> 1880. *Cyclopicera gracilicauda,* Brady, *loc. cit.,* vol. iii. p. 58, pl.
>   lxxxiii. figs. 1–10.

Habitat.—Off St Monans, Firth of Forth. Several specimens were obtained among dredged material. This was readily recognised by the slender abdomen and long caudal stylets. It appears to be a somewhat rare species.

PARASITA.

Family CHONDRACANTHIDÆ.

Chondracanthus zei, Delaroche.

> 1811. *Chondracanthus zei,* Delaroche, ' Nouv. Bull. des Sc. de la
>   ' Soc. Philm.,' vol. ii. p. 270, t. 2, fig. 2.
> 1850. *Chondracanthus zei,* Baird, ' Brit. Entom.,' p. 327, pl. xxxv.
>   fig. 1.

Habitat.—On the gills of a ' John Dory ' (*Zeus faber*), caught in the vicinity of Largo Bay, Firth of Forth. Baird (*loc. cit.*) gives a very characteristic figure of this *Chondracanthus.* The arrangement of the numerous elongate appendages (they can scarcely be called spines as Baird describes them) which surround the parasite is such as to impart to it a somewhat handsome appearance. There does not appear to be any previous record of this species from the Firth of Forth.

Chrondracanthus merluccii, Holten.

From the skin of the branchial cavity of a Hake. *Merlucius vulgaris,* landed at Newhaven, February 1885.

AMPHIPODA.

Family GAMMARIDÆ.

Cressa dubia (Spence Bate) Pl. VIII. fig. 13.

> 1855. *Montagua dubia,* Spence Bate, ' Report Brit. Assoc.'
> 1857. *Danaia dubia, idem.,* ' Ann. and Mag. Nat. Hist.,' xix. p.
>   137.
> 1870. *Cressa schiœdtei,* Boeck, ' Crust. Amph. bor et Arct.'

Habitat.—From a large ' root ' of *Laminaria* brought up in the trawl-net at Station V., Firth of Forth, in February 1892. Several specimens were obtained. From what I can learn regarding this species it appears to be somewhat rare in the British seas. It was first obtained by Spence Bate among trawl refuse ' from near the Eddystone Lighthouse.' David Robertson records having taken it ' off stones and nest of *Lima hians*

' that were dredged in 7 to 8 fathoms west of Tan Buoy, Cumbrae,' and he
adds, ' This was the only time I met with it.'*

In ' British Sessile-eyed Crustacea,' vol. i. p. 67, it is stated in the
generic description of *Danaia* that the mandibles are 'destitute of a palpi-
' form appendage.' That is not so. They possess an elongate three-jointed
palp (fig.   ) which has somehow been overlooked by the author when preparing the description of the genus. The Rev. T. R. R. Stebbing in his valuable work on the Challenger Amphipoda, referring to this genus in a foot-note
at p. 1671 (vol. xxix. of the Challenger Reports), points out that ' in 1849
' Milne Edwards and J. Haime, ("Comptes Rendus," t. xxix. p. 261), gave
' the name *Dania* to a genus of fossil Corals; this name they spell
' *Danaia* in the general index to their Monograph of the British Fossil
' Corals, Palæont. Soc. vol. for 1854, published 1865. *Danaia*, Spence
' Bate, must therefore give way to the later Cressa, Boeck, with which a
' specimen of the type species recently obtained and dissected proves it to
' be certainly synonymous.' See also a foot-note at p. 747 of Mr Stebbing's
work referred to above.

Halimedon parvimanus (Bate and Westwood).

> 1862. *Westwoodilla cæcula*, Bate, 'Cat. Amphip. Brit. Mus.,' p.
> 102.
> 1862. *Westwoodilla hyalina*, idem, *ibedem*, p. 103.
> 1863. *Œdiceros parvimanus*, Sp. Bate and Westwood, ' Brit. Sess.-
> ' eyed Crust.,' vol. i. p. 161.
> 1870. *Halimedon Mülleri*, A. Boeck, 'Crust. Amphip. bor. et
> ' Arct.,' p. 89.
> 1889. *Halimedon parvimanus*, Norman, 'Ann. and Mag.,' S. 6,
> vol. iii, p. 455, pl. xx. figs. 10-14.

Habitat.—From trawl refuse from Station V., Firth of Forth, February
1892, and on one or two previous occasions from other parts of the Forth.
This species seems to be rare in the Forth, as only one specimen at a time
has been obtained. The Forth specimens agree very closely with the
figures and description in the ' British Sessile-eyed Crustacea,' in having
the Gnathopods distinctly subchelate, the rostrum strongly produced, the
eye large and near the apex of the rostrum.

Pontocrates haplocheles (Grube).

> 1864. *Kroyeria haplocheles*, Grube, 'Die Insel Lussin und ihre
> ' Meeresfauna. Nach einen sech wochentlichen aufenthalte ges-
> 'childert, von Dr Adolph Eduard Grube.' Breslau, 1864.
> 1868. *Kroyera brevicarpa*, Bate and Westwood, ' Brit. Sess.-eyed
> ' Crust.,' vol. ii. p. 508.
> 1870. *Pontocrates haplocheles*, Boeck, 'Crust. Amphip. bor. et Arct.'

Habitat.—Largo Bay, dredged 1889. One specimen only of this apparently rare amphipod was obtained. In this species the first Gnathopods
are short and comparatively broad, and the produced part of the carpus
terminates in a distinct finger-like process. The propodos of the second
Gnathopods are long and slender : the lower angle of the carpus is very
little produced, which thus differs from other British species of *Pontocrates*
that have the lower angle of the carpus of the second Gnathopods produced
as far as, or beyond, the extremity of the propodos. In *Pontocrates haplocheles* the lower produced part of the propodos, which forms the palm of
the *chela*, consists of two distinct portions, the outer or lower is much
more slender than the other, and terminates in a slightly curved point a
little beyond the end of the *chela*. This structure, which seems to be

* Amphipoda and Isopoda of the Clyde, p. 15 (1888).

indicated by the double line in the drawing of the second Gnathopod at p. 508 of the second vol. of 'Brit. Sess.-eyed Crust.,' can only be satisfactorily observed with a moderately high power of the microscope, as a ¼ or ½ inch objective. This species has been taken at Banff by Thomas Edward.

Haustorius arenarius (Slabber).

> 1769. *Oniscus arenarius*, D. M. Slabber, 'Natuurkundige Verlus-
> 'tigingen behelzende microscopise Waarneemingen van in-en
> 'uitlandse water-en Land-Dieren, elf de Stukje,' pp. 92–96.
> Te Haarlem (1769).
> 1775. *Haustorius arenarius*, P. L. S. Müller. A Translation into
> German (with Notes) of Slabber's work. Pub. Nürnberg.
> 1818. *Lepidactylis dytiscus*, T. Say, 'An Account of the Crust
> 'of the U.S.A.' (Jour. Acad. Nat. Sc. Phil.)
> 1825. *Pterygocera arenaria*, P. A. Latreille, 'Fam. nat. Reg.
> 'Animal.,' &c.
> 1851. *Bellia arenaria*, Spence Bate, 'Ann. and Mag. Nat. Hist.,'
> ser. 2, vol. vii. pp. 318–320, pl. xi. figs. 1–8; pl. x. fig. 10.
> 1854. *Sulcator arenarius*, idem, ibidem, vol. xiii. p. 504.
> 1863. *Sulcator arenarius*, Bate and Westwood, 'Brit. Sess.-eyed
> 'Crust.,' vol. i.
> 1880. *Lepidactylis arenarius*, S. J. Smith, 'Trans. Connecticut
> 'Acad.,' vol. iv. (July 1880).
> 1888. *Haustorius arenarius*, Stebbing, 'Report on the Amphipoda
> 'of the Challenger Expedition,' vol. xxix. (text, first half),
> p. 39. (Notes on Müller's translation of Slabber's work.)

Habitat.—Sandy shore east of Burntisland. This species seems to be rather uncommon within the Forth area. There does not appear to be any previous record of it from the Forth. I obtained it by digging up the sand down to 4 or 5 inches, and then passing the sand through a fine wire sieve. It 'has been taken near Falmouth by Dr Leach; at Moray 'Firth by the Rev. G. Gordon; on the coast of Cumberland by Mr 'Albany Hancock; and in Oxwick Bay by Mr Moggridge and Dr J. 'Gwyn Jeffreys.'* Mr David Robertson of Cumbrae records it as 'moderately common all round our sandy shores near low water, and 'taken most successfully by the sieve.'†

It will be observed from the references given that this pretty Amphipod has received a considerable amount of attention from authors. Rev. Mr Stebbing remarks (*loc. cit.*): 'The excellent name *Sulcator* might well 'have been allowed to stand, but since that has been displaced on grounds 'of priority, first by *Pterygocera* and then by *Lepidactylis*, it seems only 'just to go back a step farther to Müller's *Haustorius.*' I have adopted this very reasonable proposal.

Melphidippa (?) *spinosa* (Goes).

> 1865. *Gammarus spinosus*, Goes, 'Crust. Amphip. Spitsb.'
> 1870. *Melphidippa spinosa*, Boeck, 'Crust. Amphib. bor. et Arct.'

Habitat.—Firth of Forth, west of May Island, taken with tow-net fixed to the head of the beam trawl. This appears to be a rare species in the Firth of Forth. It somewhat resembles *Dexamine* or *Atylus*. The one or two specimens I have observed in the Forth were, when captured, of a bright red colour, but from some cause none of them were perfect.

---

* Amphipoda and Isopoda of the Clyde, p. 28 (1888).
† Idem.

Gammarus marinus, Leach.

    1815. *Gammarus marinus*, Leach, 'Linn. Trans.,' vol. xi. p. 359.
    1863. *Gammarus marinus*, Bate and Westwood, 'Brit. Sess.-eyed
        'Crust.,' vol. i. p. 370.

Habitat.—Firth of Forth, inshore, in the vicinity of Culross. *Gammarus marinus* is easily distinguished by the very short inner ramus of the posterior pleiopoda. This species does not appear to have been previously recorded for the Forth, though it is not uncommon towards the head of the estuary, especially where there are sea-weeds between the tide marks.

Photis longicaudata (Bate and Westwood).

    1863. *Eiscladus longicaudatus*, Bate and Westwood, 'Brit. Sess.-
        'eyed Crust.,' vol. i. p. 412.
    1877. *Photis longicaudata*, Meinert, 'Crust. Isop. Amphip. et
        ' Decapoda Daniæ.'

Habitat.—Firth of Forth, off St Monans. Several specimens were obtained by dredging.

Family HYPERIIDÆ.

Parathemisto gracilipes (Norman).

    1868. *Hyperia oblivia*, Bate and Westwood (non Kröyer), 'Brit.
        'Sess.-eyed Crust.,' vol. ii. p. 16.
    1869. *Hyperia gracilipes*, Norman, 'Report on dredging among the
        'Shetland Islands (in Report of the 38th Meeting of the Brit.
        'Assoc., 1868.' London 1869).
    1887. *Parathemisto longipes*, Bovallius, 'Sytem. list of the Amphip.
        ' Hyperiidea ' (Behang till K. svenska Vet.-Akhad. Handling-
        ar. Band. 11, No. 16. Stockholm 1887).

Habitat.—Largo Bay. A number of specimens taken with tow-net. The specimens here ascribed to *Parathemisto gracilipes*, Norman, are small (4 to 5 mm.), apparently all ♂, and most of them with ova. They differ in several respects from *Parathemisto oblivia*, Kröyer, and particularly in the carpus of the second Gnathopods being much less produced inferiorly,—the carpal process being only about one-third of the length of the propodos ;—and in the pereiopods being less slender and not so elongate proportionally. Rev. T. R. R. Stebbing, who kindly examined one or two specimens for me, writes :—'The Small Hyperid is apparently *Parathemisto gracilipes*, Norman, wrongly described and named *Hyperia oblivia*, Kröyer, in B. and W.'

Euthemisto compressa (Goes).

    1865. *Themisto compressa*, Goes, 'φfvers. af Kgl. Svenska Vetensk
        'Akad. førhandl.,' p. 533, pl. xli. fig. 34.
    1870. *Parathemisto compressa*, Boeck, 'Crust. Amph. bor. et Arct.'
        (Særskilt aftryht af Vidensk. Selsk. Forhandlinger).
    ? 1878. *Lestrigonus spinidorsalis*, Sp. Bate, 'Ann. and Mag. Nat.
        'Hist.' (May 1878), p. 411, fig. 2.
    1890. *Euthemisto compressa*, G. O. Sars, 'Crustacea of Norway,'
        vol. i. p. 12, pl. v. fig. 2.

Habitat.—West of May Island, February 1892. This species was obtained among tow-net material collected when trawling Station V. Attention was first drawn to it by its larger size and darker colour than *Parathemisto oblivia*. It is readily distinguished by the body being much compressed, by the dorsum being distinctly keeled, and by the two last segments of the percion and the two first of the pleon being produced posteriorly in the median dorsal line into more or less sharp tooth-like processes. The posterior pleopods have the outer ramus much shorter than the inner.

The name *Euthemisto* was established by Dr Bovallius in 1887 to replace *Themisto*, Guiérin (1828), which was pre-occupied. Professor G. O. Sars describes three species of *Euthemisto* as belonging to the Norwegian Fauna.

Rev. Mr Stebbing suggests that the *Lestrigonus spinidorsalis*, Spence Bate, from the Aberdeenshire coast, and recorded in the *Annals and Magazine of Natural History* for May 1878, is the *Euthemisto compressa* (Goes). There seems to be little doubt that Mr Stebbing's suggestion is correct.

ZOANTHARIA.

Cerianthus lloydii (Gosse).

Habitat.—Off St Monans, in about 14 fathoms water; bottom clean, but not very fine sand. Fragments consisting of the head and tentacles of this sea anemone have been obtained on one or two occasions among sand dredged at the locality mentioned. I have obtained the same species at extreme low water in Rothesay Bay near the Royal Aquarium, but in this case also it was only the head part. The difficulty of capturing a whole specimen is indicated by the following remarks of Mr Robertson of Millport—'It must be approached with the greatest caution, and a spade ' or other such implement placed in the gentlest manner 4 or 5 inches ' from the spot where it is, and when all is ready, drive the spade suddenly ' in beneath it, cutting off its escape by passing through the tube. If the ' animal takes the alarm before the thrust is made, I should say, speaking ' from my own experience, that it is almost hopeless to follow up the ' pursuit.' * The one or two fragments obtained off St Monans evidently show that the dredge in passing through the sand had come upon the creatures unawares and cut their heads off.

ADDITIONAL NOTES.

Lichomolgus agilis, n. sp.

A species of *Lichomolgus*, apparently new, and of which a description with figures is being prepared for publication by my son, Andrew Scott, and myself, has been found living inside the siphons, and between the branchial folds and the body of the common cockle (*Cardium edule*). My son first discovered the *Lichomolgus* a short time ago, while examining some cockles from Morecambe Bay, Lancashire ; more recently we have obtained the same Copepod also in specimens of the common cockle from the neighbourhood of Cramond Island, Firth of Forth. This Copepod agrees with *Lichomolgus litoralis* and one or two others in having the inner branches of the first four pairs of swimming feet all three-jointed, but differs from any species known to us in several important points. It is very active (hence the specific name we have provisionally adopted) and transparent. If a cockle be opened in such a way that a portion of the contained water will remain within the hollow of the opened valves of the shell, specimens of the *Lichomolgus* may be observed darting hither and thither with great rapidity, their presence being in many cases only rendered apparent by the dark coloured line of the alimentary tract. The ovisacs are very large—about half the length of the animal ; the

* ' On the Sea Anemones of the Shores of the Cumbraes' (*Proc. Nat. Hist. Soc. of Glasgow*, vol. ii. pp. 24-30).

inner margin of the ovisacs is nearly straight, the outer margin forms a flat but more or less regular curve ; they contain numerous, moderately large ova, and, as a considerable number of the Copepods are females, we find, what in such circumstances might be expected, that the species is a comparatively common one,—we have obtained as many as sixteen specimens from a single cockle.

The fact that this *Lichomolgus* has been obtained in cockles from the coast of Lancashire and from the Firth of Forth, and that most of the cockles examined were infested with the Copepod, seems to imply that it is a generally distributed species ; if this be so, it may then be of interest to inquire further, whether (*a*) the *Lichomolgus* is found at particular seasons or all the year round, (*b*) as a semi-parasite or as commensal only, (*c*) if its presence has any connection with a healthy or unhealthy condition of the mollusc. Though Copepods, when present in more or less abundance in fresh water, may, in some cases, be rightly considered as 'danger signals,' they are in themselves innoxious, and their presence, though sometimes in considerable numbers, in the cockles, may after all be no indication of hurtful conditions.

Cyclopicera nigripes, Brady and Robertson.

This handsome species has only recently been observed within the Forth area. It was obtained by washing a quantity of Zoophytes brought up in the trawl-net while working some miles east of May Island. This is readily distinguished from other species of *Cyclopicera* by its large size and by the dark colour of the foot-jaws and swimming feet. Dr Brady records its occurrence from several places of the North East Coast of England, Shetland (Norman), the Firth of Clyde, and from Lough Swilly, Ireland. I have taken it in Cromarty Firth and in East Loch Tarbert (Loch Fyne).

Thysanoessa borealis (G. O. Sars).

This Schizopod has been taken in several parts of the Forth area. I am indebted to Rev. A. M. Norman for the name of the species. He also informs me that among a few Schizopoda sent to him, including the *Thysanoessa*, was what he considers to be a specimen of *Nematobelis megalops* (G. O. Sars.), but it wanted the long slender first pair of legs, which had become detached, and which form one of the chief distinctive characters of the species. The eyes of *Thysanoessa* and *Nematobelis* are distinctly constricted near the middle, so that they appear to consist of an under and upper eye, and this character enables them to be readily distinguished from *Boreophausia* and *Nyctiphanes.*

[DESCRIPTION OF THE PLATES.

DESCRIPTION OF THE PLATES.

PLATE VII.

Stephos minor, nov. gen. et. sp.

Fig. 1. Female, dorsal view,	.	.	.	.	magnified 80 diameters.
Fig. 2. Anterior Antenna,	.	.	.	.	,, 130 ,,
Fig. 3. Posterior Antenna,	.	.	.	.	,, 130 ,,
Fig. 4. Mandible and palp,	.	.	.	.	,, 253 ,,
Fig. 5. Anterior foot-jaw,	.	.	.	.	,, 253 ,,
Fig. 6. Posterior foot-jaw,	.	.	.	.	,, 253 ,,
Fig. 7. Foot of first pair,	.	.	.	.	,, 190 ,,
Fig. 8. Foot of second pair,	.	.	.	.	,, 190 ,,
Fig. 9. Foot of fourth pair,	.	.	.	.	,, 190 ,,
Fig. 10. Fifth pair of feet—female,	.	.	.	.	,, 380 ,,
Fig. 11. Fifth pair of feet—male,	.	.	.	.	,, 190 ,,
Fig. 12. Abdomen of female,	.	.	.	.	,, 130 ,,
Fig. 13. Abdomen of male,	.	.	.	.	,, 130 ,,

Acartia bifilosus (Giesbrecht).

Fig. 14. Foot of fifth pair—female, . . . magnified 253 diameters.

Pseudocyclopia crassicornis, nov. gen. et. sp.

Fig. 15. Female, lateral view,	.	.	.	.	magnified 80 diameters.
Fig. 16. Anterior antenna,	.	.	.	.	,, 380 ,,
Fig. 17. Posterior antenna,	.	.	.	.	,, 190 ,,
Fig. 18. Mandible,	.	.	.	.	,, 253 ,,
Fig. 19. Mandible palp,	.	.	.	.	,, 253 ,,
Fig. 20. Anterior foot-jaw,	.	.	.	.	,, 500 ,,
Fig. 21. Posterior foot-jaw.	.	.	.	.	,, 253 ,,
Fig. 22. Foot of first pair,	.	.	.	.	,, 190 ,,
Fig. 23. Foot of second pair,	.	.	.	.	,, 190 ,,
Fig. 24. Foot of third pair,	.	.	.	.	,, 190 ,,
Fig. 25. Fifth pair of feet—female,	.	.	.	.	,, 380 ,,
Fig. 26. Fifth pair of feet—male,	.	.	.	.	,, 253 ,,
Fig. 27. Abdomen of female,	.	.	.	.	,, 95 ,,
Fig. 28. Abdomen of male,	.	.	.	.	,, 95 ,,
Fig. 29. Spermatophore,	.	.	.	.	,, 190 ,,

PLATE VIII.

Pseudocyclopia minor, nov. gen. et. sp.

Fig. 1. Female, lateral view,	.	.	.	.	magnified 180 diameters.
Fig. 2. Anterior antenna,	.	.	.	.	,, 500 ,,
Fig. 3. Posterior antenna,	.	.	.	.	,, 380 ,,
Fig. 4. Foot of first pair,	.	.	.	.	,, 380 ,,
Fig. 5. Foot of third pair,	.	.	.	.	,, 380 ,,
Fig. 6. Foot of fourth pair,	.	.	.	.	,, 380 ,,
Fig. 7. Fifth pair of feet—female,	.	.	.	.	,, 760 ,,
Fig. 8. Fifth pair of feet—male,	.	.	.	.	,, 380 ,,
Fig. 9. Abdomen of female,	.	.	.	.	,, 190 ,,
Fig. 10. Abdomen of male,	.	.	.	.	,, 190 ,,

? *Tetragoniceps maleolata*, Brady.

Fig. 11. Anterior antenna—female, . . . magnified 253 diameters.
Fig. 12. Foot of fifth pair—female, . . . ,, 380 ,,

Cressa dubia (Spence Bate).

Fig. 13. Mandible and palp, . . . magnified 253 diameters.

Tachidius crassicornis, nov. sp.

Fig. 14. Female, lateral view, ... magnified 80 diameters.
Fig. 15. Male and female—lateral view, . . ,, 80 ,,
Fig. 16. Anterior antenna—female, . . . ,, 380 ,,
Fig. 17. Anterior antenna—male, . . . ,, 380 ,,
Fig. 18. Posterior antenna, ... ,, 380 ,,
Fig. 19. Mandible and palp, ... ,, 500 ,,
Fig. 20. Maxilla, ... ,, 500 ,,
Fig. 21. Anterior foot-jaw, ... ,, 760 ,,
Fig. 22. Posterior foot-jaw, ... ,, 500 ,,
Fig. 23. Foot of first pair, ... ,, 380 ,,
Fig. 24. Fifth pair of feet—female (minus lateral seta on one side), 500 ,,
Fig. 25. Foot of fifth pair, male, . . . ,, 1000 ,,
Fig. 26. Appendage to first abdominal segment—male, . ,, 1000 ,,
Fig. 27. Last abdominal segment and caudal stylets, . ,, 190 ,,

PLATE IX.

Ameira longicaudata, nov. sp.

Fig. 1. Female, lateral view, ... magnified 53 diameters
Fig. 2. Anterior antenna, female, . . . ,, 126 ,,
Fig. 3. Anterior antenna, male, . . . ,, 126 ,,
Fig. 4. Posterior antenna, ... ,, 190 ,,
Fig. 5. Mandible and palp, ... ,, 380 ,,
Fig. 6. Maxilla . . . . . ,, 380 ,,
Fig. 7. Anterior foot-jaw, ... ,, 380 ,,
Fig. 8. Posterior foot-jaw, ... ,, 400 ,,
Fig. 9. Foot of first pair, ... ,, 190 ,,
Fig. 10. Foot of third pair, ... ,, 190 ,,
Fig. 11. Foot of fourth pair, ... ,, 126 ,,
Fig. 12. Foot of fifth pair—female, . . . ,, 250 ,,
Fig. 13. Foot of fifth pair—male, . . . ,, 250 ,,
Fig. 14. Appendage of first abdominal segment—male, . ,, 250 ,,
Fig. 15. Part of abdomen, and caudal stylets, . . ,, 95 ,,
Fig. 16. Posterior margin of abdominal segment, . ,, 380 ,,
Fig. 17. Female, lateral view—variety, . . ,, 53 ,,
Fig. 18. One of the abdominal segments—variety, . ,, 190 ,,

Tetragoniceps bradyi, nov. sp.

Fig. 19. Female, lateral view, ... magnified 80 diameters.
Fig. 20. Anterior antenna, ... ,, 250 ,,
Fig. 21. Posterior antenna, ... ,, 190 ,,
Fig. 22. Mandible and palp, ... ,, 250 ,,
Fig. 23. Maxilla, ... ,, 380 ,,
Fig. 24. ? Mouth, ... ,, 190 ,,
Fig. 25. Anterior foot-jaw, ... ,, 380 ,,
Fig. 26. Posterior foot-jaw, ... ,, 380 ,,
Fig. 27. Foot of first pair, ... ,, 190 ,,
Fig. 28. Foot of third pair, ... ,, 190 ,,
Fig. 29. Foot of fourth pair, ... ,, 190 ,,
Fig. 30. Foot of fifth pair, ... ,, 190 ,,
Fig. 31. Abdomen and caudal stylets . . . ,, 90 ,,
Fig. 32. One of the caudal stylets . . . ,, 250 ,,

PLATE X.

? Lichomolgus littoralis, nov. sp. ♂

Fig. 1. Anterior antenna, ... magnified 126 diameters.
Fig. 2. Posterior antenna, ... ,, 126 ,,
Fig. 3. Mandible, ... ,, 190 ,,
Fig. 4. Maxilla. 4 a. Anterior foot-jaw, . . ,, 190 ,,
Fig. 5. Posterior foot-jaw, ... ,, 190 ,,
Fig. 6. Foot of first pair (inner branch minus last joint), ,, 126 ,,
Fig. 7. Foot of fourth pair, ... ,, 126 ,,
Fig. 8. Foot of fifth pair, ... ,, 250 ,,
Fig. 9. Abdomen and caudal stylets . . . ,, 190 ,,

Cletodes lata, nov. sp. ♀

Fig. 10.	Female, dorsal view,	.	.	.	.	magnified 80 diameters.
Fig. 11.	Anterior antenna,	.	.	.	.	,, 380 ,,
Fig. 12.	Posterior antenna,	.	.	.	.	,, 500 ,,
Fig. 13.	Mandible and palp,	.	.	.	.	,, 760 ,,
Fig. 14.	Posterior foot-jaw,	.	.	.	.	,, 500 ,,
Fig. 15.	Foot of first pair,	.	.	.	.	,, 380 ,,
Fig. 16.	Foot of third pair,	.	.	.	.	,, 380 ,,
Fig. 17.	Foot of fifth pair,	.	.	.	.	,, 250 ,,
Fig. 18.	Foot of fifth pair—variety,	.	.	.	.	,, 250 ,,

Tetragoniceps macronyx, nov. sp.

Fig. 19.	Female, lateral view,	.	.	.	magnified 160 diameters.
Fig. 20.	Anterior antenna—female,	.	.	.	,, 250 ,,
Fig. 21.	Anterior antenna—male,	.	.	.	,, 250 ,,
Fig. 22.	Posterior antenna,	.	.	.	,, 760 ,,
Fig. 23.	Anterior foot-jaw,	.	.	.	,, 500 ,,
Fig. 24.	Posterior foot-jaw,	.	.	.	,, 250 ,,
Fig. 25.	Foot of first pair,	.	.	.	,, 380 ,,
Fig. 26.	Foot of fourth pair,	.	.	.	,, 190 ,,
Fig. 27.	Foot of fifth pair—female,	.	.	.	,, 500 ,,
Fig. 28.	Foot of fifth pair—male,	.	.	.	,, 500 ,,

Plate XI.

Laophonte inopinata, nov. sp.

Fig. 1.	Female, lateral view,	.	.	.	magnified 160 diameters.
Fig. 2.	Male, dorsal view,	.	.	.	,, 160 ,,
Fig. 3.	Anterior antenna—female,	.	.	.	,, 500 ,,
Fig. 4.	Anterior antenna—male,	.	.	.	,, 500 ,,
Fig. 5.	Posterior antenna,	.	.	.	,, 500 ,,
Fig. 6.	Anterior foot-jaw,	.	.	.	,, 500 ,,
Fig. 7.	Posterior foot-jaw,	.	.	.	,, 500 ,,
Fig. 8.	Foot of first pair,	.	.	.	,, 380 ,,
Fig. 9.	Foot of third pair,	.	.	.	,, 500 ,,
Fig. 10.	Foot of fifth pair	.	.	.	,, 380 ,,
Fig. 11.	One of the caudal stylets	.	.	.	,, 250 ,,
Fig. 12.	One of the body segments,	.	.	.	,, 500 ,,

Thalestris harpactoides, Claus.

Fig. 13.	Posterior foot-jaw,	.	.	.	magnified 250 diameters.
Fig. 14.	Foot of first pair,	.	.	.	,, 190 ,,
Fig. 15.	Foot of second pair—male,	.	.	.	,, 190 ,,
Fig. 16.	Foot of fifth pair—male,	.	.	.	,, 125 ,,

Cylindropsyllus minor, nov. sp.

Fig. 17.	Female, dorsal view,	.	.	.	magnified 80 diameters.
Fig. 18.	Anterior antenna,	.	.	.	,, 250 ,,
Fig. 19.	Posterior foot-jaw,	.	.	.	,, 760 ,,
Fig. 20.	Foot of first pair,	.	.	.	,, 380 ,,
Fig. 21.	Foot of second pair,	.	.	.	,, 380 ,,
Fig. 22.	Foot of third pair,	.	.	.	,, 380 ,,
Fig. 23.	Foot of fourth pair,	.	.	.	,, 380 ,,
Fig. 24.	Foot of fifth pair,	.	.	.	,, 190 ,,

Lichomolgus concinnus, nov. sp.

Fig. 25. Female, dorsal view, ... magnified 60 diameters.
Fig. 26. Anterior antenna, . . . ,, 190 ,,
Fig. 27. Posterior antenna, . . . ,, 190 ,,
Fig. 28. Anterior foot-jaw . . . . ,, 500 ,,
Fig. 29. Posterior foot-jaw, ... ,, 500 ,,
Fig. 30. Foot of first pair, ... ,, 125 ,,
Fig. 31. Foot of fourth pair, ... ,, 125 ,,
Fig. 32. Foot of fifth pair, ... ,, 500 ,,
Fig. 33. Abdomen and caudal stylets, . . . ,, 195 ,,

Plate XII.

Tetragoniceps incertus, nov. sp.

Fig. 1. Female, lateral view, . . . magnified 80 diameters.
Fig. 2. Male, dorsal view, . . . ,, 80 ,,
Fig. 3. Anterior antenna, female, . . ,, 250 ,,
Fig. 4. Anterior antenna, male, . . ,, 250 ,,
Fig. 5. Posterior antenna, . . . ,, 760 ,,
Fig. 6. Mandible and palp, . . . ,, 380 ,,
Fig. 7. Maxilla, ... ,, 380 ,,
Fig. 8. Anterior foot-jaw . . . ,, 500 ,,
Fig. 9. Posterior foot-jaw, . . . , 500 ,,
Fig. 10. Foot of first pair, . . . ,, 380 ,,
Fig. 11. Foot of second pair—female, . . ,, 380 ,,
Fig. 12. Foot of second pair—male, . . ,, 380 ,,
Fig. 13. Foot of third pair—female, . . ,, 380 ,,
Fig. 14. Foot of third pair—male, . . ,, 380 ,,
Fig. 15. Foot of fourth pair, . . . ,, 380 ,,
Fig. 16. Fifth pair of feet—female, . . ,, 380 ,,
Fig. 17. Foot of fifth pair—male, . . ,, 380 ,,

Paramesochra dubia.

Fig. 18. Female, dorsal view, . . . magnified 125 diameters.
Fig. 19. Male, dorsal view, . . . ,, 125 ,,
Fig. 20. Anterior antenna, female, . . ,, 500 ,,
Fig. 21. Anterior antenna, male, . . ,, 500 ,,
Fig. 22. Posterior antenna, . . . ,, 760 ,,
Fig. 23. Mandible, ... ,, 760 ,,
Fig. 24. Mandible palp, . . . ,, 760 ,,
Fig. 25. Maxilla, ... ,, 500 ,,
Fig. 26. Anterior foot-jaw, . . . ,, 760 ,,
Fig. 27. Posterior foot-jaw, . . . ,, 760 ,,
Fig. 28. Foot of first pair, . . . ,, 500 ,,
Fig. 29. Foot of second pair, . . . ,, 500 ,,
Fig. 30. Foot of fourth pair, . . . ,, 500 ,,
Fig. 31. Foot of fifth pair—female, . . ,, 500 ,,
Fig. 32. Foot of fifth pair—male, . . ,, 500 ,,

Plate XIII.

Cylindropsyllus lævis, Brady.

Fig. 1. Female, lateral view, . . magnified 80 diameters.
Fig. 2. Male, lateral view, . . ,, 80 ,,
Fig. 3. Anterior antenna—female, . . ,, 250 ,,
Fig. 4. Anterior antenna—male, . . ,, 250 ,,
Fig. 5. Posterior antenna, . . . ,, 250 ,,
Fig. 6. Mandible and palp, . . . ,, 500 ,,
Fig. 7. ? Maxillæ and mouth, . . ,, 500 ,,
Fig. 8. ? Labium, . . . ,, 500 ,,
Fig. 9. Anterior foot-jaw, side view, . ,, 760 ,,
Fig. 10. ,, ,, front view, . ,, 570 ,,
Fig. 11. Posterior foot-jaw, . . ,, 500 ,,

Fig. 12. Foot of first pair, ... magnified 250 diameters.
Fig. 13. Foot of second pair—male, . . . ,, 250 ,,
Fig. 14. Foot of third pair—male, . . . ,, 250 ,,
Fig. 15. Foot of fifth pair—female, . . . ,, 500 ,,
Fig. 16. Foot of fifth pair—male, . . . ,, 380 ,,
Fig. 17. Appendage of first abdominal segment—male, . ,, 380 ,,
Fig. 18. Structure of carapace, highly magnified.

Neobradya pectinifer, nov. gen. et. sp.

Fig. 19. Male, dorsal view, ... magnified 53 diameters.
Fig. 20. Female, lateral view, . . . ,, 53 ,,
Fig. 21. Anterior antenna—female, . . . ,, 190 ,,
Fig. 22. Anterior antenna—male, . . . ,, 190 ,,
Fig. 23. Posterior antenna, ... ,, 380 ,,
Fig. 24. Mandible and palp, ... ,, 190 ,,
Fig. 25. Maxilla, ... ,, 340 ,,
Fig. 26. Anterior foot-jaw, ... ,, 340 ,,
Fig. 27. Posterior foot-jaw, ... ,, 510 ,,
Fig. 28. Foot of first pair, ... ,, 225 ,,
Fig. 29. Foot of second pair, ... ,, 225 ,,
Fig. 30. Foot of fourth pair, ... ,, 225 ,,
Fig. 31. Foot of fifth pair—female, . . . ,, 340 ,,
Fig. 32. Foot of fifth pair—male, . . . ,, 340 ,,

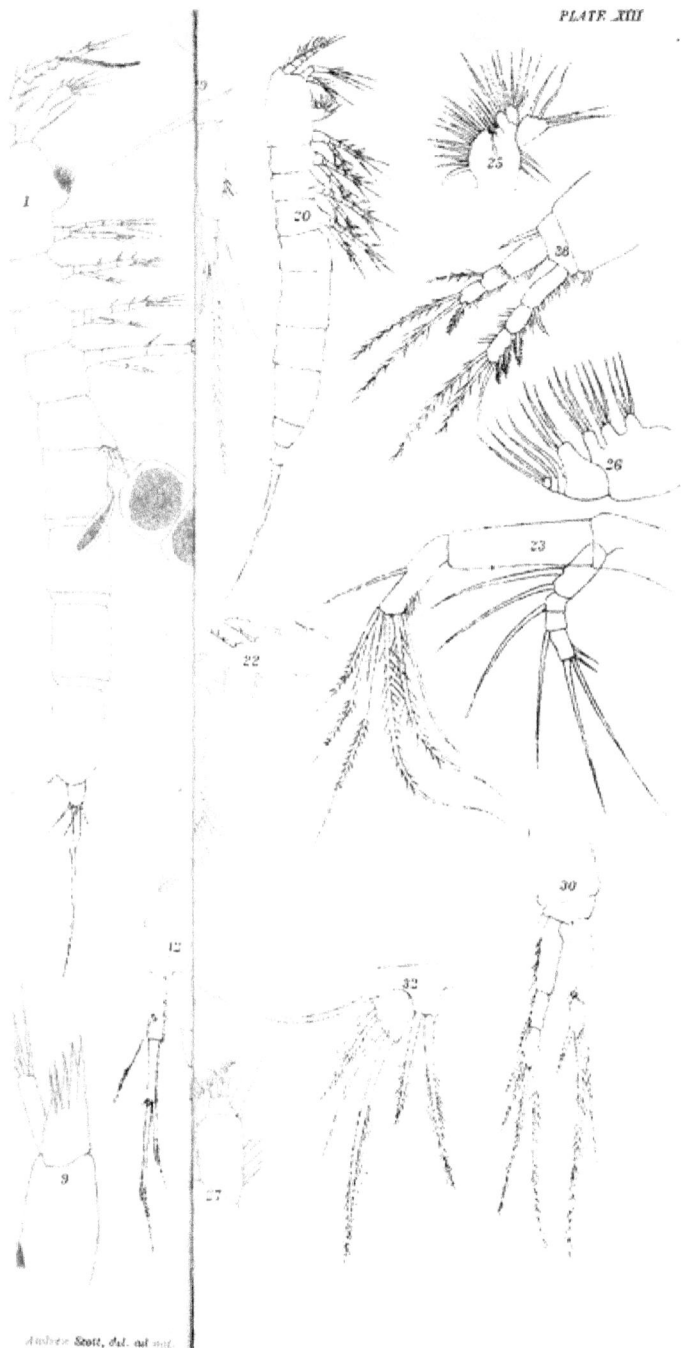

PLATE XIII

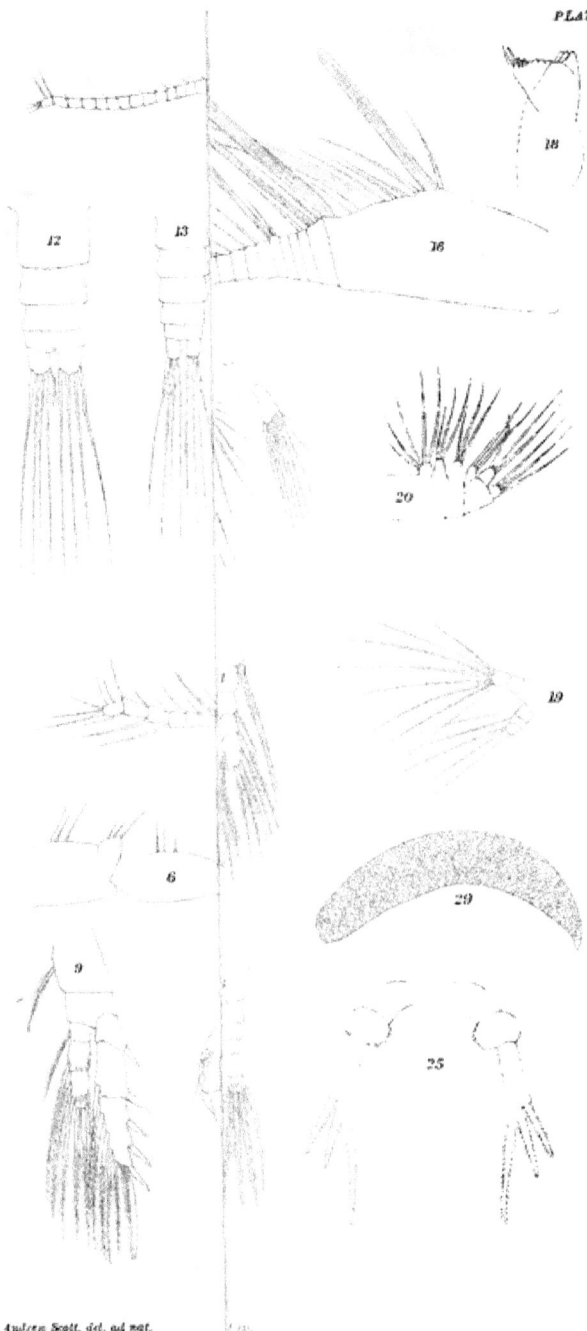

PLATE VII.

18

12   13   16

20

19

6   29

9   25

PLATE 76

PLATE VIII.

... crassicornis, nov. sp.

PLATE IX

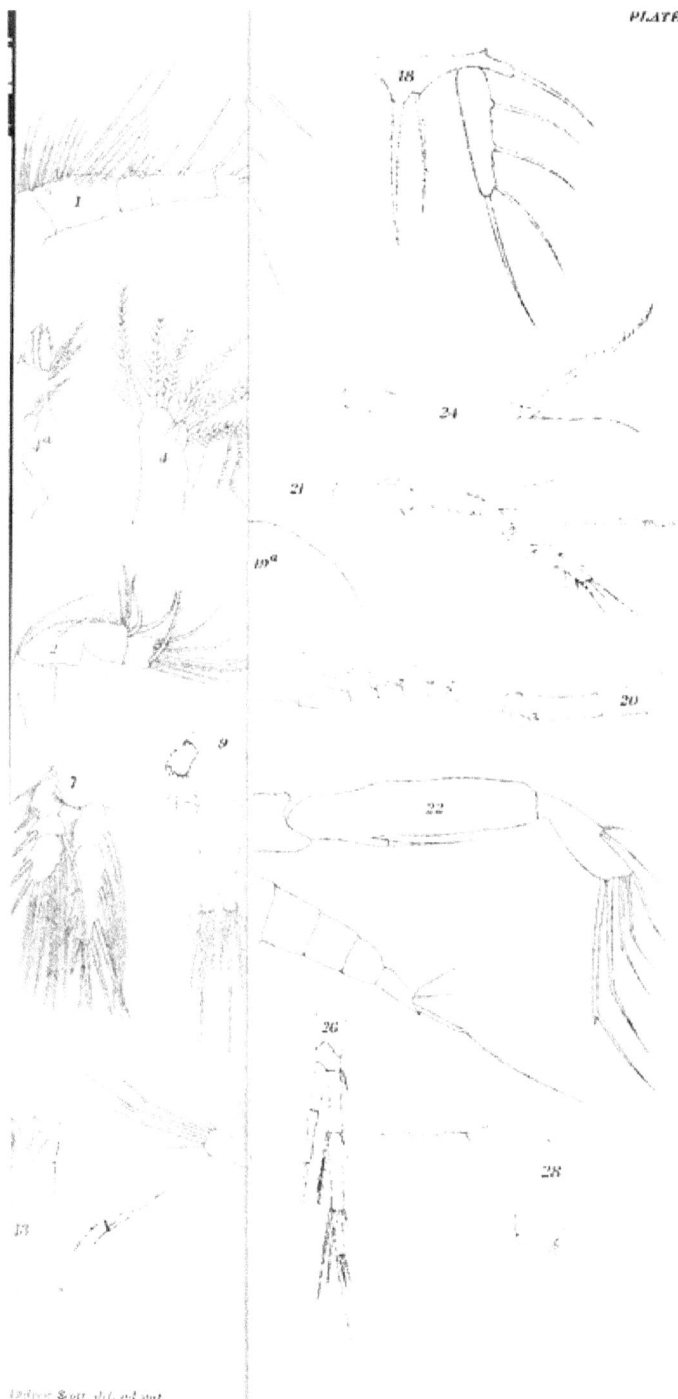

PLATE X.

18

24

21

19ᵃ

20

1

4

1ᵃ

2

9

7

22

26

28

13

Plate X

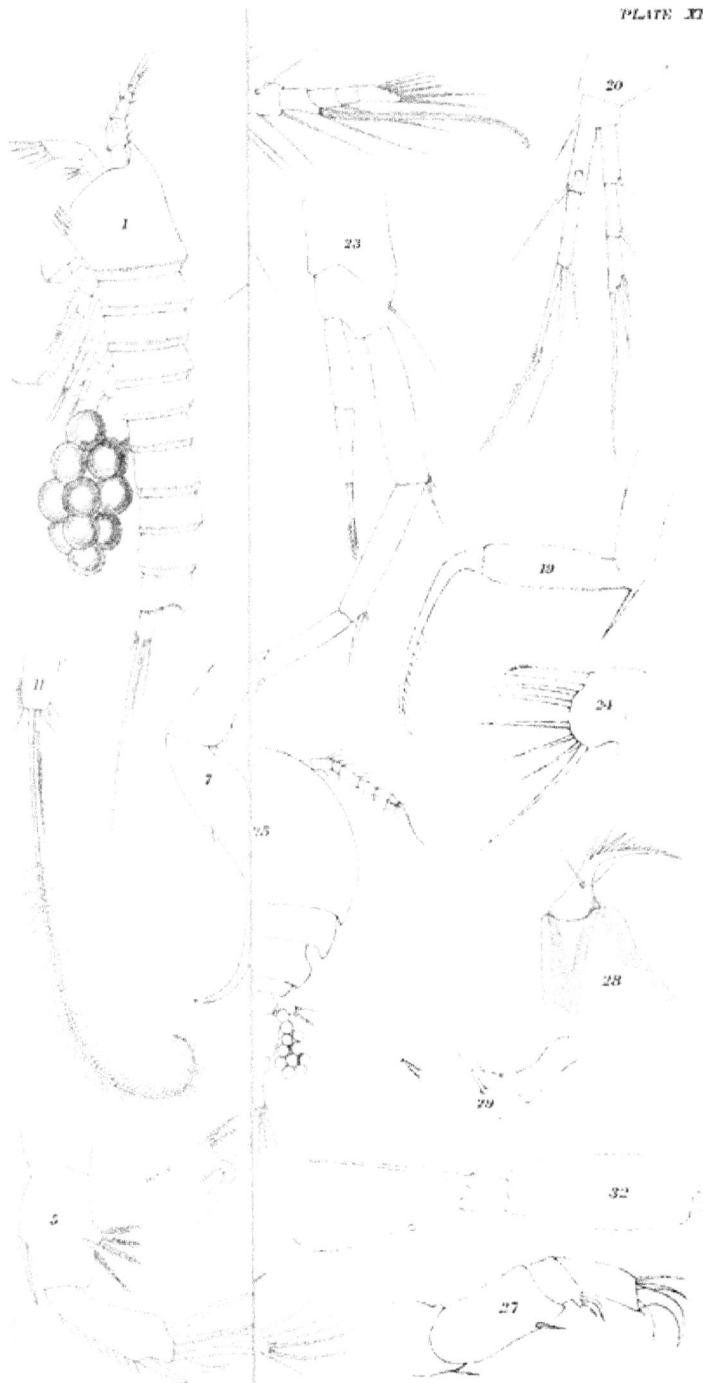

PLATE XI.

1

20

23

19

24

11

7

28

25

29

32

5

27

Figs. 4 (–5.) Lepeoptera gracious nov. sp. Figs. 11-16. Pandarus tenuiculus nov. n. nov. Figs. 17-18. Caligoenpsicus vacue nov. sp. Figs. 19-20. Lernanthropus nauclerus nov. sp.

PLATE XII.

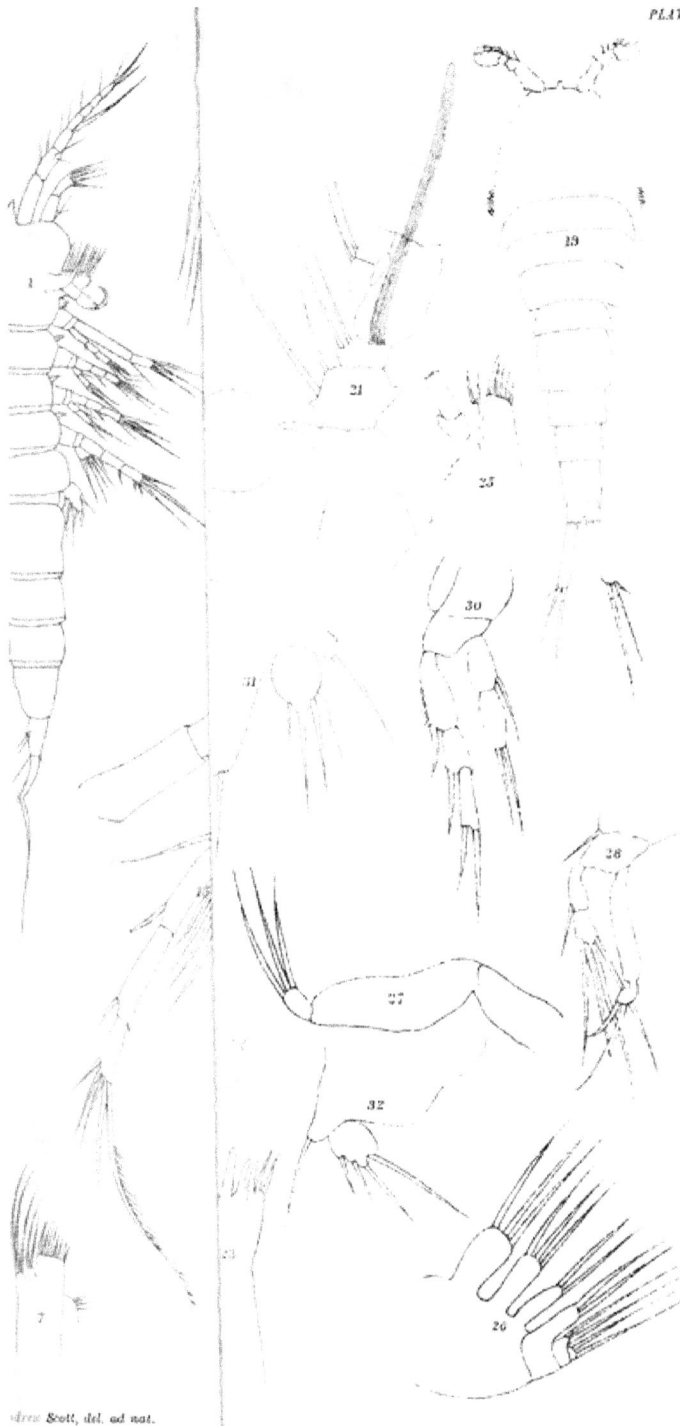

PLATE 86

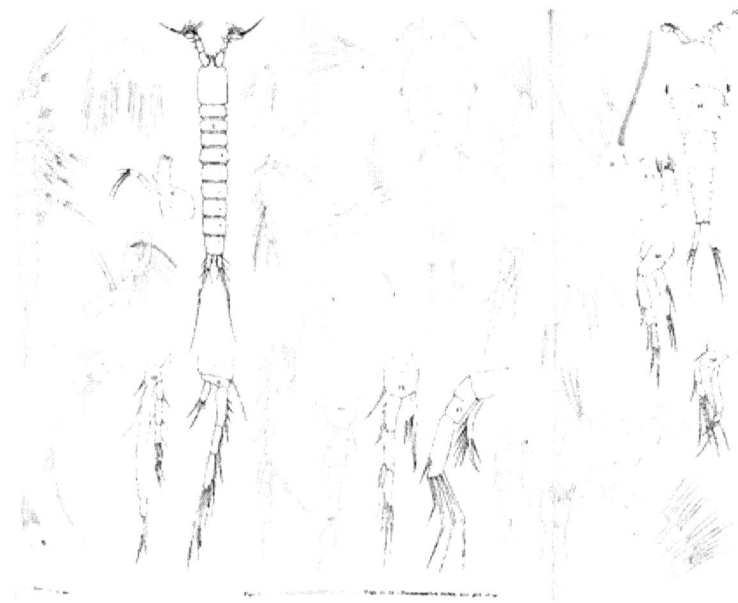

Fig. 44.—*Paramonotropa similis*, one-half size.

Longipedia coronata

minor

Canuella perplexa "

Jonesiella hyaena

Dactylopus rostratus "

Cletodes tenuiremis "

Cyclopicera purpurocincta "

" lata

Laophontodus hirsutipes "

Modiolicola insignis

Paranthessius rechardi "?

Pelmatia aemula "

Bomolochus soleae

Platychelipus littoralis

Zosime typica

II.—ADDITIONS TO THE FAUNA OF THE FIRTH OF FORTH. (Pls. II.–V.)

Part V. By Thomas Scott, F.L.S.

In this—the fifth—contribution towards a better knowledge of the invertebrate fauna of the Firth of Forth, only the lesser crustacea, and chiefly the copepoda, are dealt with. There are several points of interest in the natural history of the Forth crustacea referred to in this contribution: it is shown, for example, that the genus *Longipedia*, which has hitherto been considered to comprise only one species—*Longipedia coronata*—included not only a distinct variety (if it be not a distinct species) of the same genus, but also a supposed sexual form, for which it has been found necessary to provisionally institute a new generic name. Some of the species, though only now recorded for the first time for the Firth of Forth, were collected so long ago as 1889, but were allowed to stand over for want of sufficient information concerning them.

In preparing this paper, I have again the pleasure of thanking the kind friends mentioned in my papers of previous years for assistance and encouragement. I have also gratefully to acknowledge the kindness and sympathy of the following eminent French naturalists who have corresponded with me on matters relating to the subject of these papers :— MM. Le Baron Jules de Guerne and Jules Richard, Dr Raphael Blanchard, Dr Eugène Canu, and Dr Georges Roche. The drawings which accompany this paper are the work of my son, Mr Andrew Scott, who has also prepared the greater part of the necessary dissections. To carefully dissect a copepod, some of which are not more than the thirtieth to the fiftieth of an inch in length, requires no little dexterity ; yet a complete series of dissections are often prepared from a single specimen. It is only by such careful work that I have been enabled to add so many rare and interesting species to the number of the Forth crustacea.

Perhaps the following brief statement of the method pursued by my son may be of interest to other students :—When necessary, the specimen is left to soak for a time in caustic potash, but this, while it clears the tissues, has sometimes a tendency to cause the specimen to shrivel up more or less, and not only while it remains in the caustic potash, but also after it is mounted in glycerine jelly—which is the medium generally made use of for mounting. To overcome this difficulty, the specimen, after being removed from the caustic potash, is well washed in fresh-water, which restores it to its normal form. If a full-sized drawing is wanted, and the specimen still shows a tendency to shrink in the glycerine jelly, it is simply mounted in water under a cover-glass till the full-sized drawing is made,—three or perhaps four tiny bits of stout paper being inserted at intervals under the edge of the cover-glass to keep it from pressing on the specimen. The manner in which shrinkage usually takes place is by the contraction of the tissue between each body segment, so that the one segment is more or less drawn in under the other—telescope fashion—and the specimen becomes in this way apparently shorter and more robust than it should be ; it is seldom that mounting in glycerine jelly makes any difference on the dissected appendages of the copepoda. When the specimen is to be dissected, it is placed in a little water on a slide on the stage of the microscope—a one-inch objective is the power used—with a fine needle, the thick end of which has been pushed into a small bit of common timber for a handle, and fixed with sealing-wax ; the specimen is usually divided into two parts by an incision immediately in front of the first pair of swimming feet. The mouth appendages and antennæ, or the swimming feet, are then first carefully dissected off—each pair by itself—and at once mounted under a separate cover-glass (or bit of cover-glass). A complete series of appendages is sometimes mounted on a single slide—this is done partly for convenience, and partly to save time—and when this is done the series always begins from the same end of the slide, and each appendage or pair of appendages occupies a particular position on the slide ; and also, if very small, it is surrounded by a minute ring drawn with common writing-ink. There is thus no confusion, time is made the most of, and comparatively few slides are required.

CRUSTACEA.

COPEPODA.

Family HARPACTICIDÆ.

Genus *Longipedia*, Claus (1863).

Longipedia coronata, Claus.

As there appears to have been some misconception in regard to *Longipedia coronata*, Claus, the following description of what appears to be a typical female, and of two other and distinct forms that have probably been accidentally included with it in the same species, may be of interest.

Longipedia coronata, Claus. (Pl. II. figs. 1–13.)

  1863. *Longipedia coronata*, Claus, ' Die frei lebenden Copepoden,' p. 110, pl. xiv.
  1864. *Longipedia coronata*, Boeck, ' Oversigt Norges Copepoder,' p. 252.
  1880. *Longipedia coronata*, Brady, ' Mon. Brit. Copep.,' vol. ii. p. 6, pls. xxxiv.-xxxv. (pars.).
  1882. *Longipedia coronata*, Giesbrecht, ' Die frei lebenden Copep. ' der Kieler Föhrde,' p. 99, pls. i., iv.-xii. (? var.).
  1892. *Longipedia coronata*, Canu, ' Les Copep. du Boulon.,' p. 146.
  1893. *Longipedia coronata*, T. and A. Scott, ' Ann. Scot. Nat. ' Hist.,' vol. ii. pt. 2, p. 91, pl. ii. figs. 4–6.

Length from apex of rostrum to end of caudal stylets (exclusive of setæ) 1·5 mm. ($\frac{1}{16}$th of an inch). Rostrum broad, with a bluntly rounded apex. The form of the animal seen from above is narrow, elongate, and tapering gradually to the end of the abdomen ; the whole length is about equal to four times the breadth at the posterior end of the first cephalothoracic segment. Anterior antennæ short, stout, and curved ; densely setiferous, most of the setæ being plumose in varying degrees ; an elongate sensory filament springs from the third joint, and two similar but short filaments from the extremity of the antennæ. Primary branch of the posterior antennæ, three-jointed. The secondary branch is rather longer than the primary and six-jointed ; the joints become longer and narrower towards the apex, so that while the length of the first joint is equal to little more than half the breadth, the last is about one and a half times longer than broad. The posterior antennæ bear numerous plumose setæ. The mandibles have the masticatory portion broad, the truncate end of which is armed both with papilliform and pointed teeth. The two branches of the palp arise from a dilated base ; one of the branches is stout and apparently one-jointed ; the other is more slender and two-jointed—both branches strongly setiferous. The maxillæ are very similar to those of the *Calanidæ* (fig. 5). Anterior foot-jaws furnished with several marginal setiferous processes; the last process armed with a strong spine in addition to three aculeate setæ; terminal joints small, furnished with a number of slender setæ. Posterior foot-jaws somewhat rudimentary, bearing numerous marginal and delicate plumose setæ. A long delicately plumose hair springs from the lateral aspect and near the base of the foot-jaw, and a few similar hairs from near the middle and distal extremity (fig. 7). The first pair of swimming feet are comparatively short; both branches are of about equal length, and composed of three sub-equal joints. The second joint of the outer branch bears a long, curved, spiniform, and blunt-pointed marginal seta. The last joint is armed with three spiniform marginal setæ, but these are considerably shorter than that on the second joint; both branches otherwise more or less setiferous. The second pair of swimming feet have the inner branches greatly elongated—being equal to about three times the length of the outer branches (the third joint alone is nearly equal to twice the length of the outer branch). The end of the outer branch extends to a little beyond the second joint of the inner one. The long joint of the inner branch is armed on the outer aspect and near the middle—but nearer the proximal than the distal end—with a large aculeate spine, and with two similar but smaller spines on the inner aspect—one near the middle of the joint, but posterior to the large outer spine, and one about one-fourth of the length of the joint from the distal end. This joint is also furnished with three stout and moderately long terminal spines. Both branches of the third and fourth pairs of feet are nearly equal and similar to each other, except that the fourth is a little smaller (figs. 10, 11). Fifth pair small, basal joint scarcely developed, produced exteriorly into an elongate digitiform process, bearing a single apical seta, and furnished interiorly with a very long and curved aculeate seta, the proximal part of which is stout. The secondary branch (or joint) is foliaceous, spathulate in form, the greatest breadth being rather less than half the length; the inner margin nearly straight, bearing several minute teeth; outer margin and end sinuate and provided with a number of setæ, the innermost being nearly three times the length of the joint from which it springs (fig. 12). The postero-lateral angles of the first abdominal segment are produced into tooth-like processes. The postero-lateral angles of all the thoracic and abdominal segments are acutely angular—the last abdominal segment is very short—with the dorsal part of the posterior margin spiniferous, the central spine being large and prominent, the others small. Caudal stylets short; the longest of the caudal setæ are scarcely equal to twice the length of the abdomen and caudal stylets combined. One ovisac.

No males of this form have yet been observed in the Firth of Forth, but females are frequent.

This appears to be the form described in the monograph of the British Copepoda as the 'male' of *Longipedia coronata*. It also closely agrees with the description and figures of Claus, both as regards its size and structural details, with the exception of the arrangement of the spines on the long joint of the inner branches of the second pair of swimming feet. Claus's figure agrees with that of Giesbrecht in this respect. But the form now described does not agree with that described by Giesbrecht in size, in the form of the second and fifth pairs of thoracic feet, and in some other important points. That described by Giesbrecht, on the other hand, agrees perfectly with a form recently discovered by my son, Andrew Scott, while examining some dredged material from the Firth of Forth, and which is provisionally described as *Longipedia coronata*, var. *minor*. The following is a description of this variety:—

Longipedia coronata, var. *minor*, T. and A. Scott. (Pl. II. figs. 14–20.)

1893. *Longipedia coronata*, var. *minor*, T. and A. Scott, 'Ann.' Scot. Nat. Hist.,' vol. ii. pt. 2, p. 93.

1882. (?) *Longipedia coronata*, Giesbrecht, *loc. cit.*

Length from apex of rostrum to end of caudal stylets, exclusive of tail setæ, ·82 mm. (one-thirtieth of an inch). Female—anterior and posterior antennæ and mouth organs similar to those of *Longipedia coronata*, but smaller. Male—anterior antennæ short and robust, much less setiferous than those of the female, hinged between the third and fourth joints, terminal joint forming a comparatively small and curved claw-like appendage (fig. 15). First pair of swimming feet somewhat similar to those of *Longipedia coronata*, but smaller. The second pair in the female differ from those of *Longipedia coronata* in the following particulars : —The outer branch is considerably longer in proportion to the inner branch, the end of the second joint in the one reaching to about the end of the second joint in the other ; and the long third joint of the inner branches is only about one and a half times the length of the outer branches ; and the spine on the outer aspect of the long third joint is situated between the two spines on the inner aspect, but nearer to the proximal one (fig. 16). The exterior spine on the long third joint is wanting in the male ; the outer branch of the second pair of the male is also somewhat proportionally shorter than in the female (fig. 17). The other swimming feet are nearly similar to those of *Longipedia coronata*, but are rather smaller. The fifth pair in the female differ considerably from those of *Longipedia coronata* in the form of the secondary branch (or joint). This branch is elongate and narrow (fig. 18), the greatest breadth being only equal to about one-fourth of the length ; and also, though the var. *minor* is little more than half the length of the (supposed) typical form, the length of the secondary branch of its fifth foot is greater than that of the other. This difference may be indicated in another way —*i.e.*, in *Longipedia coronata* the length of the secondary branch of the fifth foot is scarcely equal to one-fourteenth of the entire length of the animal, but in *Longipedia coronata*, var. *minor*, the length of the secondary branch is equal to one-eighth of the length of the animal. In the male the fifth pair are small, and the basal joint is proportionally rather more developed than in the female ; the secondary branch is rather broader and shorter ; and the interior spiniform basal seta is straight, and considerably shorter than in the female (fig. 20). The appendages of the first abdominal segment in the male are nearly as large as the basal part of the fifth feet, and are furnished with three setæ. The central one of the three marginal spines of the last abdominal segment is not so large proportionally, and the apical setæ of the caudal stylets are much longer than in *Longipedia coronata* (fig. 19). One ovisac. Frequent in dredged material from Largo Bay and off Musselburgh. This is certainly not the form described as *Longipedia coronata* (male) in the monograph of the British Copepoda ; neither does it appear to agree with that described by Claus, except in the arrangement of the spines on the last elongate joint of the inner branch of the second pair of swimming feet, but it agrees in size and in structural details with that described by Giesbrecht in his account of the free living Copepoda of Kiel fiord.

Genus *Canuella*, T. and A. Scott.
'Ann. Scot. Nat. Hist.,' 1893.
Longipedia, Brady (in part).

Somewhat like *Longipedia coronata*, Claus ; but the inner branches of the second pair of swimming feet are not longer than the outer branches.

Fifth pair of feet in both sexes rudimentary ; last abdominal segment not spiniferous ; ovisacs two.

Canuella perplexa, T. and A. Scott. (Pl. II. figs. 21–35.)
       1893. *Canuella perplexa*, T. and A. Scott ,' Ann. Scot. Nat. Hist.,'
          vol. ii. p. 92, pl. ii. figs. 1–3.
       1880. *Longipedia coronata*, Brady (in part), 'Mon. Brit. Copep.,'
          vol. ii. p. 6, pl. xxxiv. figs. 3, 9 ; pl. xxxv. figs. 1, 3, 9.
       1867. (?) *Sunaristes paguri*, Hesse, ' Ann. des Sci. Nat.,' 5th ser.
          (Zool.), vol. vii. p. 205, pl.   .
       1884. (?) *Longipedia paguri*, W. Müller, ' Archiv. für Naturg.,'
          Jahrgang 50, 1st Band, p. 19, pl. 13.

Length (exclusive of caudal setæ) 1·4 mm. ($\frac{1}{18}$th of an inch). Body seen from above elongate, nearly cylindrical, tapering slightly to the posterior end ; forehead produced into a stout and somewhat conical rostrum. Anterior antennæ of the female stout, especially the basal portion, curved, and indistinctly five-jointed, furnished with numerous elongate setæ, most of which are plumose ; two sensory filaments spring from the third joint. In the male the middle joints of the anterior antennæ are narrower than the preceding joints, or than the one immediately following. This joint, which is the penultimate one, is dilated, while the last is small and hook-like—the two forming together an efficient grasping organ. Primary branch of posterior antennæ three-jointed, the middle joint short ; secondary branch rather stouter than the other and six-jointed, tapering slightly towards the posterior end ; the third joint is somewhat longer than the other five ; the breadth of the first joint is equal to about three times the length, and of the last to about twice the length (fig. 23). Mouth organs nearly as in *Longipedia coronata*. First pair of swimming feet also similar to those of *Longipedia coronata*, but shorter, more robust, and armed with stouter spines and longer plumose setæ. The outer margin of the first joint of the outer branch bears a pectinate fringe of spine-like setæ immediately anterior to the large spine, and the outer margins of the second and third joints of the inner branch are similarly fringed. Both branches of the second pair of swimming feet are of about equal length. First joint of the inner branch very short, and armed on the lateral aspect with a stout conical and tooth-like process, which reaches slightly beyond the end of the next joint ; the third joint in both branches is rather longer than the combined length of the other two (fig. 28). The third and fourth pairs are somewhat similar to those of *Longipedia coronata*. Fifth pair in both sexes rudimentary, and consisting of a very small basal joint bearing four setæ ; the second seta from the inside is longer than the others and plumose. Caudal stylets strongly divergent ; length equal to rather more than twice the breadth ; caudal setæ comparatively short, scarcely equal to three-fifths the length of the animal. Ovisacs broadly ovate ; their transverse diameter about three-fifths of the length.

This species is of frequent occurrence, especially in material dredged off Musselburgh and in Largo Bay, but specimens with ovisacs are scarce.

Canuella perplexa appears to be the form described in the monograph of the British Copepoda as the female of *Longipedia coronata*.

In 1867 Hesse described* a copepod he had discovered living in the same shell with a *Pagurus*, and to which he gave the name of *Sunaristes paguri*. This copepod resembles *Canuella perplexa* in some respects, but the difference in habitat, the difference in size (Hesse states, *loc. cit.*, that *Sunaristes* is 5 mm. long, but this probably includes the tail setæ), and the difference in important structural details is so great that it seems

---

* *Ann. des Sc. Nat.*, 5th ser. (Zoology), vol. vii. p. 205.

scarcely possible they can be the same species. In *Sunaristes* the genital segment of the female abdomen is equal to the entire length of the other abdominal segments, the ovisacs are elongate ovate, somewhat pointed at both ends, and reach to the end of the abdomen ; the buccal appendages and swimming feet also differ. Further, when describing the *habitat* of *Sunaristes* he says, *loc. cit.*, 'Sont les compagnons intimes des Pagures, 'c'est avec la plus grande peine qu'on peut les en séparer, non qu'ils ' soient fixés sur eux comme le sont leurs parasites, mais par leur adresse ' à ce cacher dans l'intérieure ou en dessous des coquilles que ceuxci ' habitent.' But *Canuella* is free-living like *Longipedia*. Specimens both of *Paguri* and their shells have been examined without obtaining a single specimen of *Canuella ;* all our specimens have been obtained in dredged material, or with hand-net, along with *Longipedia* and other free-living species. In 1884 Dr Wilhelm Müller described * a large copepod he also had obtained living as a messmate with a species of *Paguri* [*Pagurus (Eupagurus) bernhardus*], and which he named *Longipedina paguri*. This may be the same species as that described by Hesse as *Sunaristes paguri*, but if so the description and figures of the one certainly differ very widely from those of the other.

Longipedina paguri, W. Müller, has even a closer resemblance to *Canuella* than *Sunaristes* has, but there are still important differences— *Longipedina* is twice the size of *Canuella*, its length, exclusive of tail setæ, being, as stated by Müller, 2·7 mm. It is a messmate with *Pagurus bernhardus*, while *Canuella* is free-living. The second pair of swimming feet in the male are different from those of the female, but in *Canuella* they are alike in both sexes.

'After a careful study of the descriptions and figures of *Sunaristes* and ' *Longipedina*, we find that ' . . . . . . 'the difference, both in respect ' of structure and *habitat*, between each of these and the species described ' by us, is apparently so great that we prefer for the present to consider ' the Forth species as distinct.' †

Zosime, Boeck (1872).

Zosime typica, Boeck. (Pl. V. figs. 14–17.)
        1872. *Zosime typica*, Boeck, 'Nye Slægter og Arter af Saltvands- 'Copepodar,' p. 14.
        1880. *Zosime typica*, Brady, ' Mon. Brit. Cop.,' vol. ii. p. 15, pl. xxxix. figs. 1–12.

Habitat.—In material dredged off Musselburgh, frequent. The Forth specimens of this interesting and well-marked species agree thoroughly in structural details with the description and figures in 'British Copepoda.' The structure and armature of the first pair of swimming feet and of the female fifth pair are characters by which the species is readily distinguished.

Genus *Jonesiella*, Brady (1880).

Jonesiella hyænæ, I. C. Thompson. (Pl. III. figs. 1–6).
        1889. *Jonesiella hyænæ*, I. C. Thompson, 'Proc. Biol. Soc. 'Liverpool,' vol. viii. p. 193, pl. ix. figs. 1–10.

This rather peculiar and interesting species was obtained among dredged material collected near Eyebrough Rock—a short distance west of Fidra

        * *Archiv für Natur.*, jahrgang 50, 1st Band, p. 19.
        † *Ann. Scot. Nat. Hist.* (April 1893), p. 94.

Lighthouse, Firth of Forth, during February this year. A number of specimens were obtained. The anterior antennæ are six-jointed, short, stout, and furnished with several strongly plumose setæ ; the first joint is large and robust, the second, third, and fifth are short, while the fourth and the last joints are very small. The basal joint projects almost straightforward from the head, the remaining joints curve outwards so that the last three are nearly at right angles to the basal joint ; two stout spines, setiferous on the upper margin, spring from the upper part of the distal end of the third joint, and an elongate sensory filament and a very long slender and plain seta from the upper distal angle of the same joint ; the fifth joint is also furnished with a stout terminal spine similar to those on the end of the third joint (fig. 2). Rostrum prominent, moderately broad; extremity rounded, and extending beyond the geniculate anterior antennæ. Secondary branch of the posterior antennæ well developed, reaching beyond the end of the primary branch ; three-jointed, middle joint very small (fig. 3). Mandible palp distinctly two-branched, one branch long and narrow, the other short, and each furnished with several apical setæ. Foot-jaws and swimming feet as described and figured by Mr Thompson. Fifth pair large, foliaceous ; internal portion of the basal joint well developed, sub-triangular ; apex broadly rounded ; exterior portion of basal joint forming a small rounded process at the base of the secondary joint ; secondary joint broadly ovate, scarcely reaching to the apex of the inner portion of the basal joint ; lateral margins of both joints ciliate, distal margins furnished with several 'spear-shaped' setæ (fig. 6).

Jonesiella hyænæ differs from other British species of *Jonesiella* by having the inner branches of the first pair of swimming feet three-jointed, and also by the secondary branch of the posterior antennæ being three-jointed : * these differences may render it necessary to modify the generic description, or to remove *Jonesiella hyænæ* to another genus. There can be no doubt, however, that the general structure of this species agrees fairly well with the more typical members of the genus in which it is placed.

Genus Delavalia, Brady (1868).

Delavalia palustris, Brady.

> 1868. *Delavalia palustris,* Brady, Nat. Hist. Trans. Northumb. 'and Durham,' vol. iii. p. 134, pl. v. figs. 10–15.
> 1880. *Delavalia palustris, idem,* 'Mon. Brit. Copep.,' vol. ii. p. 43, pl. l. figs. 1–8.

Habitat.—Estuary of the Forth, in the vicinity of Culross, rather scarce. Common in brackish water pools at the mouth of the Peffer Burn, Aberlady Bay, ♀ with ovisacs. *Delavalia palustris* appears to be restricted to localities where the water is more or less brackish. In the monograph of the 'British Copepoda' it is recorded from only one locality—the mouth of the Seaton Burn, Northumberland.

It may be of interest to enumerate some of the Copepoda that have been found associated together in the upper reaches of the Forth estuary, as *Eurytemora affinis,* Poppe ; *Tachidius crassicornis,* Scott ; *Delavalia palustris,* Brady ; *Thalestris harpactoides,* Claus ; *Platychelipus littoralis,* Brady ; *Hersiliodes littoralis* (Scott) ; *Acartia longiremis,* Lilljeborg ; *Temora longicornis,* Müller, &c. A few of these, as *Eurytemora affinis, Tachidius crassicornis, Delavalia palustris,* and *Hersiliodes littoralis,* have

---

* We have recently ascertained that *Jonesiella spinulosa* has the secondary branch also three-jointed, the intermediate joint being very small.

not been observed in the Forth beyond the area referred to, while others are not so restricted, and have been obtained in different localities—seaward as well as inshore.

Delavalia æmula,[*] sp. n. (provisional name).   (Pl. IV. figs. 36–47.)

Length, ·73 mm. Similar in form to *Delavalia reflexa*, Brady and Robertson. Anterior antennæ of the female nearly as in *Delavalia robusta*, Brady and Robertson; male anterior antennæ nine-jointed, hinged between the sixth and seventh joints; a sensory filament springs from the end of the fourth joint in both sexes, but that of the female is shorter. The formula shows approximately the relative length of the joints of the antennæ in both male and female :—

$$\text{Female antennæ,} \quad \frac{20 \cdot 10 \cdot 8 \cdot \; 6 \cdot 4 \cdot \; 8 \cdot 6 \cdot 4 \cdot}{1 \cdot 2 \cdot 3 \cdot \; 4 \cdot 5 \cdot \; 6 \cdot 7 \cdot 8 \cdot 9 \cdot}$$

$$\text{Male antennæ,} \quad 24 \cdot 11 \cdot 5 \cdot 13 \cdot 5 \cdot 10 \cdot 8 \cdot 5 \cdot 7 \cdot$$

Secondary branch of posterior antennæ three-jointed, middle joint very small. Basal joint of mandible palp—elongate—about three times longer than broad; the branches, which are subequal, are less than half the length of the basal joint. Posterior foot-jaws nearly as in *Delavalia robusta*, the last joint short, narrow, and curved, and appearing to be merely a continuation of the base of the terminal claw (fig. 41). Both branches of all the four pairs of swimming feet—except the inner branches of the second pair in the male—three-jointed; the inner branches of the first pair are rather longer than the outer, and the first joint is about one and a half times longer than the next one (fig. 42). The second joint of the inner branch of the second pair in the male is nearly twice the length of the first joint, and probably consists of two coalesced joints, rather slender, and with the inner margin of the proximal half strong, gibbous, and furnished with a plain, flexuous, terminal seta (fig. 43). Fifth pair in both sexes nearly as in *Delavalia reflexa*, except that the basal joint carries one very small and four elongate stout plumose setæ, and the outer joint six plain setæ. Caudal stylets about equal in length to the last abdominal segment. The inner of the two principal caudal setæ very long, equal to the combined length of the abdomen and caudal stylets.

Habitat.—Largo Bay.   Not rare.

Delavalia æmula differs especially in the inner branches of the first pair of swimming feet being three-jointed, and seems otherwise to combine characters belonging to all the other three British genera.

Genus *Cletodes*, Brady (1872).

(?) *Cletodes tenuiremis*, sp. n. (provisional name).   (Pl. III. figs. 21–28.)

Animal resembling *Cletodes linearis* (Claus).   Length, ·96 mm. Anterior antennæ seven-jointed, basal joints robust; all the joints, with the exception of the first, subequal in length. The approximate length of the joints are shown by the formula—

$$\frac{14 \cdot 8 \cdot 7 \cdot 7 \cdot 5 \cdot 5 \cdot 6 \cdot}{1 \cdot 2 \cdot 3 \cdot 4 \cdot 5 \cdot 6 \cdot 7 \cdot}$$

Secondary branch of posterior antennæ small, uniarticulate, and bearing two short terminal setæ. Mandible palp consisting of a small one-jointed branch (fig. 24). The first pair of swimming feet have the inner branches

---

[*] *Æmulus*, an imitator.

short, two-jointed, the outer three-jointed branches also short, but rather longer than the inner branches; exterior marginal spines of the outer branches elongate, slender; inner and outer branches provided with very long, slender, and blunt-pointed terminal filaments (fig. 25). Inner branches of second, third, and fourth pairs very short, one-jointed; the intermediate terminal seta of both branches of the second and third pairs very long and sparingly plumose (fig. 26). Intermediate terminal seta of the inner branches of fourth pair not reaching much beyond the end of the outer branches. Fifth pair very small, almost rudimentary; basal joint furnished with four subequal plain setæ, and the secondary joint with five setæ of unequal length (fig. 28). Caudal stylets short, their breadth equal to about two-thirds the length; primary caudal seta equal to the combined length of the last abdominal segment and stylets, the other setæ minute. The last four abdominal segments are adorned with three to four transverse rows of minute cilia, and the posterior margins of thoracic and abdominal segments are fringed with aculeate setæ, as shown in figure .

Habitat.—Vicinity of Inchkeith. February 1893. Dredged; rare.

This species, which is doubtfully referred to *Cletodes*, differs from that genus in having the inner branches of all but the first pair of the swimming feet one-jointed, but as it agrees with *Cletodes* in most of the other important characters, and as no male has yet been obtained the structure of which might have assisted in more satisfactorily indicating the affinities of the species, it seems better in the meantime to place it in the genus *Cletodes*. One peculiar character of the species that distinguishes it from almost all others of the genus to which it is referred, is the long terminal filaments of the first pair of feet; they do not appear to be hairs or setæ in the proper sense, but have rather the appearance of filamentous conferva. There can be no doubt, however, that they are organically connected with the first feet, and are not accidental parasitic growths.

Genus, *Platychelipus*, Brady (1880).

Platychelipus littoralis, Brady. (Pl. V. figs. 11–13.)

    1880. *Platychelipus littoralis*, Brady, 'Mon. Brit. Copep.,' vol. ii. p. 103, pl. lxxix. figs. 20–23; pl. lxxx. figs. 15–19.

Habitat.—Forth estuary, near Culross, and also off Musselburgh. This well marked species was of frequent occurrence in the material collected off Musselburgh. The long, curved, claw-like, and spiniform terminal seta of the inner branches of the first pair of swimming feet (fig. 12), together with the *Enhydrosoma*-like form of the animal, enable the species to be distinguished almost at first sight. *Platychelipus* is well described and figured by Dr Brady in the 'Monograph of the British 'Copepoda.' I do not know of any previous record of it from the Scotch coasts.

Genus, *Dactylopus*, Claus (1863).

Dactylopus rostratus, sp. n. (provisional name). (Pl. III. figs. 7–20.)

This copepod, for which I propose the provisional name *Dactylopus rostratus*, closely resembles *Dactylopus flavus*, Claus, in some structural details, and may be a large variety of that species. (*Dactylopus rostratus* is 1 mm. in length, whereas *Dactylopus flavus* is little more than half that size.) But besides being nearly double the size of *Dactylopus flavus*, it is readily distinguished from that species by the prominent and bluntly-rounded rostrum, which, like *Delavalia palustris*, is provided with two

O

minute lateral setæ. The anterior antennæ are short and seven-jointed (fig. 8). The formula shows the relative length of the joints :—

$$\frac{10 \cdot 12 \cdot 10 \cdot 10 \cdot 8 \cdot 5 \cdot 7 \cdot}{1 \cdot 2 \cdot 3 \cdot 4 \cdot 5 \cdot 6 \cdot 7 \cdot}$$

The inner branch of the posterior antennæ is two-jointed (fig. 10). The mandibles are well developed ; apex of biting part broad, truncate, and armed with a row of stout blunt-pointed teeth, and a marginal divergent seta ; there is also a stout tooth, larger than the others, arising from the lateral distal aspect of the mandible. Posterior foot-jaws stout, with a long slender terminal claw ; the inner margin of the second joint has an intermediate fringe of cilia, and two small setæ near the distal end ; a prominent setose spine springs from the inner distal angle of the first joint, and immediately behind the seta is a transverse row of small hairs. The first pair of swimming feet, which somewhat resemble those of *Dactylopus flavus*, differ in the spines on the exterior distal angles of the first and second joints of the outer branches being not larger than the marginal spine of the third joint, and in the apical setæ of the third joint being non-geniculate ; the outer margins of all the three joints are strongly setiferous ; the inner branch, which has also the outer margin of all the joints fringed with small setæ, is armed with a stout, moderately long, and straight terminal spine and two setæ (fig. 14). The inner branches of the second pair in the male terminate in a stout, slightly curved, conical spine as long as the third joint (fig. 15). The fourth pair resemble those of *Dactylopus flavus*, but the outer margins of both branches are setiferous, and the elongate setæ on the inner margins are plumose (fig. 16). The inner segment of the basal joint of the fifth pair is slightly produced and rounded, and provided with four marginal setæ, one being greatly elongate ; the secondary joint is short, obliquely truncate at the end, and furnished with four unequal terminal setæ ; fifth pair in the male smaller, the inner portion of the basal joint less produced, and armed with two spine-like setæ of unequal length, the longer one being plumose (fig. 18). Abdominal segments strongly ciliate. Caudal stylets short and stout ; the inner one of the two caudal setæ much longer than the other, and equal to half the length of the entire animal. Spermatophore broadly ovate (fig. 20).

Several specimens of this species were obtained by washing some shells inhabited by *Pagurus bernhardus*, and collected west of Inchkeith. The copepods may only have been accidentally harbouring about the shells; they have scarcely the appearance of ‘commensals’ or ‘ messmates.’

Family SAPHIRINIDÆ, Thorell.

Genus *Lichomolgus.*

Lichomolgus hirsutipes, sp. n. (provisional name). (Pl. IV. figs. 1–12.)

Length 1·4 mm. ($\frac{1}{18}$th of an inch). Seen from above, the first four segments of the cephalo-thorax are together broadly ovate, the first segment has a shield-like form, the fifth segment is narrow, and the proximal rather narrower than the distal end. Anterior antennæ seven-jointed. The relative length of the joints is shown by the formula—

$$\frac{19 \cdot 44 \cdot 8 \cdot 19 \cdot 20 \cdot 15 \cdot 11 \cdot}{1 \cdot 2 \cdot 3 \cdot 4 \cdot 5 \cdot 6 \cdot 7 \cdot}$$

Posterior antennæ four-jointed, third joint short, last joint bearing two stout and hooked terminal claws. Mandibles nearly as in *Licho-*

molgus poucheti, Canu ; the broad anterior margin of the basal portion is distinctly furrowed, the furrows extend to the edge, and give to it a crenate appearance (fig. 4), maxilla (*a*) with three terminal setæ. The moderately long slender extremity of the anterior foot-jaws is strongly pectinate on the upper edge, but plumose towards the end ; a stout and strongly plumose seta, equal to about a third of the length of the slender terminal part of the foot-jaw, springs from the inner edge of the base, where there is also a small seta on the lateral aspect. Posterior foot-jaws of the female three-jointed ; the first and second joints, which are sub-equal in length, are of moderate size, but the last joint is small ; two setæ spring from near the middle of the second joint ; the third joint, besides being furnished with a stout terminal seta, is armed at the apex with a robust conical spine, somewhat longer than the joint to which it is articulated (fig. 6). The posterior foot-jaw in the male is armed with a long, curved, terminal claw, fully twice the length of the joint from which it springs (fig. 7). In the female the fifth pair of feet are elongate ; they reach to nearly the end of the genital segment of the abdomen ; the whole of the upper portion of each foot is clothed with short aculeate setæ ; the fifth pair in the male are much shorter, and without setæ. Female abdomen slender, and composed of five segments. Genital segment comparatively small, but fully twice the length of the next ; breadth about equal to the length ; the remaining four segments are nearly equal to each other in length. Caudal stylets fully one and a half times longer than the last abdominal segment, and provided with one subterminal and four apical setæ ; the second seta from the inside is considerably longer than the others, being equal to the entire length of the abdomen. Genital segment of the male abdomen large, subquadrate, the sides slightly convex, the distal end truncate, and bearing two small setæ on each side of the following segment ; the length of the genital segment is about equal to that of the other four segments and the caudal stylets combined. The caudal stylets of the male are somewhat shorter than those of the female (fig. 12).

The fifth pair of feet in the female of this species form a well-marked and distinctive character.

Habitat.—'Rath Ground,' a short distance north of the Bass Rock, Firth of Forth.

Genus *Modiolicola*, Aurivillius.

Modiolicola insignis, Aurivillius. (Pl. IV. figs. 13–24.)

1883. *Modiolicola insignis*, Aurivillius, ' Akademisk Afhandling,' Stockholm (1883), p. 10, t. ii., figs. 1–10 ; t. iv., figs. 1–8.

1885. *Lichomolgus insignis*, Raffaele é Monticelli, ' Mem. d. R. 'Accad. d. Lincei,' ser. 4, vol. i. p. 302, figs. 13–16.

1892. *Modiolicola insignis*, Canu, ' Les Copep. du Boulon.,' p. 238, pl. xxiv. figs. 14–20.

Length 1·2 mm. ($\frac{1}{20}$th of an inch). Cephalo-thorax broadly ovate ; abdomen elongate, narrow, and equal to about two-thirds the length of the cephalo-thorax. First cephalo-thoracic segment subtriangular, and equal to the combined length of the other four thoracic and first abdominal segments. Forehead narrowly rounded. Anterior antennæ short, scarcely half the length of the first bony segment, seven-jointed, the second joint much longer than any of the others. The subjoined formula exhibits the relative length of the joints :—

$$\frac{8 \cdot 17 \cdot 4 \cdot 8 \cdot 10 \cdot 7 \cdot 4}{1 \cdot 2 \cdot 3 \cdot 4 \cdot 5 \cdot 6 \cdot 7}$$

Posterior antennæ four-jointed, joints subequal in length, but the last two are more slender than the first and second ; the apex is furnished with three claw-like and hooked spines, and also a seta of about the same length as the spines.   Mandibles nearly as in *Lichomolgus forficula*, Thorell.   Basal part of the anterior foot-jaws stout, terminal part extremely long and slender, gradually tapering to a setiform extremity, and bent at nearly right angles to the basal part ; the upper edge of the terminal part is furnished with a fringe of cilia which extend from the geniculation, where they are stout and setiform, but rapidly decrease in size towards the extremity.   Posterior footjaws in the female rudimentary, three-jointed, the last very small, and without any spines or setæ ; the posterior foot-jaws in the male form powerful grasping organs, composed of two stout joints, and a very long falciform terminal claw, which carries a small seta at its base ; the second joint has the distal half of the inner edge fringed with cilia (fig. 19).   Both branches of the first four pairs of swimming feet three-jointed, and of nearly equal length ; the first basal joint of the first pair is furnished interiorly with an elongate and stout plumose seta, the inner margins of both branches are also provided with long plumose setæ ; the armature of the inner branch comprises one seta on the distal end of the inner margin of the first joint, two on the second, and two on the third joint ; there are also four short dagger-shaped spines round the end of the third joint, and the outer distal angles of the first and second joints form tooth-like processes ; there is no seta on the inner edge of the first joint of the outer branch, one on the second and five on the third joint ; the third joint is also armed on the outer margin and end with four dagger-shaped spines—the terminal one being the largest ; the second also bears one dagger-shaped spine on the outer distal angle, and the first joint one.   The armature of the fourth pair differs from that of the first in the following manner—there is no seta on the interior edge of the second basal joint, the first and second joints of the inner branch bear each a long plumose seta on the distal end of the inner margin, there is no seta on the third joint ; this joint is truncate at the end, and armed with two stout dagger-shaped terminal spines ; the outer distal angle of the second joint forms a bifid toothed process, and there is a small tooth on distal end of the outer margin of the first joint ; the second joint of the outer branch bears one long seta on the inner margin, and the third joint five setæ ; the third joint is also armed at the extremity with a long sabre-like spine, ciliate along the inner edge, and with two short dagger-like spines on the outer margin ; and the first and second joints are each provided with a similar spine on the outer distal angle (fig. 21).   The fifth pair are small and one-jointed, the apex truncate, and furnished with one long, stout, and slightly curved spine, and a small spiniform seta, both of which are plain (fig. 22), fifth pair alike in both sexes. First abdominal segment in both sexes considerably dilated ; that of the female has the sides rounded, is widest across the middle, and furnished on each side, on the ventral aspect, with a small setiferous appendage ; that of the male is widest across the distal end, the distal angles are somewhat produced, and provided with three small setæ.   The second, third, and fourth abdominal segments have the posterior margins in both sexes strongly setose.   Caudal stylets about equal to twice the length of the last abdominal segment, and furnished with a small seta on the proximal half of the outer margin, and with four apical setæ ; the male abdomen is rather smaller than that of the female.   Ovisacs two, large. Colour of the animal, including ovisacs, usually brilliant red.

Habitat.—Living as a messmate within the shell of the ' horse mussel,' *Mytilus modiolus.*   Frequent in the Firth of Forth.   I have also obtained *Modiolicola* in *Mytilus modiolus*, both in the Moray Firth, on the East

Coast, and in the vicinity of Mull on the West Coast. It frequents the branchial lamellæ of the mollusc. A considerable number of specimens may sometimes be obtained in a single mussel, while in others it may be rare or altogether absent.

Family ARTOTROGIDÆ, Brady.

Genus *Cyclopicera*, Brady.

Cyclopicera purpurocincta, sp. n. (provisional name). (Pl. III. figs. 29–40.)

Length, exclusive of caudal seta, 1 mm. ($\frac{1}{25}$th of an inch). Seen from above, the cephalo-thorax is broadly ovate; the first segment is large, and equal to twice the combined length of the second, third, and fourth segments; the fifth thoracic segment is of about equal breadth with the narrow elongate abdomen; the colour of the second, third, and fourth segments is dark purple, and seems to be very little affected by a lengthened immersion in methylated spirit—the specimen figured was obtained in 1889, and though it has been in spirit since then, no perceptible change has taken place in the colour of these segments; the posterolateral angles of the second and third segments are produced into toothlike processes. Anterior antennæ slender, sixteen-jointed, sparingly setiferous; a sensory filament springs from the end of the third last joint. The subjoined formula shows the relative length of the joints:—.

$$\frac{11 \cdot 9 \cdot 3 \cdot 3 \cdot 4 \cdot 5 \cdot 4 \cdot 4 \cdot 3 \cdot 4 \cdot 5 \cdot 5 \cdot 5 \cdot 8 \cdot 3 \cdot 11}{1 \cdot 2 \cdot 3 \cdot 4 \cdot 5 \cdot 6 \cdot 7 \cdot 8 \cdot 9 \cdot 10 \cdot 11 \cdot 12 \cdot 13 \cdot 14 \cdot 15 \cdot 16}$$

Posterior antennæ nearly as in *Cyclopicera gracilicauda*, Brady. Mandibles also nearly as in that species; mandible palp slender, bearing two apical setæ—one very long and plumose, and scarcely half as long, plain, and very slender (fig. 32). The two simple branches of the maxillæ are about equal in length, but one branch is more slender than the other, and bears three apical setæ, while the stout branch bears five setæ at the apex (fig. 33). The anterior foot-jaws are furnished with a long, slender, and curved terminal claw. Posterior foot-jaws four-jointed, elongate, slender, resembling those of *Cyclopicera nigripes* Brady and Robertson (fig. 35). The inner portion of the second basal joint of the first pair of swimming feet is considerably delated, and the inner branch is attached to this part, while the outer branch is attached to the very reduced exterior portion, so that though the two branches are of about equal length, the outer branch does not extend much beyond the second joint of the inner one; the joints of the inner branches are subequal. In the second, third, and fourth pairs the inner portion of the second basal joint is not so enlarged as in the first pair; the outer branches are considerably longer than the outer, and the first joint of the inner branches is much shorter than the second or third joints (figs. 36, 37). Fifth pair small, two-jointed; the breadth of the first joint is greater than the length, and the length of the second joint, which is narrower than the first, is greater than the breadth; a small seta springs from the anterior distal angle of the first joint, and also from each of the later angles of the truncate apex of the second joint, which is also armed with a large dagger-shaped apical spine, intermediate between the small angular setæ. Abdomen slender, composed of four segments of nearly equal breadth; genital segment nearly as long as the next three together; and the second segment is nearly equal to the combined length of the third and fourth. Caudal stylets slender, rather longer than the two last abdominal segments; the inner margin of each stylet is ciliated, the outer margin plain; terminal setæ, four unequal, the two intermediate densely plumose, and slightly thickened in the middle.

Habitat.—'Rath ground,' north of the Bass Rock, Firth of Forth, rare, in material dredged November 20th, 1889. One specimen was also obtained in some material dredged in 1892 off the south end of the Island of Mull.

Cyclopicera purpurocincta appears to be intermediate between *Cyclopicera gracilicauda* and *Cyclopicera nigripes*, but is at once distinguished from either by the colour of the second, third, and fourth thoracic segments.

Cyclopicera lata, Brady.   (Pl. III. figs. 41, 42.)

> 1872. *Cyclopicera lata*, Brady, ' Nat. Hist. Trans. Northumb. and
>   ' Durham,' vol. iv. p. 433, pl. xviii. figs. 3–8.
> 1868. *Ascomyzon echinicola*, Norman, ' Brit. Assoc. Report,' p.
>   300.
> 1880. *Cyclopicera lata*, Brady, ' Mon. Brit. Copep.,' vol. iii. p. 56,
>   pl. lxxxix. fig. 12 ; pl. xc. figs. 11–14.

Habitat.—West of Gullane Ness, Firth of Forth. Washed from sponges, 1889. Several specimens were obtained. *Cyclopicera lata* closely resembles *Artotrogus boeckii*, Brady, which I have also obtained by washing *Chalina oculata* (a kind of sponge), and for this reason I had some doubts as to its being distinct, and deferred recording its occurrence ; but having recently been enabled to make a careful examination of its structure, I have now no doubt that it is the species described as *Cyclopicera lata* in the Monograph of the British Copepoda : the structure of the anterior and posterior antennæ is the same, the mandible palp, which is very small, bears two apical setæ, one very long, slender, and sparsely plumose, and one very short (fig. 41). The abdomen is less robust, and the caudal stylets distinctly more elongate than in *Artotrogus boeckii.*

Genus *Parartotrogus*, T. and A. Scott (1893).

Parartotrogus richardi, T. and A. Scott.   (Pl. IV. figs. 25–35.)

> 1893. *Parartotrogus richardi*, T. and A. Scott, ' Ann. and Mag.
>   ' Nat. Hist.,' ser. 6, vol. xi. p. 210, pl. vii. figs. 1–11.

Habitat.—Near Fidra, Largo Bay, the ' Fluke Hole ' (off St Monans), and other parts of the Forth between Inchkeith and May Island.

This species, which is only about the one-fiftieth of an inch in length, is readily distinguished by the peculiar subrhomboidal form of the cephalo-thorax (fig. 25). Seen from above, the sides of the cephalo-thorax diverge rapidly from the broad, almost truncate rostrum to about the middle of the first segment, where they form bluntly-rounded angles by again tapering quickly towards the last segment ; the greatest breadth of the first segment is about equal to three-fifths of the entire length of the animal ; the abdomen is moderately stout, and equal to about three-sevenths of the length of the cephalo-thorax. Anterior antenna, nine-jointed, sparingly setiferous, the third, fourth, and fifth joints much shorter than any of the others. The relative length of the joints is shown by the formula—

$$\frac{12 \cdot 12 \cdot 6 \cdot 5 \cdot 6 \cdot 10 \cdot 12 \cdot 13 \cdot 4}{1 \cdot 2 \cdot 3 \cdot 4 \cdot 5 \cdot 6 \cdot 7 \cdot 8 \cdot 9}$$

Posterior antennæ four-jointed, and terminating in a stout hooked claw ; a small curved and stout spine springs from near the middle of the last joint; secondary branch small, one-jointed, arising from the middle of the second joint, and bearing four slender apical setæ. Mandibles stylet-shaped, produced at the base into a barb-like process (fig. 28). Maxillæ two-branched, one branch stout, and bearing three apical spines and two plumose setæ, the other branch small, with three apical setæ (fig. 29). The first joint of the anterior foot-jaws stout, second joint elongate and slender, and terminating in a moderately long curved claw and a small spine (fig. 30). Posterior foot-jaw four-jointed; second joint about three times longer than broad, and bearing a small seta near the middle of the inner margin; third and fourth joints narrow, and together scarcely equal in length to two-thirds of the second joint; terminal claw rather longer than the two preceding joints, and provided with a small seta near the middle of the inner margin. Both branches of the first pair of swimming feet two-jointed, the second joint of the inner branch large and foliaceous (fig. 32). ?Third and fourth pairs somewhat similar to those of *Lichomolgus fucicolus*, Brady; slender, and with the marginal and terminal spines of the outer branches broadly dagger-shaped (figs. 33, 34). Fifth pair rudimentary, bilobed, with about three apical setæ (fig. 35, *a*). Genital segment of the abdomen considerably dilated; second and third segments short, and together scarcely equal in length to the first; last segment equal to twice the length of the preceding one; caudal stylets rather shorter than the last abdominal segment, furnished with five apical setæ, the second seta from the inside being considerably longer than the others, and about equal in length to the last three abdominal segments. Ovisacs two, large.

This species has been known to us since 1889, but, because of some uncertainty as to whether the first specimens obtained were mature, it was considered expedient to defer recording it. A few months ago a specimen with ovisacs was obtained, and quite recently several others, also with ovisacs, have been secured. By the discovery of these specimens our uncertainty as to the maturity of those previously obtained has in a great measure been set at rest.

Parartotrogus richardi resembles in some respects a curious parasitic copepod, described by Sir John Dalyell in 1851[*] under the name of *Cancerilla tubulata*, and which was discovered by him adhering to the base of one of the arms of a species of *Amphiura* (a kind of starfish). The same copepod has since been obtained on the coasts of France, and is described and figured by Dr Canu in his work 'Les Copepodes du 'Boulonnais.' But in *Cancerilla* the cephalo-thorax is greatly dilated, the abdomen is very short, the anterior antennæ are only six-jointed, and the first pair of swimming feet are more rudimentary than in *Parartotrogus*; and a further difference of considerable importance is, *Cancerilla* has only been obtained as a parasite, whereas all the specimens of *Parartotrogus* that have yet been obtained were free.

Genus *Bomolchus*, Nordman (1832).

Mikrograph. Beit. zur Naturg., II. Heft., s. 135–137.

Animal somewhat like *Lichomolgus* in form. Anterior antennæ seven-jointed. Posterior antennæ three-jointed. (?) Mandibles stylet-shaped. Maxillæ simple, the apex truncate, and provided with two broadly ovate appendages. Foot-jaws rudimentary; (?) anterior footjaw, consisting of a simple stylet-shaped joint, bearing a single plumose seta. Both branches of all the swimming feet three-jointed, first pair short, foliaceous, furnished with spathulate and densely-plumose setæ; the

[*] *The Powers of the Creator*, vol. i. p. 233, pl. lxii. figs. 1–5.

second, third, and fourth pairs rather longer, the outer branches armed with hooked marginal spines. Fifth pair nearly as in *Lichomolgus.*

Bomolchus soleœ, Claus. (Pl. V. figs. 1–13.)
     1864. *Bomolchus soleœ,* Claus, Zeitschrift fur Wissenschaft zool. vol. xiv. p. 374, pl. 35, figs. 16–20.

Length, exclusive of tail setæ, 1·3 mm. Anterior antennæ furnished with numerous moderately long and densely plumose seta ; the relative length of the joints are shown by the formula—

$$\frac{25 \cdot 15 \cdot 15 \cdot 17 \cdot 15 \cdot 11 \cdot 13 \cdot}{1 \cdot 2 \cdot 3 \cdot 4 \cdot 5 \cdot 6 \cdot 7.}$$

Middle joint of posterior antennæ short, bearing a single small hair, the last joint covered with small prickles and furnished with three pectinate setiferous appendages and three apical setæ (fig. 3). The mandibles (?) have the basal portion considerably dilated, d the terminal portion curved and stylet-shaped (fig. 4). First pair of swimming feet short, broadly foliaceous, somewhat distorted ; joints of inner branch subequal in length ; the middle joint of the outer branch very short (fig. 7). The second and third pairs longer and much narrower comparatively than the first ; inner margins furnished with elongate, densely plumose setæ ; the exterior margin and end of the outer branches armed with stout spines bearing terminal hook-like processes ; joints of the inner branches subequal ; middle joint of outer branches shorter than either of the other two (fig. 8). In the fourth pair the inner branches are rather longer than the outer, but otherwise this pair is similar to the second and third. The fifth pair consist each of a single two-jointed branch ; the first joint is very short, the second is about four times the length of the first, and is provided with three terminal and one marginal setæ. Abdomen short, tapering from the somewhat stout genital segment ; the third, fourth, and fifth segments shorter than the preceding one. Caudal stylets rather longer than the last abdominal segment ; the principal caudal seta is about one and a half times longer than the abdomen ; other tail setæ short.

Habitat.—" Fluke Hole " off St Monans, Firth of Forth. This curious copepod seems to be closely allied to the *Saphirinidæ*, and probably belongs to that group. The peculiar structure of the first pair of swimming feet give it a somewhat abnormal character. New to Britain.

Besides the *Copepoda* now described, there are still some others that do not apparently agree with known species, and which are held over for further study.

AMPHIPODA

GAMMARIDÆ.

Genus *Anonyx,* Kroyer (1838).

Anonyx nugax, Phipps. (Pl. V. figs. 18–21.)
     Cancer nugax, Phipps, ' Voyage au Pole boréale,' p. 192, pl. 12, fig. 8.
     Anonyx ampulla, Kroyer (not Phipps).
     1891. *Anonyx nugax,* G. O. Sars, ' Crust. of Norway,' vol. i. p. 88, pl. 31.

Several specimens of this fine species were obtained in February 1889, near May Island, Firth of Forth, but were not then recorded ; it was only when I read the description in G. O. Sars' excellent work, the 'Crustacea ' of Norway,' that the species was recognised. The largest of the Forth specimens measure 20 mm. (fully three-quarters of an inch) in length.

The eyes are lageniform, and, being large and black, give a marked character to the species.

Genus BATHYPOREIA, Lindström (1855).

Bathyporeia norvegica, G. O. Sars.   (Pl. V. fig. 22.)
    Bathyporeia pilosa, Boeck (not Lindström).
    1891. *Bathyporeia norvegica*, G. O. Sars, *loc. cit.*, p. 128, pl. 43.
    1892. *Bathyporeia norvegica*, T. and A. Scott, 'Ann. and Mag.
    'Nat. Hist.,' s. 6, vol. x. p. 205.

Specimens of this species are occasionally obtained in various parts of the Firth of Forth. It may be distinguished by its larger size, and especially by the tooth-like form of the postero-lateral angles of the epimeral plates of the third metasome.

Bathyporeia pelagica, Sp. Bate.   (Pl. V. figs. 23–25.)
    1862. *Bathyporeia pelagica*, Sp. Bate, 'Cat. Amphip. Brit. Mus.,'
    p. 174, pl. xxxi. fig. 6.
    1891. *Bathyporeia pelagica*, G. O. Sars, *op. cit.*, p. 129, pl. 44,
    fig. 1.

After a careful examination of a considerable number of Forth specimens of *Bathyporeia*, I have been enabled, with the assistance of Prof. G. O. Sars' 'Crustacea of Norway,' to distinguish this and the following species. The exceedingly long flagellum of the posterior antennæ of the adult male enables *Bathyporeia pelagica* to be readily distinguished except from *Bathyporeia norvegica;* but in this case, not only is there a difference in size, there is also the difference in the form of the epimeral plates of the third metasome, which, in all but *Bathyporeia norvegica*, have postero-lateral angles more or less rounded. In *Bathyporeia pelagica* the dorsal flexure of the first urosome is very marked, and the four small submedian setæ on its lower convex portion are present on all the specimens observed in the Forth ; the two lower ones are short, stout, and distinctly spiniform.

Bathyporeia robertsoni, Sp. Bate.   (Pl. V. figs. 26–29.)
    1862. *Bathyporeia robertsoni*, Sp. Bate, 'Cat. Amphip. Brit. Mus.,'
    p. 173, pl. xxxi. fig. 5.
    1891. *Bathyporeia robertsoni*, G. O. Sars, *op. cit.*, p. 131, pl. 44,
    fig. 2.

In this species the flagellum of the adult male posterior antennæ is comparatively short, and the first urosome wants the two posterior submedian dorsal spiniform setæ ; it also differs in having the last joint of the last pair of uropods very small. In the 'Revised List of the Crus' taceæ of the Firth of Forth,' *Bathyporeia pelagica* and *Bathyporeia robertsoni* are included under *Bathyporeia pilosa*, Lindström, as forms of that species.

Genus *Argissa*, Boeck (1870).

Argissa hamatipes (Norman).   (Pl. V. figs. 30, 31.)
    1868. *Syrrhoe hamatipes*, Norman, Report on Dredging among
    the Shetland Islands (in 'Report of the 38th Meeting of the
    'British Association, 1868,' London, 1868).
    1870. *Argissa typica*, Boeck, 'Crust. Amphip. bor. et arct.,' p. 45.
    1891. *Argissa typica*, G. O. Sars, *op. cit.*, p. 141, pl. 48.

This curious and somewhat anomalous Amphipod has been observed during the past autumn and winter in various parts of the Forth. It has also been obtained in Aberdeen Bay. One of its peculiar characteristics is the structure of the eyes, which consist of four pairs of small lenses about equidistant from each other, and arranged round the circumference of a nearly circular patch of pigment. In the male the second segment of the urosome has the dorsal part produced backward in the form of a free tooth-like process, that extends to nearly the end of the next segment.

Argissa hamatipes has been recorded for Shetland (Norman) and for the Clyde district (Robertson).

Genus Ampelisca, Kroyer (1842).

Ampelisca assimilis, Boeck. (Pl. V. figs. 32–35.)

> 1870. *Ampelisca assimilis*, Boeck, 'Crust. Amphip. bor. et arct.,'
> p. 142.
> 1891. *Ampelisca assimilis*, G. O. Sars, *op. cit.*, p. 168, pl. lviii.
> fig. 2.

Habitat.—Vicinity of May Island. Apparently scarce.

This, which is one of the smaller species, is distinguished by the anterior antennæ reaching very little beyond the end of the basal joints of the posterior antennæ ; the two last basal joints of the posterior antennæ are of about equal length. The epimeral plates of the last segment of the mesosome have the postero-lateral angles broadly rounded. The telson is rather longer than broad ; sides of the proximal half straight and parallel ; of the distal half slightly rounded, and converging to the blunt-pointed apex, and bearing a few very small marginal setæ.

Mr Robertson records this species from the Clyde.

Ampelisca lævigata, Lilljeborg. (Pl. V. figs. 36, 37.)

> 1855. *Ampelisca lævigata*, Lilljeborg, 'Ofv. af. Kgl. Vet. Akad.
> 'Förh.,' p. 123.
> 1862. *Ampelisca Belliana*, Sp. Bate, 'Cat. Amphip. Crust. Brit.
> 'Mus.,' p. 93, pl. xv. fig. 3.
> 1891. *Ampelisca lævigata*, G. O. Sars, *op. cit.*, p. 169, pl. lii.
> fig. 1.

Habitat.—Largo Bay and one or two other places in the Firth of Forth, but not common.

In this species the anterior antennæ scarcely reach to the end of the basal joints of the posterior antennæ. The penultimate basal joint of the posterior antennæ is considerably longer than the next, and the flagellum is comparatively short, being less than twice the length of the peduncle. The posterior pair of pereiopoda are robust ; the meral joint is produced exteriorly, and forms a lobe-like process as long as the next joint, and densely fringed with cilia. The postero-lateral margins of the epimeral plates of the last segment of the mesosome are doubly and strongly sinuate, the postero-lateral angles being produced into acutely slender teeth. Telson moderately broad, with a pointed apex.

According to Professor G. O. Sars, this species is the *Ampelisca belliana* of Sp. Bate, and *Ampelisca lævigata*, Sp. Bate, is the same as *Ampelisca tenuicornis*, Lilljeborg, which was described by Lilljeborg in 1855.

Ampelisca spinipes, Boeck. (Pl. V. figs. 38–40.)

> 1870. *Ampelisca spinipes*, Boeck, 'Crust. Amphip. bor. et arct.,'
> p. 143.
> 1891. *Ampelisca spinipes*, G. O. Sars, *op. cit.*, p. 173, pl. 60,
> fig. 2.

Habitat.—Vicinity of May Island, Firth of Forth. Not common.

The superior antennæ of *Ampelisca spinipes* are proportionally much longer than in the last species, as they extend considerably beyond the end of the peduncle of posterior antenna. The terminal joints of the last pair of pereiopods are comparatively narrow. The posterior margin of the expanded plate of the basal joint bends obliquely upwards at a moderately acute angle. The postero-lateral angles of the epimeral plates of the last segment of the mesosome are nearly rectangular. The telson is comparatively narrow. Length of Forth specimens, 15 mm.

Genus *Amphilochoides*, G. O. Sars (1892).

Amphilochoides odontonyx, Boeck.   (Pl. V. figs. 41, 42.)

> 1870. *Amphilochoides odontonyx*, Boeck, 'Crust. Amphip. bor. et 'arct.,' p. 51.
> 1892. *Amphilochoides odontonyx*, G. O. Sars., *op. cit.*, p. 221, pl. lxxv. fig 2.

Habitat.—Vicinity of Fidra Island, Firth of Forth.

This, though a small species, is quite easily distinguished from *Amphilochus manudens*, which it resembles in size and form by the structure of the gnathopods. The inner edge of the dactylus of the first and second gnathopods, but especially of the second pair, is produced near the hinge into a small but quite distinct blunt-pointed tooth, that interlocks into an opposing notch on the palm. It is further distinguished from *Amphilochoides pusillus*, G. O. Sars, which has the claws of second pair of gnathopods similarly toothed, by the postero-lateral margins of the epimeral plates of the third segment of the mesosome being sinuate, and the angles slightly produced and tooth-like. Professor Sars does not give (*loc. cit.*) any British locality for *Amphilochoides odontonyx*, but, on the authority of Meinert, records its occurrence in the Kattegat and Skagerak. He has also obtained it in a few places off the west coast of Norway. Mr David Robertson, of Millport, records its occurrence in the Clyde district

Genus *Cerapus*, Say (1817).

Cerapus crassicornis (Spence, Bate).

> 1855. *Siphonœcetes crassicornis*, Sp. Bate, 'Rep. Brit. Assoc.,' p. 59.
> 1857. *Siphonœcetes crassicornis*, White, 'Pop. Hist. Brit. Crust.,' p. 197.

This species was obtained among some material collected by a tow-net fixed to the head of the beam-trawl. It occupied a tube a little longer than itself, composed of blackish mud, held together by some kind of glutinous substance. It was able to move freely about with its tube, and to withdraw itself at pleasure. Only one specimen has been obtained.

CUMACEA.

Genus *Petalomera*, Stimpson (1858).

Petalomera declivis, G. O. Sars.   (Pl. V. fig. 43.)

> 1892. *Petalomera declivis*, T. and A. Scott, 'Ann. and Mag. Nat. Hist.,' ser. vi., vol. x. p. 206.

This Cumacean was obtained in Largo Bay in 1892. It appears to be a rare species in the Firth of Forth. More recently (April 1893) several specimens were obtained in bottom material collected a few miles east of May Island.

EXPLANATION OF THE PLATES.

PLATE II.

Longipedia coronata, Claus (♀).

Fig. 1. Female, dorsal view,   .   .   ×   32
Fig. 2. Anterior antenna,   .   .   × 152
Fig. 3. Posterior antenna,   .   .   × 152
Fig. 4. Mandible and palp,   .   .   × 190
Fig. 5. Maxilla, ...   × 127
Fig. 6. Anterior foot-jaw,   .   .   × 190
Fig. 7. Posterior foot-jaw,   .   .   × 152
Fig. 8. Foot of first pair of swimming feet,   × 100
Fig. 9. Foot of second pair,   .   .   ×   95
Fig. 10. Foot of third pair,   .   .   × 100
Fig. 11. Foot of fourth pair,   .   .   × 100
Fig. 12. Foot of fifth pair,   .   .   × 190
Fig. 13. Abdomen and caudal stylets,   .   ×   53

Longipedia coronata, var. *minor*, T. and A. Scott.

Fig. 14. Female, dorsal view,   .   .   .   ×   40
Fig. 15. Anterior antenna—male,   .   .   ,   × 190
Fig. 16. Foot of second pair of swimming feet—female,   .   × 127
Fig. 17. Foot of second pair of do.—male,   .   × 127
Fig. 18. Foot of fifth pair—female,   .   .   × 190
Fig. 19. Last abdominal segment and caudal stylets,   .   ×   80
Fig. 20. Portion of last thoracic segment and first abdominal segment—
            male (*a.* fifth foot ; *b.* abdominal appendage),   .   ×   23

Canuella perplexa, T. and A. Scott.

Fig. 21. Female, dorsal view,   .   .   .   ×   32
Fig. 22. Anterior antenna—female,   .   .   × 152
Fig. 23. Anterior antenna—male,   .   .   .   × 152
Fig. 24. Posterior antenna,   .   .   .   × 190
Fig. 25. Mandible, ...   × 190
Fig. 26. Maxilla, ...   × 190
Fig. 27. Anterior foot-jaw,   .   .   .   × 190
Fig. 28. Posterior foot-jaw,   .   .   .   × 190
Fig. 29. Foot of first pair of swimming feet,   .   .   × 152
Fig. 30. Foot of second pair,   .   .   .   × 100
Fig. 31. Foot of third pair,   .   .   .   × 100
Fig. 32. Foot of fourth pair,   .   .   .   ×   84
Fig. 33. Fifth pair of thoracic feet,   .   .   × 506
Fig. 34. Genital segment, ventral view—female,   .   × 127
Fig. 35. Genital segment, ventral view—male,   .   × 127

PLATE III.

Jonesiella hyaenae, I. C. Thompson (♀).

Fig. 1. Female, dorsal view,   .   .   .   ×   64
Fig. 2. Anterior antenna,   .   .   .   × 380
Fig. 3. Posterior antenna,   .   .   .   × 390
Fig. 4. Mandible palp,   .   .   .   × 380
Fig. 5. Foot of first pair of swimming feet,   .   .   × 200
Fig. 6. Foot of fifth pair,   .   .   .   × 300

Modiolicola insignis, Aurivillius.

Fig. 13.	Female, dorsal view,		× 54
Fig. 14.	Anterior antenna,		× 253
Fig. 15.	Posterior antenna,		× 170
Fig. 16.	Mandible and palp,		× 455
Fig. 17.	Anterior foot-jaw,		× 253
Fig. 18.	Posterior foot-jaw—female,		× 380
Fig. 19.	Posterior foot-jaw—male,		× 300
Fig. 20.	Foot of first pair of swimming feet,		× 152
Fig. 21.	Foot of fourth pair,		× 127
Fig. 22.	Foot of fifth pair,		× 380
Fig. 23.	Abdomen and caudal stylets—female,		× 63
Fig. 24.	Abdomen and caudal stylets—male,		× 115

Parartotrogus richardi, T. and A. Scott.

Fig. 25.	Female, dorsal view,		× 126
Fig. 26.	Anterior antenna,		× 253
Fig. 27.	Posterior antenna,		× 253
Fig. 28.	Mandible,		× 460
Fig. 29.	Maxilla,		× 253
Fig. 30.	Anterior foot-jaw,		× 253
Fig. 31.	Posterior foot-jaw,		× 253
Fig. 32.	Foot of first pair of swimming feet,		× 253
Fig. 33.	Foot of third pair,		× 253
Fig. 34.	Foot of fourth part,		× 253
Fig. 35.	Abdomen and caudal stylets (*a*. fifth foot),		× 170

Delavalia œmula, sp. nov.

Fig. 36.	Female, lateral view,		× 80
Fig. 37.	Anterior antenna—female,		× 253
Fig. 38.	Anterior antenna—male,		× 325
Fig. 39.	Posterior antenna,		× 200
Fig. 40.	Mandible and palp,		× 335
Fig. 41.	Posterior foot-jaw,		× 506
Fig. 42.	Foot of first pair of swimming feet,		× 253
Fig. 43.	Foot of second pair—male,		× 300
Fig. 44.	Foot of fourth pair,		× 152
Fig. 45.	Foot of fifth pair—female,		× 380
Fig. 46.	Foot of fifth pair—male,		× 300
Fig. 47.	Last abdominal segment and caudal stylets,		× 152

PLATE V.

Bomolochus soleæ, Claus.

Fig. 1.	Female, dorsal view,		× 40
Fig. 2.	Anterior antenna,		× 190
Fig. 3.	Posterior antenna,		× 127
Fig. 4.	Mandible,		× 300
Fig. 5.	Maxilla,		× 380
Fig. 6.	(?) Anterior foot-jaw,		× 380
Fig. 7.	Foot of first pair of swimming feet,		× 163
Fig. 8.	Foot of second pair,		× 127
Fig. 9.	Foot of fourth pair,		× 127
Fig. 10.	Foot of fifth pair,		× 127

Platychelipus littoralis, Brady.

Fig. 11.	Female, dorsal view,		× 70
Fig. 12.	Foot of first pair of swimming feet,		× 253
Fig. 13.	Foot of fifth pair,		× 169

Zosime typica, Boeck.

Fig. 14. Female, dorsal view, ... × 70
Fig. 15. Foot of first pair of swimming feet, . . × 253
Fig. 16. Foot of fifth pair, ... × 338
Fig. 17. Second-last abdominal segment, . . . × 380

Anonyx nugax (Phipps).

Fig. 18. Cephalon with superior and inferior antennæ, . × 7·6
Fig. 19. Mandible and palp, ... × 20
Fig. 20. One of the last epimeral plates of metasome, . × 9
Fig. 21. Telson, ... × 26

Bathyporeia norvegica, G. O. Sars.

Fig. 22. One of the last epimeral plates of metasome, × 13

Bathyporeia pelagica, Sp. Bate.

Fig. 23. One of the last epimeral plates of metasome, . . × 26
Fig. 24. Dorsum of first segment of urosome, . . . × 26
Fig. 25. One of the last pair of uropods, ... × 26

Bathyporeia robertsoni, Sp. Bate.

Fig. 26. Cephalon with superior and inferior antennæ, . . × 21
Fig. 27. One of the last epimeral plates of metasome, . . × 16
Fig. 28. Dorsum of first segment of urosome, . . . × 40
Fig. 29. One of the last pair of uropods, . . . × 20

Argissa hamatipes (Norman).

Fig. 30. Cephalon with superior and inferior antennæ (a. eye), . . × 24
Fig. 31. Urosome, ... × 21

Ampelisca assimilis, Boeck.

Fig. 32. Cephalon with superior and inferior antennæ, . . × 26
Fig. 33. One of the epimeral plates of metasome, . . × 13
Fig. 34. Telson, ... × 40
Fig. 35. One of the last pair of pereiopods, . . . × 13

Ampelisca lævigata, G. O. Sars.

Fig. 36. One of the last epimeral plates of metasome, . . × 21
Fig. 37. One of the last pair of pereiopods, . . . × 20

Ampelisca spinipes, Boeck.

Fig. 38. Cephalon with superior and inferior antennæ, . × 21
Fig. 39. One of the last epimeral plates of metasome, . × 8
Fig. 40. Telson, ... × 13

Amphilochoides odontonyx (Boeck).

Fig. 41. One of the second gnathopods, . . . × 53
Fig. 42. One of the last epimeral plates of metasome, . . × 18

Petalomera declivis, G. O. Sars.

Fig. 43. One of the thoracic appendages, . . . × 40

PLATE II.

PLATE 2

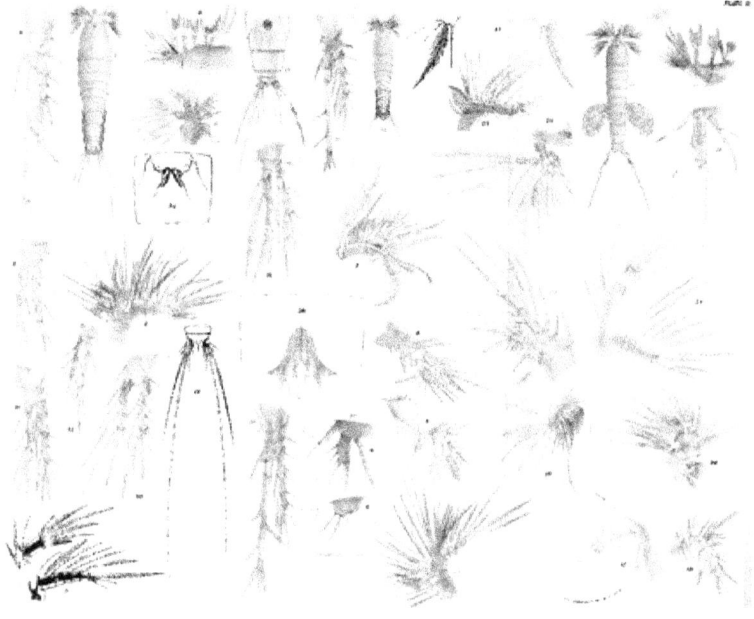

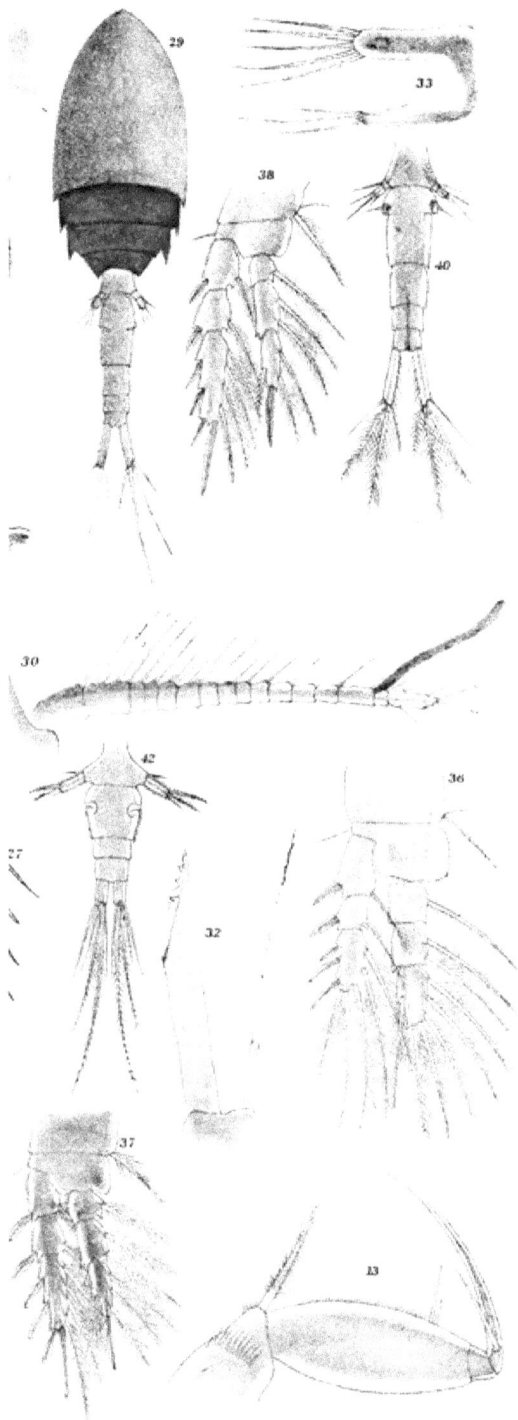

—Cyclopicera purpurocincta, sp. n.

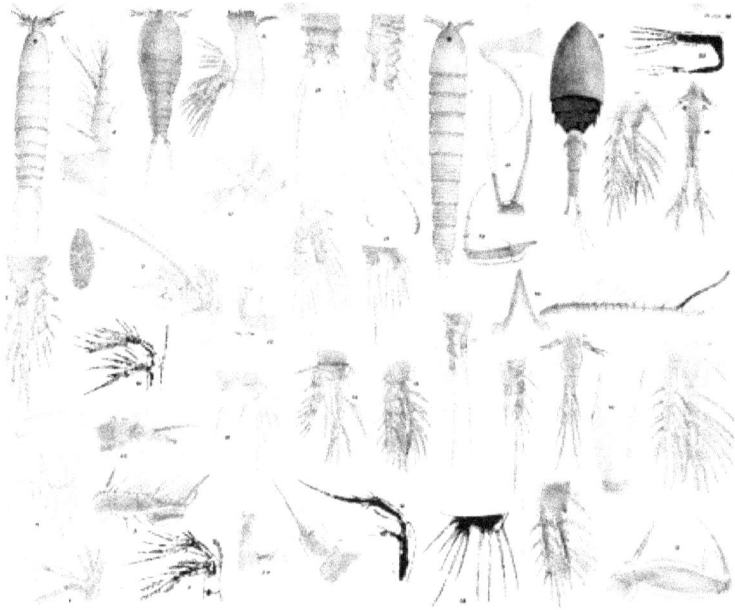

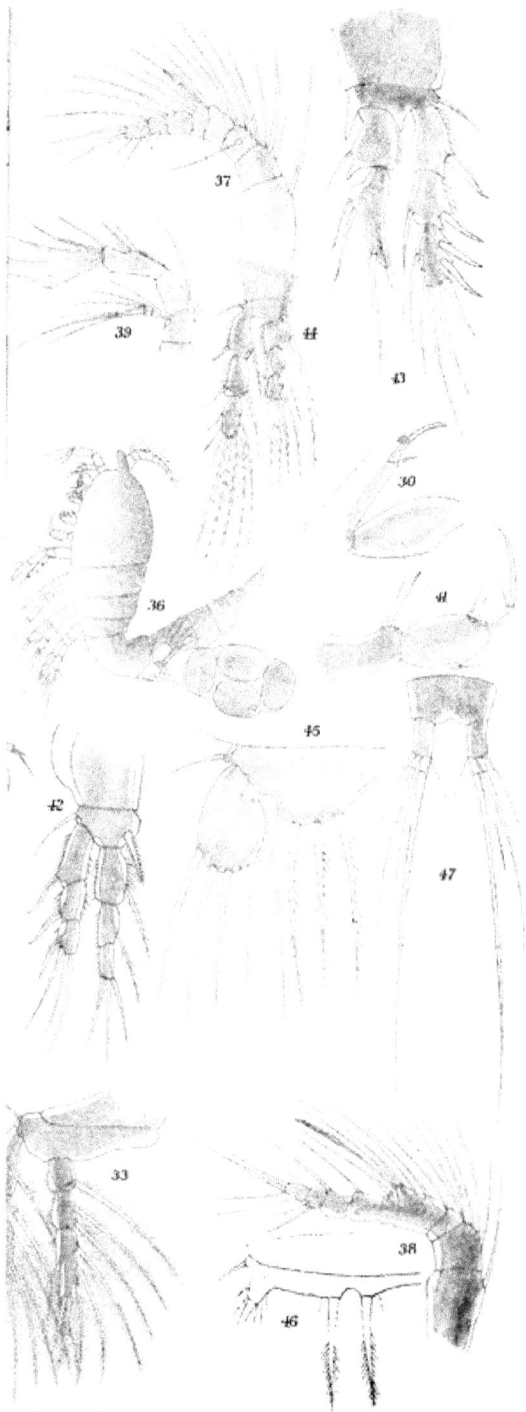

37

39

41

43

30

36

41

45

42

47

33

38

46

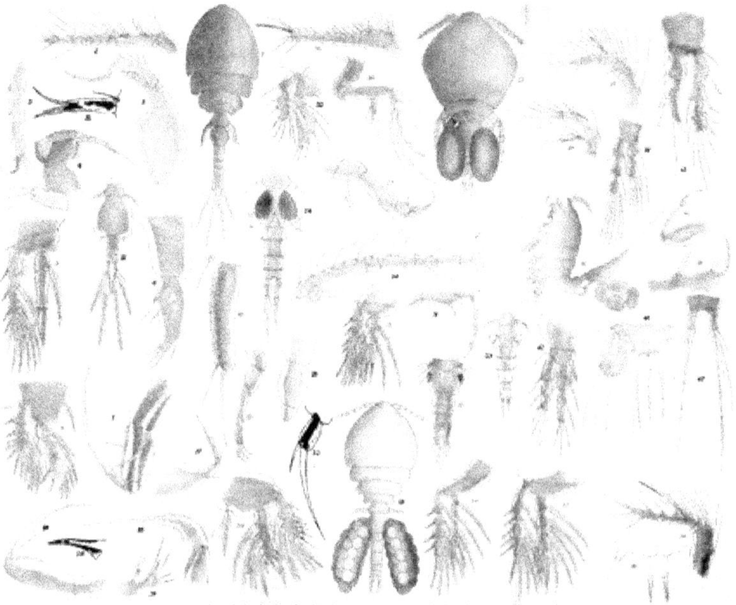

Figs. 1—11. Ameiridae brevicaudata, n.sp. Figs. 12—24. Nitocrella amara, Berlanova. Figs. 25—37. Psychrocamptus arcticus, J. and A. Scott. Figs. 38—45. Bradyella arctica, sp. n.

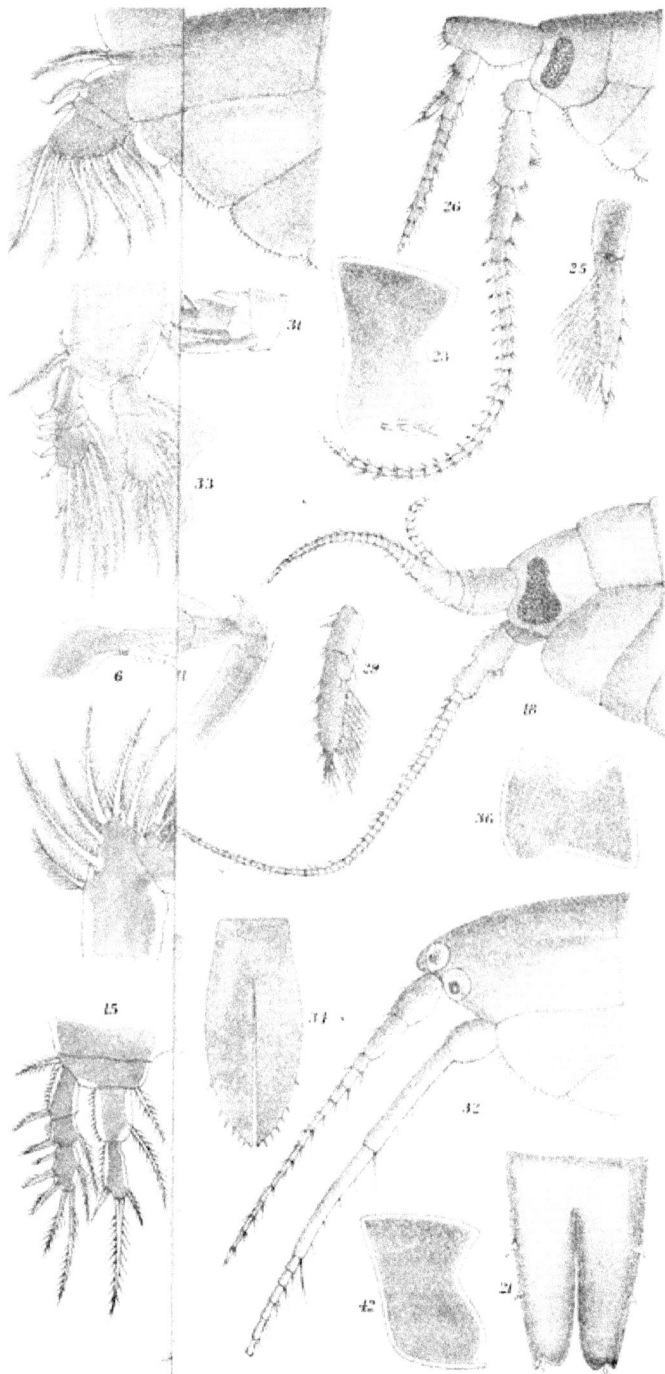

Chilochondus odontonyx (*Boeck*).

Andrew Scott, del.

III.—THE INVERTEBRATE FAUNA OF THE INLAND WATERS OF SCOTLAND. PART III. By Thomas Scott, F.L.S. (Plates VI., VII.)

Loch Morar, Inverness-shire.

Introduction.

I was requested to visit this loch in September last year (1892), to make an investigation of its invertebrate fauna, and to obtain some information concerning the depth of the loch, the specific gravity of its bottom water, &c.

Mr Anderson Smith, one of the members of the Fishery Board, who was at this time engaged on board the fishery steamer 'Garland,' in carrying on an extensive series of observations—biological and physical—among the islands and fiords of the West of Scotland, assisted me very materially in carrying out the work that I had been requested to take in hand.

I was conveyed on board the 'Garland' to Tarbet, Loch Nevis, where there is a nice sheltered little bay, and where there is a small inn, at which boats and men for Loch Morar may be obtained or arranged for. There is a fairly good road leading from Tarbet Inn to Loch Morar, which is only about three-quarters of a mile, if so much, distant. The inn is nearly midway between the upper and lower ends of the loch, and forms an excellent rendezvous for anglers and others who desire to work the upper as well as the lower ends. Unfortunately, the inn is not readily accessible for visitors, being much out of the way of steamers and coaches. The nearest port of call for steamers is Arisaig, which is, therefore, the usual landing-place for visitors to this district.

The road leading from Tarbet Inn to Loch Morar is formed through what appears to be a 'fault' in the schistoze rocks that form the barrier between Loch Nevis and Loch Morar; this road gradually rises to a height of about 200 feet, then descending towards Loch Morar, it continues westward along the north side and round the lower or west end of the loch, and thence to Arisaig. This is the road by which the postman from Arisaig reaches Tarbet Inn. I understand he comes three times a week to the inn during the summer, and twice a week during the winter months. Tarbet Inn is thus the postal terminus in this district. A little sheltered bay in Loch Morar, opposite Tarbet Bay, Loch Nevis, is called South Tarbet Bay. Along the north side of Loch Morar, from South Tarbet Bay eastward, there is no road deserving the name: there is, indeed, a kind of pathway, but of such a break-neck character that it requires some experience and 'nerve' to traverse it safely. There appears to be no road along the whole of the south side, with the exception of a pathway similar to that described.

Loch Morar is justly described as 'a large and beautiful loch amid ' magnificent scenery ; ' * it lies nearly due east and west, between stupendous ridges of rocks, sparsely clothed with vegetation, and which, in many places, rise precipitously from the water's edge to summits of over 1000 feet in height, and in one or two instances to summits of over 2000 feet in height, at a distance of less than a half to three-quarters of a mile from the edge of the loch. Towards the lower end the sides, though rugged, are less steep and precipitous. At the upper end three streams

---

* *Sportsman's Guide to the Rivers, Lochs, Moors, and Deer Forests of Scotland*, September 1890, p. 220.

enter the loch from the westward ; one of these issues from a deep, gloomy ravine or cañon, which terminates abruptly in bold headlands. The loch is shallow at its extreme upper or east end, and there is a considerable extent of nearly level marsh and meadow land that has probably been formed by the silting up of this part of the loch. The length of Loch Morar, as shown on the 'one-inch' Ordnance Survey Map, is a little over 11½ statute miles, but in the *Sportsman's Guide* it is stated to be 'about ' 14 miles long ;' * the narrowest part (about 3¼ furlongs) is near the east end, at Sron an Drutain, and the greatest width (1 mile 4¾ furlongs) is at the west end, near Bracorina. Several streams flow into the loch along both the north and south sides : the largest is the River Meoble, formed by the overflow waters of Loch Beoraid (a small loch about 3 miles south of Loch Morar), and augmented by several rivulets in its course north-wards, to where it falls into Loch Morar, near Camas Luinge, nearly opposite, but a little west of, South Tarbet Bay. The overflow waters of Loch Morar form the River Morar, which, flowing from the west end, and after a tortuous course of about 1¾ miles, falls into the Atlantic. The normal height of Loch Morar is about 31 feet above sea-level, but the height varies considerably according as the season may be wet or dry.

'Loch Morar belongs to Lord Lovat, Mr J. A. Macdonald of Glenala-'dale, Mrs Campbell of Callert, and Mrs Nicholson of Arisaig and Morar ; ' it contains salmon, sea-trout, and loch-trout in great abundance, and yields ' good sport to the angler. The loch fishes well all the summer months, ' but salmon never rise well.' † It may be remarked here, by way of com-parison, that angling for salmon in Loch Ness—also a deep loch—is usually not very successful, though these fish abound in that loch also.

Depth of Loch Morar.

Loch Morar was sounded at four different places, as near as possible in mid-channel, and here referred to as Stations I., II., III., and IV. The extreme upper end of the loch is shallow, but, proceeding westward from the head of the loch, the depth rapidly increases, so that at Station I. (about 7 furlongs from the head of the loch) the sounding-lead touched bottom at a depth of 59⅔ fathoms (358 feet) ; the depth continued still further to increase, so that at Station II., a little over a mile from Station I., the sounding-lead touched bottom at 100 fathoms (600 feet). At Station III. (a little west of Aron Mhor, or a little over 3¼ miles from the head of the loch) we obtained 103⅔ fathoms (622 feet) ; the bottom at Stations I. to III. appeared to be rock or gravel. Three soundings were taken in the vicinity of Station IV. (about 3 furlongs west of South Tarbet Bay). The first was with the dredge, which touched bottom, — the line being straight up and down,—at 170 fathoms (1020 feet) ; when hauled up the dredge-bag contained a quantity of fine mud. The second sounding was taken with the sounding-lead (with reversing thermometers attached) ; the bottom was touched at 156½ fathoms (941 feet). The third sounding was taken with the self-locking water-bottle, let down for a sample of bottom water, and which when hauled up contained a mixture of water and impalpable mud ; the water-bottle touched bottom at 145 fathoms (870 feet). The difference in depth shown by these three separate soundings may, I think, be easily accounted for, in this way : the distance of Station IV. from either side of the loch, opposite to where the soundings were taken, is under half a mile ; the sides must therefore dip rapidly downward to form a deep, precipitous, sub-aqueous ravine. The drifting of the boat, though but for a short distance, by winds or currents during the time intervening between the one sounding and the other, would thus

* *Loc. cit.*‡ † *Loc. cit.*

P

be quite sufficient to account for the difference in depth between the
different soundings.   The depth of the west portion of the loch was not
tested, but the west end is probably comparatively shallow, if one may
judge by the numerous islets at that end, and by the contour of the land
contiguous to this part of the loch.

Specific Gravity and Temperature of the Bottom Water.

The self-locking water-bottle,* which was let down and touched bottom
at 145 fathoms, as already stated, was when hauled up found to be perfectly closed, and filled with water mixed with impalpable mud.   This
mixture was emptied into a sufficiently large vessel, and when the mud
had subsided the supernatant water was run off and tested by one of the
set of hydrometers used on board the 'Garland.'   The specific gravity
registered by this hydrometer was 1007·9 ; temperature of water at the
time the specific gravity was taken, 13·9° C. (57·2° F.)   The temperature
of the water at the bottom of the loch, taken with a 'Negretti-Zambra'
deep-sea reversing thermometer, was 42° F. ; at 100 fathoms the temperature was 49·6° F. ; and at the surface, 57·7° F.   The water, as in
the case of many Highland lochs, was of a brown peaty colour.

Invertebrate Fauna.

The bottom mud brought up by the dredge and by the water-bottle at
Station IV. was carefully examined, with the result that no living organism
could be detected in it ; a considerable portion of it consisted of fragments
of Entomostracan tests that formed a distinct layer on the surface of the
mud after it had been allowed to subside : it thus resembled very closely
the mud from the bottom of Loch Ness.   The loch was tow-netted in
various places from Brinacory eastward to the head of the loch.   The
tow-net collections were chiefly made at the surface or at a depth of a few
fathoms.   On one or two occasions, however, the tow-net was attached to
the dredge and lowered to a depth of from 50 to 100 fathoms.   Entomostraca were found to be abundant at and near the surface, but were
much fewer at 50 and 100 fathoms: the grouping of the species,
however, appeared to vary very little in the surface and deep-water
gatherings.   A few gatherings were made by dragging the tow-net
through the weeds in the shallow water at the head of the loch, and also
close inshore at Brinacory—on the north side of the loch, near Brinacory
island.   Two days were devoted to the examination of Loch Morar, but
the weather was not very favourable.   On the first day there was almost continuous rain ; the first part of the second day was rather better, but a
smart westerly breeze sprang up shortly after noon, which caused the
water to become choppy, and made rowing more exhaustive and difficult.
Mr Anderson Smith accompanied us the second day, and but for his
assistance and encouragement our success, small though it may be, would
have been less.

Pelagic Entomostraca, though abundant in Loch Morar at the time of
my visit, included comparatively few species, viz., two (or three) species
of *Copepoda* and five of *Cladocera;* on the other hand, the bottom material,
though comparatively poor in number of specimens, was rich in species,
especially of *Cladocera,*—the number obtained in the bottom gatherings
include eight species of *Copepoda,* five of *Ostracoda,* and sixteen of
Cladocera.   The apparent scarcity of the *Ostracod* group is noteworthy,
and is probably attributable to non-suitability of habitat.   The physical

* For description of this water-bottle, see *Sixth Annual Report of the Fishery
Board for Scotland,* part iii. p. 346, pl. x. (1888).

conditions of fresh-water lochs—as Loch Ness and Loch Morar, where the water is of great depth, and where the shores dip rapidly, leaving very little or no shallow margin—do not seem to favour the development of the *Ostracoda*, and it is interesting to compare in this respect the distribution of the marine species of the same group, which extends from the shore down to many hundreds of fathoms. Another point of interest may be noted here, viz., though several species of Copepoda were obtained, including one or two rare Harpactids, not a single *Canthocamptus* was observed,—not even *Canthocamptus minutus*. *Gammarus pulex*—the common fresh-water amphipod—was also conspicuous by its absence.

Mollusca appeared to be very scarce: the only species observed were *Limnæa peregra* and *Pisidium pusillum*.

LIST AND DESCRIPTION OF SPECIES.

CRUSTACEA.

COPEPODA.

Family CALANIDÆ.

Genus *Diaptomus*, Westwood.

Diaptomus gracilis (G. O. Sars).

      1863. *Diaptomus gracilis*, G. O. Sars, Oversigt af de indenlandske Ferskvands copepoder, p. 9.
      1891. *Diaptomus gracilis*, Brady, Revis. Brit. Cyclop. and Calan., p. 29, Pl. XII. figs. 7–9 ; Pl. XII. figs. 1–8.

Of common occurrence all over the loch, but comparatively few of the specimens mature.

Family CYCLOPIDÆ.

Genus *Cyclops*, Müller.

Cyclops viridis (Jurine).

      1820. *Monoculus quadricornis viridis*, Jurine, Hist. de Monocles, p. 46, Pl. III. fig. 1.
      1891. *Cyclops viridis*, Brady, *op. cit.*, p. 17, Pl. V. figs. 6–10.

Common all over the loch.

Cyclops serrulatus, Fischer.

      1851. *Cyclops serrulatus*, Fischer, Bull. Soc. Imp. Moscow, vol. xxiv. p. 423, t. x. figs. 22, 23, 26–31.
      1891. *Cyclops serrulatus*, Brady, *op. cit.*, p. 18, Pl. VII. fig. 1.

Frequent in bottom material collected at the head of the loch and at Brinacory.

Cyclops ewarti, Brady. (Pl. VI. figs. 1–10).

      1888. *Cyclops ewarti*, Brady, Sixth Annual Report of the Fishery Board for Scotland, Part III. p. 232, Pl. VIII. figs. 1–10.
      1891. *Cyclops ewarti*, Brady, Revis. Brit. Cyclop. and Calan., p. 22, Pl. VII. figs. 4–7.

Length 1·5 mm. ($\frac{1}{17}$th of an inch). Seen from the dorsal view, body elongate, pyriform, tapering gradually to the abdomen ; first cephalo-

thoracic segment equal in breadth to about four-fifths the length, and
rather longer than the entire length of the other thoracic segments ; forehead moderately broad, somewhat produced and truncate.   Anterior
antennæ slightly longer than the first thoracic segment, eleven-jointed,
moderately stout, and sparingly setiferous ; the fifth joint distinctly
shorter, the first and seventh subequal and longer than any of the others.
The formula shows very nearly the comparative length of the joints :—

$$\frac{23 \cdot 7 \cdot 12 \cdot 8 \cdot 5 \cdot 11 \cdot 22 \cdot 18 \cdot 11 \cdot 16 \cdot 19}{1 \cdot 2 \cdot 3 \cdot 4 \cdot 5 \cdot 6 \cdot 7 \cdot 8 \cdot 9 \cdot 10 \cdot 11} \cdot$$

Posterior antennæ four-jointed, second and third joints smaller than the
first and last.   A long plumose seta springs from the lower distal extremity
of the first joint ; opposite to this, on the upper edge, are two short plain
setæ : there is a small seta near the middle of the upper margin of the
second joint, and on the upper margin of the third joint there is a row of
small spiniform setæ arranged in a pectinate manner,—the setæ at the
distal being much longer than the posterior ones ; the extremity of the
posterior antennæ bears a fascicle of setæ, one of which is longer and
plumose ; the second, third, and last joints are ciliate on the lower
margin.   The base of the mandible is provided with two long plumose
and one very small plain setæ (fig. 4).   Anterior footjaws short, four-jointed,
the first being equal in length to the other three ; the last is very small ;
the first joint bears three setæ, the second two, and the third and fourth
one each.   These setæ are furnished with short bristle-like cilia arranged
somewhat widely apart ; the fourth joint also bears two small terminal
plain setæ (fig. 5).   Posterior footjaw stout, well developed, four-jointed ;
the first joint bears two plumose setæ that arise from a papilliform process
near the upper distal margin ; a seta with bristle-like cilia springs from
a conical base of the upper margin of the second joint, at the distal end of
the same joint, two similar setæ arise from an elongate digitiform process ;
the upper part of the distal end of the third joint is produced so as to
extend fully beyond the last two small joints, and terminates in a seta
similar to those just described and a stout elongate spine (fig. 6).   The
basal joint of the fifth pair of feet is furnished with one stout seta
plumose of the distal half ; a long stout seta also plumose on the distal
half springs from the end of the second joint ; on the upper margin of
the same joint, and near the base of the terminal seta, there is a fringe of
stout cilia ; and a short, moderately-stout spine, serrate on the interior
edge, springs from the lower margin.   The abdomen tapers very gradually ;
the posterior margins of the abdominal segments are fringed with small
teeth ; the last segment is considerably longer than the preceding one.
Caudal stylets nearly one and a half times the length of the last abdominal
segment.   There is a small toothed notch on the outer margin near the
middle of the proximal half of each stylet, and a small seta springs from
near the middle of the lower half ; the terminal setæ are plumose and of
moderate length.

This species was first obtained in the upper reaches of the Firth of
Forth in Nov. 1887 ; but though thus apparently a marine species, Dr Brady
suggested that its real habitat might be ' in some of the streams or ponds
' whose contents find their way into the Forth.'   The discovery of this
species in Loch Morar—the second time it has been observed in
Scotland—shows that the explanation of its occurrence in the Forth
estuary suggested by Dr Brady is probably correct.   *Cyclops ewarti* was
of frequent occurrence in Loch Morar.

? *Cyclops Kaufmanni*, Uljanin.

> 1875. *Cyclops Kaufmanni*, Uljanin, Reise in Turkestan (Crust.),
>   p. 38, t. xii., figs. 2-4.
> 1891. *Cyclops Kaufmanni*, Brady, Revision, p. 24, Pl. VII. fig. 3.

Several specimens of what appears to be this species were obtained in Loch Morar.

Cyclops fimbriatus, Fischer.

    ? 1785. *Cyclops crassicornis*, Müller, Entomostraca, p. 113, Pl. XVIII. figs. 15–17.

    1853. *Cyclops fimbriatus*, Fischer, Bull. Soc. Imp. Moscow, p. 94, Pl. III. figs. 19–28, 30.

    1891. *Cyclops fimbriatus*, Brady, *op. cit.*, p. 25, Pl. IX. fig. 1.

This pretty *Cyclops* was moderately frequent in the bottom material collected at the head of the loch.

Family HARPACTICIDÆ.

Genus *Attheyella*, Brady. (See note at p. 235.)

Attheyella spinosa, Brady. (Pl. VI. figs. 11–20.)

    1880. *Attheyella spinosa*, Brady, Mon. Brit. Copep., vol. ii. p. 58, Pl. XLIII. figs. 15–18; Pl. XLVI. figs. 13–18.

Female as described by Dr Brady (*op. cit.*).

Male.—The third joint of anterior antennæ dilated exteriorly so as to form a large lobe-like process (fig. 23). Mouth organs and first and second pairs of swimming feet as in the female. Both branches of third pair of swimming feet three-jointed ; the first two joints of the outer branches are broad and of about equal length, the second joint is armed on the outer half of the distal end with a strong conical spine slightly curved outward at the extremity, third joint narrow and as long as the other two together ; the first two joints of the inner branch are very short, the second being armed with a moderately long spiniform process, the two edges of which instead of being straight are curved in a somewhat irregular zigzag manner ; the last joint, which is nearly twice as long as the first two together, is furnished with a long plain terminal seta, besides a short and plumose subterminal one. Fourth pair as in the female. Fifth pair small ; basal joint moderately broad, somewhat rounded, but not much produced posteriorly, and provided with two stout plumose setæ ; secondary joint elongate-ovate, furnished with several setæ, as shown in the figure (fig. 18). Caudal stylets rather longer than, and not so much dilated as those of the female.

Attheyella spinosa was frequent in the bottom material from the head of the loch, and also in that from Brinacory. The large spines on the outer branches of the third swimming feet of the male form a striking character.

Attheyella cryptorum, Brady. (Pl. VI. figs. 21–31).

    1868. *Canthocamptus cryptorum*, Brady, Jour. of Micros. Sci., vol. ix., Pl. VI. figs. 1–10.

    1880. *Attheyella cryptorum*, Brady, Mon. Brit. Copep., vol. ii. p. 60, Pl. LII. figs. 1–18.

Length, exclusive of caudal setæ, ·66 mm. ($\frac{1}{38}$th of an inch). The general form is something like that of *Canthocamptus*. Anterior antennæ, of the female, short, moderately stout, eight-jointed. A sensory filament springs from the upper distal angle of the fourth joint. The proportional length of the joints are as follows :—

$$\frac{11 \cdot 11 \cdot 7 \cdot 7 \cdot 6 \cdot 8 \cdot 7 \cdot 10.}{1 \cdot 2 \cdot 3 \cdot 4 \cdot 5 \cdot 6 \cdot 7 \cdot 8.}$$

The male antennæ are considerably dilated. There is a constriction between the third and fourth joints ; the articulation between these forms a hinge to enable the antennæ to be used for grasping ; the fifth joint is small, and bears a short stout spine on its upper distal angle, while the sixth joint is furnished with a sensory filament. First pair of swimming feet short, the outer branch three-jointed, the inner two-jointed, both, branches of about equal length, the basal joint of the inner branch stouter and rather shorter than the other ; the margins of both joints are fringed with short spiniform setæ ; the last joint bears at its distal extremity a moderately short plumose spine, and one short and one long bent subterminal seta ; each of the three joints of the outer branch is armed exteriorly with a subterminal and somewhat stout plumose spine, the second joint carries a moderately long seta on the inner distal angle, while the last joint carries two terminal and one long bent subterminal setæ (fig. 24). Inner branches of the third pair scarcely longer than the first two joints of the outer branch, first joint very short ; in the male the first joint bears a long stout setaform appendage, which is at least two and a half times the length of the inner branch ; there is a small seta at the base of the appendage ; the only armature of the second joint—which is somewhat lageniform—consists of two terminal plumose setæ, one being much longer than the other (fig. 27). Inner branches of the fourth pair shorter than the first two joints of the outer branch, first joint very small ; the armature of both branches somewhat similar to that of the third pair (female). Fifth pair small, foliaceous ; in the female the basal joint bears four long and two short plumose setæ ; the long setæ are arranged as follows :—one on the inner and one on the outer margin near the somewhat truncate extremity, and two of them terminal, one of which is considerably longer than either of the other three ; the two short setæ spring from the inner margin, as shown in the figure (fig. 29) ; the second joint is smaller, obliquely truncate, and provided with five setæ, as shown in the figure. In the male the basal joint is boldly convex, and bears two short plumose terminal setæ, second joint small subovate, furnished with one stout, coarsely plumose, and moderately long terminal seta, and two small setæ on the inner and three on the outer margin (fig. 30). Abdominal segments ornamented with two concentric rows of small prickles—one row near the lower and one near the upper margin, but the last row is not so conspicuous as the other. The opercular plate has the edge strongly aculeate. Caudal stylets short, somewhat dilated, and provided with two long terminal unequal setæ, one being about twice the length of the other, and a few short hairs (fig. 31).

This species was described by Dr Brady from specimens obtained by Mr Thomas Atthey, amongst gelatinous algæ growing on the root of the pit-workings of the low main West Cramlington Colliery, near Newcastle, in 1868 : so far as I can learn, it was not known to occur anywhere else till it was obtained in Loch Maree during our examination of that loch. Specimens were sent to Dr Brady, and he was able to recognise them as belonging to the species he had described from the West Cramlington Colliery.

It may be of interest to note here, that quite recently, when examining some material collected by hand-net in the vicinity of Harelaw Dam, Balerno, near Edinburgh, in August 1880, and which had not been examined previously for want of time, several specimens of this interesting species were observed, thus indicating that its distribution may not be so restricted as had been supposed. Its occurrence among the gelatinous algæ on the roof of the pit-workings at West Cramlington is very curious and interesting, and opens up questions respecting distribution of species which, like that of the 'toad in the rock,' may not be easily solved.

Attheyella propinqua sp. n. (provisional name), (Pl. VII. figs. 1–11).

Length, exclusive of caudal setæ, ·7 mm. ($\frac{1}{36}$th of an inch), anterior antennæ rather longer and more slender than those of *Attheyella cryptorum*, but resembling in this respect those of *Attheyella spinosa*; the relative length of the joints is also somewhat different, as shown by the formula—

$$\frac{7 \cdot 7 \cdot 6 \cdot 7 \cdot 6 \cdot 7 \cdot 7 \cdot 8.}{1 \cdot 2 \cdot 3 \cdot 4 \cdot 5 \cdot 6 \cdot 7 \cdot 8.}$$

The secondary branch of the posterior antennæ is fairly well developed and two-jointed; the first joint bears one terminal seta, the second joint bears two plumose terminal setæ and a marginal one. The mandible, which is elongate and of moderate breadth, has a small two-jointed palp bearing a few setæ (fig. 5). Posterior footjaws two-jointed and armed with a slender terminal claw; the first joint is provided, at the inner distal angle, with a small spiniform seta. First pair of swimming feet short, inner branch rather longer than the outer, and composed of two nearly equal joints (fig. 7). The second, third, and fourth pairs similar to those of *Attheyella spinosa*. Fifth pair, especially in the female, somewhat like those of *Attheyella cryptorum*; those of the female rather larger than in that species; the basal joint is broader and subquadrate, and provided with shorter setæ on the subtruncate end; the second joint is broadly subovate, the outer margin bearing a fringe of cilia in addition to the five terminal and subterminal setæ. Fifth pair in the male smaller than in the male of *Attheyella cryptorum*; the extremity of the produced inner part of the basal joint is subtruncate and provided with two small plumose setæ, second joint ovate with one moderately long plumose and one short plain terminal setæ; there are also two short plumose setæ on the inner margin, and three short setæ on the outer margin. The second, third, and fourth abdominal segments are each furnished with one concentric fringe of short prickles close to the posterior margin, instead of two as in *Attheyella cryptorum*. Seen from the side, the abdomen terminates abruptly, with the fringed opercular plate projecting slightly upward and backward. The male differs little from the female; the anterior antennæ are not so stout as in *Attheyella cryptorum*, and do not possess the strongly developed lobe-like process, which is such a marked character of the male antennæ of *Attheyella spinosa*; the third pair of swimming feet are provided with shorter spines than the third pair of the male of *Attheyella cryptorum*, but are otherwise somewhat similar to those of that species; caudal stylets very short. In bottom material both from the head of Loch Morar and from Brinacory.

The species now described is somewhat intermediate between *Attheyella spinosa* and *Attheyella cryptorum*, but seems more closely allied to the latter; the three forms when placed together are readily distinguished from each other by the form of the abdomen and stylets. In the species now described, the abdomen, viewed laterally, is seen to end abruptly, with the opercular plate showing as a distinct tooth-like process; and the abdominal segments have only one fringe of setæ. In *Attheyella cryptorum*, the abdomen is more slender and ends much less abruptly than in the last species, the opercular plate is less prominent, and there are two rows of setæ on each of the abdominal segments; moreover, the ovisac is somewhat larger, so that the setæ of the fifth feet do not reach to the end of it as in *Attheyella propinqua*. In *Attheyella spinosa* the form of the stylets, which are larger than in either of the other two species, are very characteristic, as are also the curled setæ of the fifth feet. Dissection brings out other more or less important differences as shown by the description and figures.

Genus *Moraria*,* T. and A. Scott.

Ann. and Mag. Nat. Hist., ser. 6, vol. xi. p. 213, March 1893.

Anterior antennæ short, seven-jointed; inner branch of posterior antennæ, small, one-jointed; mandible palp, one-branched, two-jointed; posterior footjaws, three-jointed, provided with a prehensile terminal claw; first pair of swimming feet short; inner branches, two-jointed, not longer than the outer branches, joints subequal; inner branches of the second, third, and fourth pairs much shorter than the outer branches, two-jointed, joints subequal.

This genus somewhat resembles *Cylindropsyllus*, Brady, in form and in some structural details. In the form and structure of the posterior antennæ, mandibles, and first pair of swimming feet it resembles both *Attheyella cryptorum* and *Mesochra robertsoni*, Brady, but it differs very markedly from both *Attheyella* and *Mesochra* in the structure of the second, third, and fourth pairs.

Moraria anderson-smithi, T. and A. Scott. (Pl. VII. figs. 12–26).

1893. *Moraria anderson-smithi*, T. and A. Scott, Ann. and Mag. Nat. Hist., ser. 6, vol. xi. p. 213, Pl. VIII. figs. 1–14.

Length, exclusive of caudal setæ, ·62 mm. ($\frac{1}{40}$th of an inch). Forehead produced into a short rostrum. The anterior antennæ in both male and female are short, stout, and seven-jointed, the male antennæ being hinged between the fourth and fifth joints, and adapted for grasping. The relative length of the joints of the female antennæ are as follows :—

$$\frac{6 \cdot 8 \cdot 6 \cdot 5 \cdot 5 \cdot 7 \cdot 10 \cdot}{1 \cdot 2 \cdot 3 \cdot 4 \cdot 5 \cdot 6 \cdot 7 \cdot}$$

A stout sensory filament springs from the upper distal angle of the fourth joint in the female and the third in the male. From the same joint in the male there springs a curved spine-like process from a produced basal part. Posterior antennæ short, stout, three-jointed; a small one-jointed secondary branch springs from near the middle of the second joint. Mandibles well developed, having a broad biting part and a one-branched two-jointed palp. The maxillæ consist of a broad masticatory portion and a two-branched appendage; the truncate end of the masticatory portion is armed with five moderately long spine-like teeth; there is also a small seta on the exterior margin; the outer of the two branches of the appendage terminates in a comparatively stout spine, plumose on the outer half, and is also furnished with four small marginal setæ; the inner branch is small, and terminates in a small spine. Anterior footjaws stout, terminating in a stout clawed spine bearing a small seta on its outer margin near the base of the claw; slightly anterior to the claw are two digitiform processes, each of which terminates in a moderately stout spine and an inward curved process not so long as the spine, and bearing several small setæ on its interior distal margin. Posterior footjaws three-jointed; last joint small, bearing a moderately slender, but not very long terminal claw. A stout plumose seta springs from the interior distal angle of the first joint; inner margin of the second joint fringed with cilia. First pair of swimming feet nearly as in *Attheyella cryptorum*, but the last joint of the inner branch is shorter than the first joint, being scarcely three-fourths its length; it is also narrower, and bears a stout terminal spine and two subterminal setæ—one very short and one very long and slender. The second and third pairs of feet in the female are somewhat similar to those of *Mesochra lilljeborgii*, but the first joint is more dilated

* From Loch Morar.

and rather longer than the last, and the whole inner branch is rather shorter than the first two joints of the outer branch. In the male the last joint of the inner branch of the second pair is much smaller than the first, and is armed with a long terminal spine and a subterminal seta ; the first joint bears a stout conical tooth-like process, slightly hooked at the end on the outer aspect, and a seta on the inner margin. The fourth pair have the inner branches scarcely reaching to the middle of the second joint of the outer branches ; the spines of the first and second joints are much longer than the joints to which they are attached ; the inner of the three terminal spines of the last joint, is of considerable length, and, like the middle spine, is ciliate along one side. Fifth pair in both sexes somewhat similar to *Attheyella*, but the armature is shorter and stouter (figs. 25, 26). Caudal stylets about equal in length to the last abdominal segment. The abdomen in the male consists of five, in the female of four segments, the first and second segments in the female being coalesced.

Moraria anderson-smithi was obtained in gatherings of bottom material both from the head of the loch and from Brinacory ; it appeared to be a rather scarce species in both gatherings, but was, if anything, more frequent in the last.

During our study of the closely allied genera,—*Attheyella*, *Moraria*, and *Mesochra*,—it seemed to us that the first and last of these, included species that differed considerably from each other. For example, *Attheyella spinosa* (including the male form) appears to belong to quite a different type from that of *Attheyella cryptorum*; *Attheyella spinosa* has both branches of the first pair of swimming feet three-jointed and of moderate length, while in *Attheyella cryptorum* the first pair is short, with the inner branches composed of two nearly equal joints. Moreover, in *Attheyella spinosa*, the male antennæ and the inner branches of the male third pair of feet, are strikingly different from those of *Attheyella cryptorum*. On the other hand, *Mesochra lilljeborgii* has the inner branches of the first pair elongate and two-jointed, the first joint being longer than the entire length of the outer branch ; while *Mesochra robertsoni* has the first pair short, both branches of about equal length, and with the inner branches composed of two subequal joints, thus differing very distinctly from *Mesochra lilljeborgii*, but agreeing very closely with *Attheyella cryptorum* and with *Attheyella propinqua*. After summarising all the points in which the various species referred to agree and in which they differ, we are inclined to think that, while *Attheyella spinosa* and *Mesochra lilljeborgii* may be considered as the types of the genera to which they belong, the others, *Attheyella cryptorum*, *Attheyella propinqua*, and *Mesochra robertsoni*, form a distinct and closely related group, which if removed from the genera where they are at present located and included in a separate genus or subgenus, the systematic study of these various forms would be somewhat simplified.

OSTRACODA.

PODOCOPA.

Family CYPRIDIDÆ.

Genus *Cypria*, Zenker.

Cypria serena (Koch).

1838. *Cypris serena*, Koch, Deutchlands Crustaceen, H. xxi. 22.

    1868. *Cypris lævis*, Brady, Mon. Rec. Brit. Ostrac., p. 374, Pl. XXIV. figs. 6–8.

    1889. *Cypria serena*, Brady and Norman, 'Mon. M. and F. W. Ostrac. of the N. Atlantic and N.W. Europe, p. 70.

In bottom material from the head of the loch, rather scarce.

Cypria ophthalmica (Jurine).

    1820. *Monoculus ophthalmicus*, Jurine, Hist. des Monocles, p. 178, Pl. XIX. figs. 16, 17.

    1868. *Cypris compressa*, Brady, *op. cit.*, p. 372, Pl. XXIV. figs. 1–5; Pl. XXXVI. fig. 6.

    1889. *Cypria ophthalmica*, Brady and Norman, *op. cit.*, p. 69, Pl. XI. figs. 5–9.

In the same material with the last, rather scarce.

Genus *Cyclocypris*, Brady and Norman.

Cyclocypris globosa (G. O. Sars).

    1863. *Cypris globosa*, G. O. Sars, Om en i Sommeren 1862 foretagen zoologisk Reise i Christianias og Trondhjems Stifter, p. 27.

    1868. *Cypris cinerea*, Brady, *op. cit.*, p. 374, Pl. XXIV. figs. 39–42; Pl. XXXVI. fig. 7.

    1889. *Cypris globosa*, Brady and Norman, *op. cit.*, p. 71, Pl. XIV. figs. 1, 2; Pl. XI. figs. 10–18.

In bottom material collected at Brinacory, scarce.

Genus *Candona*, Baird.

Candona candida (Müller).

    1785. *Cypris candida*, Müller, Entomostraca, p. 62, Pl. VI. figs. 7–9.

    1889. *Candona candida*, Brady and Norman, p. 98, Pl. X. figs. 1, 2, 14–23.

In bottom material from the head of the loch, scarce.

Candona kingsleii (Brady and Robertson).

    1870. *Candona kingsleii*, Brady and Robertson, Ann. and Mag. Nat. Hist., ser. iv. vol. vi., Pl. V.

    1889. *Candona kingsleii*, Brady and Norman, *op. cit.*, p. 102, Pl. IX. figs. 19–22; Pl. XIII. fig. 19.

Also in bottom material from the head of the loch.

CLADOCERA.

CALYPTOMERA.

Family SIDIDÆ.

Genus *Latona*, Straus.

Latona setifera (Müller).   (Pl. VII. fig. 27.)

    1821. *Latona setifera*, Straus, Mém. Mus. d'Hist. Nat.

This curious Cladoceran was obtained very sparingly in a bottom gathering from close inshore at Brinacory. *Latona setifera* was first recorded as British by Conrad Beck in the *Journal of the Microscopical Society* for December 1883, and is there described and figured. Herrick, in *Crustacea of Minnesota* (1884), states that it 'is not yet recognised in 'Minnesota, but was found by Professor Birge in Lake Michigan.'

Latona is distinguished from other British Cladocera by, among other characters, the structure of the posterior antennæ. The inner branches of these appendages are two-jointed; the distal portion of the basal joint next the outer branch, is continued forwards in a leaf-like process to fully half the length of the second joint, as shown in the figure. The Loch Morar specimens were small—under 1 mm.—but the size appears to vary. Mr Conrad Beck states that specimens 2·5 mm. ($\frac{1}{10}$th of an inch) are often obtained.*

Family HOLOPEDIDÆ.

Genus *Holopedium*, Zaddach.

Holopedium gibberum (Zaddach).
>   1855. *Holopedium gibberum*, Zaddach, Wiegman's Archiv. für **Naturges.**, Bd. xxi. p. 159, Pl. VIII. fig. 9.

This remarkable species was common in Loch Morar, and especially so near the surface of the water. It appears to be widely, but perhaps not generally, distributed. In the Fishery Board's Ninth Annual Report there are records of its occurrence in Loch Ness, Loch Oich, and Loch Lochy, Inverness-shire. It is distinguished by the greatly elevated and hood-like form of the brood-pouch. They frequently measure over an eighth of an inch from the apex of the brood-pouch to the under side of the body. When seen adhering to the tow-net after being removed from the water, they closely resemble at first sight the little medusiform gonophores sometimes so abundant in the sea.

Family DAPHNIDÆ.

It is noteworthy that no species belonging to this family were obtained in Loch Morar at the time of our visit.

Family BOSMINIDÆ.

Genus *Bosmina*, Baird.

Bosmina longispina, Leydig.
>   1860. *Bosmina longispina*, Leydig.

This species was very abundant in Loch Morar. It is readily distinguished from *Bosmina longirostris* by its much greater size and by the larger postero-ventral spine-like processes of the carapace.

Family MACROTHRICIDÆ.

Genus *Macrothrix*, Baird.

Macrothrix laticornis (Jurine). (Pl. VII. figs. 28, 29.)
>   1820. *Monoculus laticornis*, Jurine, Hist. des Monocles, p. 151, Pl. XV. figs. 6, 7.
>   1850. *Macrothrix laticornis*, Baird, Brit. Entom., p. 103, Pl. XV. fig. 2.

* C. Beck, *op. cit.* p. 757.

This is a small species, being only about ·5 mm. ($\frac{1}{50}$th of an inch). It was obtained in the gathering of bottom material from Brinacory; only a few specimens were obtained.   The figure represents an apparently adult form, and the post abdomen.

Family LYNCODAPHNIA, G. O. Sars (1861).

Genus *Ilyocryptus*, G. O. Sars.

Ilyocryptus sordidus (Lievin).
>   1858. *Acanthocercus sordidus*, Lievin.
>   1861. *Ilyocryptus sordidus*, G. O. Sars, Om de i Omeg. af Chris-
>   tiania forekom. blad., p. 12.

A few specimens of *Ilyocryptus sordidus* were obtained in a gathering of bottom material from the shallow water at the head of the loch.   It appears to be a moderately common species where the conditions are favourable , it was very common in June 1890 among vegetable mud on the south shore of Loch Leven, Kinross-shire ; and in July 1892 by the side of Lochend Loch, near Edinburgh.

Family LYNCEIDÆ.

Genus *Eurycercus*, Baird.

Eurycercus lamellatus (Müller).
>   1776. *Lynceus lamellatus*, Müller, Zool. Dan. Prod., No. 3396.
>   1850. *Eurycercus lamellatus*, Baird, Brit. Entom., p. 124, Pl.
>   XV. fig. 1.

A few specimens of this large and well-known species were obtained in the same gathering with the last.

Genus *Acroperus*, Baird.

Acroperus harpæ, Baird.
>   1835. *Lynceus harpæ*, Baird, Trans. Berw. Nat. Club, vol. i.
>   p. 100, Pl. II. fig. 17.
>   1850. *Acroperus harpæ*, idem, Brit. Entom., p. 129, Pl. XVI.
>   fig. 5.

This is not unlike *Camptocercus macrourus* in form, but much smaller. A few specimens only were obtained.

Genus *Camptocercus*, Baird.

Camptocercus macrourus (Müller).
>   1776. *Lynceus macrourus*, Müller, Zool. Dan. Prod., No. 2397.
>   1850. *Camptocercus macrourus*, Baird, Brit. Entom., p. 128,
>   Pl. XVI. fig. 9.

This *Lynceid* is readily distinguished by the elongated narrow post-abdomen and *Acroperus*-like form of the carapace.   Several specimens were obtained in bottom gatherings, both from the head of the loch and from the shore at Brinacory.

Genus *Alonopsis*, G. O. Sars.

Alonopsis elongata, G. O. Sars.
>   1862. *Alonopsis elongata*, G. O. Sars, Om de i Omeg. af Christ.
>   forekom. blad., Andet Bidrag., p. 41.

Frequent in the same gatherings with the last.

Genus *Graptoleberis*, G. O. Sars.

Graptoleberis testudinarius (Fischer).
>   1851. *Lynceus testudinarius*, Fischer, Mem. de Lav. étrangers,
>   St Petersbourg, vol. vi. p. 191, Pl. IX. figs. 3–6.
>   1862. *Graptoleberis reticulata*, G. O. Sars, op. cit., Andet Bedrag.,
>   p. 41.
>   1884. *Graptoleberis testudinarius*, Herrick, Crust. of Minnesota,
>   p. 90.

The peculiar hood-like portion of the carapace that forms the head imparts a distinctive appearance to this species. It was very scarce in the Loch Morar gatherings, one or two specimens only being obtained in bottom material from the head of the loch.

Genus *Alona*, Baird.

Alona guttata, G. O. Sars.
>   1862. *Alona guttata*, G. O. Sars, op. cit., Andet Bidrag., p. 38.

This *Alona* was one of the more common of the smaller species in the gatherings of bottom material. The carapace viewed laterally is sub-quadrilateral in form, and the surface of the carapace is usually ornamented by being thickly covered with puncture-like markings. By means of the form and sculpture of the carapace the species is readily distinguishable among its more common associates.

Alona costata, G. O. Sars.
>   1862. *Alona costata*, G. O. Sars, op. cit., Andet Bidrag., p. 38.

In the same material with the last, but much scarcer. This species, though somewhat like the last, wants the peculiar puncture-like markings; it is rather longer in proportion to the breadth, and there are usually impressed parallel longitudinal lines observable on the carapace.

Alona quadrangularis (Müller).
>   1776. *Lynceus quadrangularis*, Müller, Zool. Dan. Prod., p. 199,
>   No. 2393.
>   1850. *Alona quadrangularis*, Baird, Brit. Entom., p. 131, Pl.
>   XVI. fig. 4.

In the gatherings of bottom material, but not very common.

Genus *Alonella*, G. O. Sars.

Alonella exigua (Lilljeborg).
>   1853. *Lynceus exiguus*, Lilljeborg, De Crust. in Scenia, p. 79, Pl.
>   VII. figs. 9–10.

This is a very small species—rather less than the $\frac{1}{10}$th of an inch—but distinguishable by the broadly ovate form of the carapace, and the distinctly toothed posterior angle of the front margin. It appeared to be a scarce species.

Alonella nana (Baird).
>   1843. *Acroperus nanus*, Baird, Ann. and Mag. Nat. Hist., vol. ii.
>   p. 92, Pl. III. fig. 8.
>   1862. *Alonella pygmæa*, G. O. Sars, op. cit., Andet Bidrag., p. 52.

This was also apparently a rare species, but from its small size—less than the $\frac{1}{100}$th of an inch—it is easily overlooked. It is readily distinguished by the beautiful sculpture of the shell.

Genus *Peracantha*, Baird.

Peracantha truncata (Müller).

    1781. *Lynceus truncatus*, Müller, Entom., p. 75, Pl. 11. figs. 4–6.
    1850. *Peracantha truncata*, Baird, Brit. Entom., p. 136, Pl. XVI.
        fig. 1.

This also was one of the rarer species, and, like the two previous forms, was only obtained in the gatherings of bottom material.

Genus *Leptorhynchus*, Herrick.

Leptorhynchus falcatus (G. O. Sars).   (Pl. VII. fig. 30.)

    1861. *Alona falcata*, G. O. Sars, *op. cit.*, p. 20.
    1862. *Harporhynchus falcatus*, idem, ibidem, Andet. Bidrag.,
        p. 41.
    1884. *Leptorhynchus falcatus*, Herrick, Crust. of Minnesota, p. 114.

The long slender curved beak of this species (fig. 30) enables it to be readily distinguishable. It was of frequent occurrence in the bottom material only; both males and females were obtained. The beak of *Leptorhynchus falcatus* is said to exceed in length that of any known species of Lynceid. Norman and Brady* record it from 'Sweet-hope and Green Lee Lochs, Northumberland ; Lochend Loch, Kirkcudbright-shire ; and Lochmaben Castle Loch, Dumfriesshire ;' and add, in ' all these localities it was numerically scarce.'

Genus *Chydorus*, Leach (1816).

Chydorus sphæricus (Müller).

    1776. *Lynceus sphæricus*, Müller, Zool. Dan. Prod., No. 2932.
    1816. *Chydorus milleri*, Leach, Encyclop. Brit., Supp. Art.
        Annulosa.
    1863. *Chydorus sphæricus*, Baird, Ann. and Mag. Nat. Hist., vol.
        ii. p. 89, t. 2, figs. 11–13.

Frequent, and only in gatherings of bottom material, both from the head of the loch and from Brinacory.

GYMNOMERA.

Family POLYPHEMIDÆ, Baird, 1845.

Genus *polyphemus*, Müller.

Polyphemus pediculus (Linné).

    1746. *Monoculus pediculus*, Linné, Faun. Suec. No. 2048.
    1776. *Polyphemus oculus*, Müller, *op. cit.*, No. 2417.
    1850. *Polyphemus pediculus*, Baird, Brit. Entom., p. iii. t. xvii.,
        fig. 1.

Taken in abundance with the tow-net, near the surface of the water, all over the portion of the loch examined.

Genus *Bythotrephes*.

Bythotrephes longimanus, Leydig.

This species was frequent in the surface tow-net gatherings ; it was also

  * A Mon. of the Brit. Bosmin., Macroth., and Lynceidæ, p. 36.

obtained, though sparingly, in gatherings from fifty and one hundred fathoms, but probably some of the specimens may not have been obtained at the depths stated, but may have entered the net when it was being hauled up.

Family LEPTODORIDÆ.

Genus *Leptodora*, Lilljeborg.

Leptodora hyalina, Lilljeborg.

This beautiful and interesting Cladoceran was frequent in some of the surface tow-net gatherings, but its extreme transparency when living enables it usually to escape observation, except by those who are more or less accustomed with the method of hunting for such organisms.

It will be observed from what has been stated in the foregoing notes on the results of my examination of Loch Morar, that this loch contains an abundant and interesting crustacean fauna which is well suited as a food supply for its numerous finny inhabitants. In this respect Loch Morar compares very favourably with Loch Ness, that somewhat resembles Loch Morar both in size and physical conditions. A partial examination of Loch Ness was made on two occasions during the summer of 1890, * when *Entomostraca* were found to be much less numerous than they were in Loch Morar in 1892. On the other hand, Loch Leven, Kinross-shire, which is a comparatively shallow loch, though covering a considerable area, and which was examined also in 1890, * closely resembled Loch Morar in the abundance of its crustacean fauna. There was of course some difference in the proportion and in the kinds of associated species, but this difference is of comparatively small importance when the organisms are considered from the utilitarian point of view of their value as a food supply for fishes, as almost all the micro-crustacea are of equal importance in this respect. Moreover, Loch Morar, like Loch Leven, is reputed to be a loch where the angler is almost certain of a good and successful day's fishing; but being so much out the way and so difficult of access, comparatively few anglers visit Loch Morar.

NOTE ON *Attheyella*, Brady (see page 225).

The following information was obtained after the preceding notes were in the press :—

G. O. Sars, in a paper entitled 'Oversigt af de indenlandske Fersk-'vands-copepoder,' published in Christiania by Brøgger & Christie in 1863, described among other things three species of *Canthocamptus*, new to science, viz., *Canthocamptus crassus*, *Canthocamptus pygmæus*, and *Canthocamptus brevipes*, but without illustrative figures.

In 1883 Professor Lilljeborg exhibited a collection of crustacea at the International Fisheries Exhibition held that year in London. This collection, which included the three species of *Canthocamptus* here referred to, was secured by the Rev. A. M. Norman, F.R.S., and added to his extensive museum.

Through the kindness of Dr Norman, my son, Mr Andrew Scott, was recently favoured with an opportunity of examining Lilljeborg's specimens of Sars' *Canthocamptus*, and of comparing them with the description of

* See Part III. of the Ninth Annual Report of the Fishery Board for Scotland.

the species contained in the work of that author ; and while doing so, he recognised a close resemblance between *Canthocamptus crassus*, G. O. Sars, and *Attheyella spinosa*, Brady, and between *Canthocamptus pygmæus*, G. O. Sars, and *Attheyella cryptorum*, Brady ; a further examination tended still more to confirm the identity of the two species of G. O. Sars with the corresponding species described by Dr Brady.

Should my son's identification prove to be correct, the specific names adopted by Professor Brady will necessarily give place to those of Professor G. O. Sars, and in that case the nomenclature of the two species will stand thus :— *Attheyella* (*Canthocamptus*) *crassa* (G. O. Sars) = *Attheyella spinosa*, Brady. *Attheyella* (*Canthocamptus*) *pygmæa* (G. O. Sars) = *Attheyella cryptorum*, Brady.

The following is Professor G. O. Sars' description of the species referred to :—

(1) *Canthocamptus crassus*, G. O. Sars (*loc. cit.*, p. 23).

' Corpus quam in speciebus ceteris robustus segmentis abdominalibus ' postice attenuatis in margine postico subtus et ad latera pilis vel aculeis ' sat longis pectenatim ornatis. Rami caudalis forma singulari fere ovati et ' ad basim valde constricti setis apicalibus brevibus er valde divergentibus ; ' setæ intermediæ duæ insolito modo flexuosæ in medio sparsim ciliatæ ' exteriore quam interiore triplo breviore. Antennæ 1 mi paris sat crassæ ' setis longis dense obsitæ. Ramus interior pedum 1 mi paris 3-articulatis ' exterior aliquanto longior in paribus sequentibus 3st brevissimus et ' biarticulatus. Pedum 5 ti paris articulus basilis introrsum parvum dila- tatus et ut articulus ultimus setis longissimus præditus. Color albidus, longit circit ¾ mm.'

(2) *Canthocamptus pygmæus*, G. O. Sars (*loc. cit.*, p. 21).

' Corpus postice parum attenuatum segmentis abdominalibus sat crassis ' versus marginem posteriorem ad latera et subtus serie transversa aculeorum ' instructis. Rami caudales brevissimis latioris quam longiores setis ' majoribus apicalibus duabus sat divergentibus in medio aculeatis, ' exteriore dimidiam longitudinem interioris æquante. Operculum anale ' dentatum, dentibus majusculis. Antennæ 1 mi paris brevissimæ pedun- ' culo (articulis basilibus 4-composito) crasso flagello vero valde attenuato. ' Ramus interior pedum 1 mi paris exteriore paullo brevior ut in paribus ' sequentibus duobus biarticulatus, in pari 4 to unarticulatus minimus ' articulo 1 mo rami exterioris vix longior. Pedum 5 ti paris biarticula- ' torum, articulus 1 mus introrsum in processum foliiformen magnum et ' setiferum articulatum ultimum minimum et rotundatum longe superantem ' exit. Oculus minimus a margine antico capitis remotus saccus oviferus ' elongato-ovatus. Animal sat pellucidum colore plerumque albido interdum ' leviter rubicundo longit vix ½ mm.'

The species provisionally described in this paper on the fauna of the Inland Waters of Scotland under the name of *Attheyella propinqua* may probably be identical with Sars' *Canthocamptus brevipes;* but as there are apparently one or two important differences between them, it seems better for the present to retain the specific name adopted in this paper.

EXPLANATION OF THE PLATES.

Plate VI.

Cyclops ewarti, Brady.

Fig.	1.	Adult female, dorsal view,			×	50
Fig.	2.	Anterior antenna,			×	168
Fig.	3.	Posterior antenna,			×	126
Fig.	4.	Mandible and palp,			×	168
Fig.	5.	Anterior footjaw,			×	126
Fig.	6.	Posterior footjaw,			×	126
Fig.	7.	Foot of first pair,			×	168
Fig.	8.	Foot of fourth pair,			×	168
Fig.	9.	Foot of fifth pair,			×	380
Fig.	10.	Abdomen and caudal stylets,			×	96

Attheyella spinosa, Brady.

Fig.	11.	Adult female, side view,	×	46
Fig.	12.	Anterior antenna, female,	×	336
Fig.	13.	Anterior antenna, male,	×	336
Fig.	14.	Foot of first pair,	×	253
Fig.	15.	Foot of third pair, male,	×	253
Fig.	16.	Foot of fourth pair,	×	253
Fig.	17.	Foot of fifth pair, female,	×	253
Fig.	18.	Foot of fifth pair, male,	×	336
Fig.	19.	Abdomen and caudal stylets, female,	×	126
Fig.	20.	Abdomen and caudal stylets, male,	×	126

Attheyella cryptorum, Brady.

Fig.	21.	Adult female, side view,	×	80
Fig.	22.	Anterior antenna, female,	×	506
Fig.	23.	Anterior antenna, male,	×	380
Fig.	24.	Foot of first pair,	×	253
Fig.	25.	Foot of second pair, male,	×	253
Fig.	26.	Foot of third pair, female,	×	253
Fig.	27.	Foot of third pair, male,	×	253
Fig.	28.	Foot of fourth pair,	×	253
Fig.	29.	Foot of fifth pair, female,	×	253
Fig.	30.	Foot of fifth pair, male,	×	380
Fig.	31.	Abdomen and caudal stylets,	×	100

Plate VII.

Attheyella propinqua, n. sp.

Fig.	1.	Adult female, side view,		×	80
Fig.	2.	Anterior antenna, female,		×	253
Fig.	3.	Anterior antenna, male,		×	253
Fig.	4.	Posterior antenna,		×	506
Fig.	5.	Mandible and palp,		×	760
Fig.	6.	Posterior footjaw,		×	760
Fig.	7.	Foot of first pair,		×	336
Fig.	8.	Foot of third pair, male,		×	336

Q

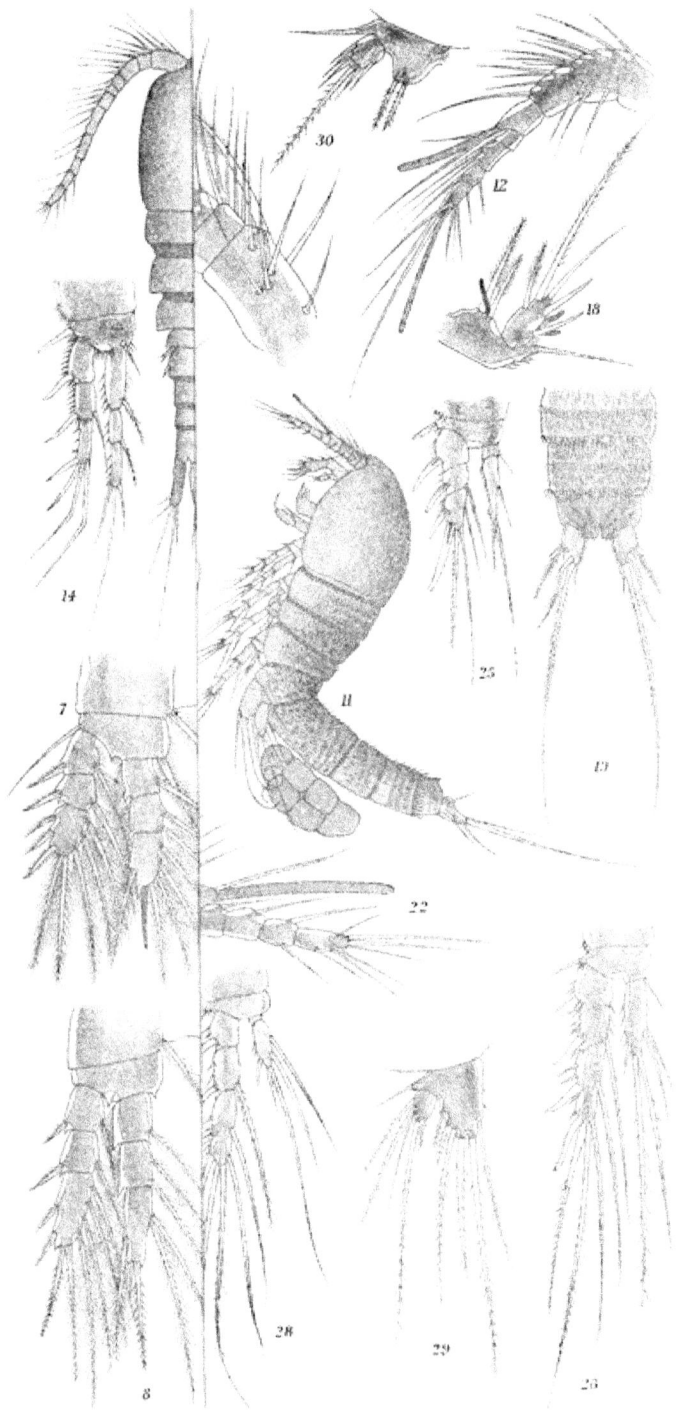

30

12

18

14

25

13

7

11

22

28

29

26

8

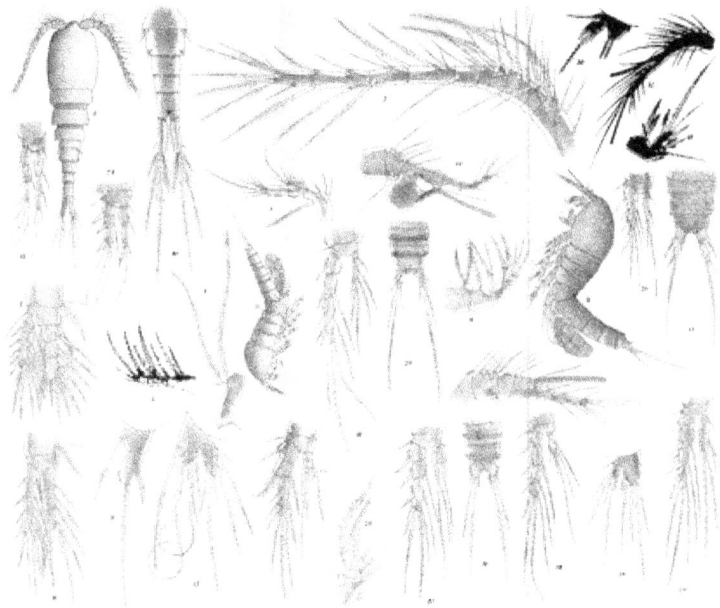

PLATE VII.

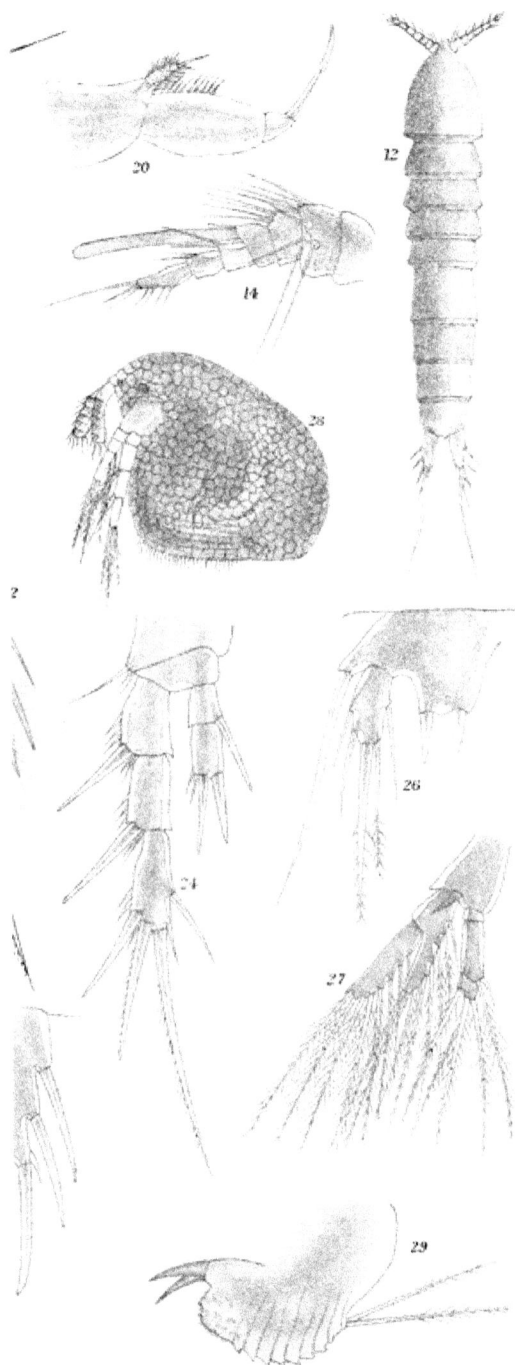

laticornis (Jarine). Fig. 30.—Leptorhynchus falcatus (G. O. Sars).

PLATE 14

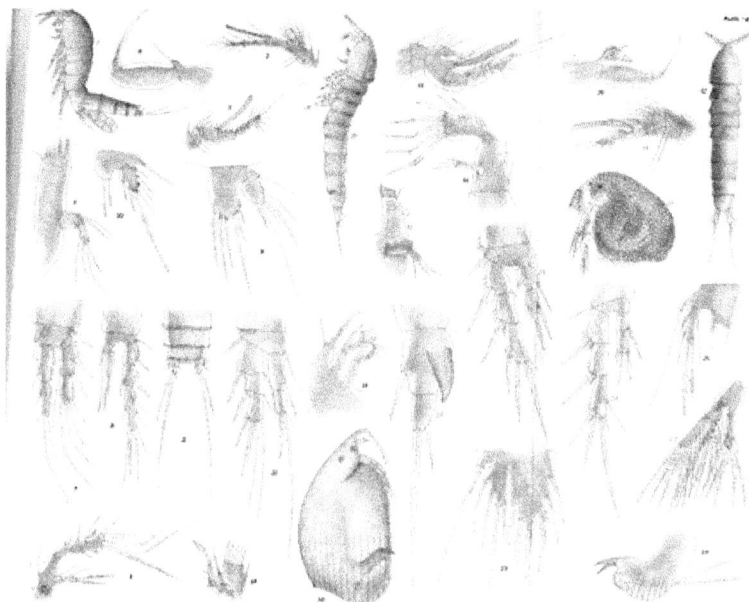

(II.)—ADDITIONS TO THE FAUNA OF THE FIRTH OF FORTH.

PART VI. By Thomas Scott, F.L.S. (Plates V.-X.)

In my paper—' Additions to the Fauna of the Firth of Forth,' Part V. published in the *Eleventh Annual Report of the Fishery Board for Scotland* (1893), it is stated that some of the species that are there described for the first time had been known for some years previously, but had been allowed to stand over for want of sufficient information concerning them ; the same statement may be made now, as not a few of the species recorded in the sequel have been in my possession since 1889, 1890, and 1891. There is nothing unusual in this, because an accurate knowledge of such small organisms, as many of the *Copepoda* are, and even of larger forms, can only be acquired by careful study extending over a lengthened period. Though much has now been done to throw light on the distribution of the Crustacean fauna of the Firth of Forth, my experience leads me to believe that much still remains to be done ere our knowledge of this group—a group that forms an important part of fish-food—attains to anything approaching completeness. In the present paper, the additions to the Forth fauna include 43 species of *Copepoda*, 1 species of *Ostracoda*, 10 species of *Amphipoda*, 1 species of *Annelida*, and 1 species of *Mollusca*,—in other words, 56 species of invertebrates, of which, so far as I know, no previous records of their occurrence in the Forth have been published, now fall to be added to those recorded in preceding papers. I again find it necessary to leave over several species of *Copepoda* for further study ; the *Amphipoda* also require further investigation ; and, thanks to the excellent work of Professor G. O. Sars of Norway, the study of this difficult group will, ere long, become comparatively easy. Very much, also, still remains to be done among the *Annelida* of the Forth. One interesting group—the *Nemertians* —has scarcely yet been touched ; but the sea is a boundless storehouse whose treasures will never be exhausted.

Seventeen species of the *Copepoda* now recorded are described and figured here for the first time ; preliminary descriptions of other 7 species new to science, and 1 new to Britain, have been published in the *Annals and Magazine of Natural History* for October 1893, and February 1894, and 1 species, new to Britain, is recorded here. A few of the others are new to the East Coast.* In my paper ' Additions to the Forth Fauna,' Part III. (*Ninth Annual Report of the Fishery Board for Scotland*, 1891), the following statement occurs :—' I venture to predict that when ' the Firth of Forth becomes more thoroughly and systematically worked ' up, the number of Crustacea will be little, if at all, short of 500 species.' In connection with this statement it may be of interest to mention that the number of species of Forth Crustacea, including those recorded in the present paper, now amounts to over 480.

My son, Mr Andrew Scott, has prepared all the drawings, and a large part of the necessary dissections required in the preparation of this paper. I may also say that, as in former years, the great interest taken in all our work, by Dr Fulton, has been a source of much encouragement. By his assistance, ever willingly given, I have been enabled to have the

---

* Several rare and new species have also recently been added to the British fauna from the Moray Firth District (see the parts of the *Annals and Magazine of Natural History* already referred to).

privilege of consulting literature that would otherwise have been difficult to get hold of. I have also to acknowledge my indebtedness to Mr Webster of the University Library for valuable assistance in this respect.

Before proceeding to describe the animals that have been added to the Forth fauna during the past year, it will perhaps be useful and interesting if in some measure, at least, I try to show what progress has now been made towards the attainment of a better knowledge of the Invertebrate fauna of the Firth of Forth through the investigations that have been carried on in the estuary under the direction of the Fishery Board for Scotland, and also the amount of success that has attended these investigations. For the present I can only indicate what has been accomplished in the study of the *Copepoda*, leaving the other groups to be dealt with later on. In this attempt to summarise the results of the work that has been done during the last few years among this important group of the Crustacea, I propose, in the first place, to give a list of all those species obtained in the Firth of Forth that have already been described and figured in the 'Monograph of the British Copepoda,' published in 1880 ; this will form a basis for, and an incitement to, further work by showing what blanks have yet to be filled up ere this part of the list can be considered complete. Second, I propose to give a list of those species obtained in the Firth of Forth which are not included in that Monograph, but have been added to the British fauna by other investigators since the publication of that work. A third list will include those species that have been added to the British fauna as the direct result of the Scottish Fishery Board's Investigations. And a fourth list will contain those species, apparently new to science, discovered during the progress of the investigations referred to.

The species are arranged in alphabetical order, and those described in this report are included with the others in the various lists.

List No. I.

The Copepods contained in this list include only those species that are described in the 'Monograph of British Copepoda,' by Dr Brady, published in 1878–80. (Fresh-water species are not included in the list.)

Acartia longiremis (Lillj.).
Acontiophorus scutatus (B. and R.).
Alteutha depressa, Baird.
　　,,　　*interrupta* (Goodsir).
Ameira longipes, Boeck.
Amymone sphaerica, Claus.
Anomalocera patersonii, Templeton.
Artotrogus boeckii, G. S. Brady.
　　,,　　*magniceps*, G. S. Brady.
Ascidicola rosea, Thorell.
Calanus finmarchicus (Gunner).
Candace pectinata, G. S. Brady.
Canthocamptus palustris, Brady.
Centropages hamatus (Lillj.).
　　,,　　*typicus*, Kroyer.
Cletodes limicola, Brady.
　　,,　　var. *gracilis*, Brady.
　　,,　　*longicaudata*, B. and R.
　　,,　　*propinqua*, B. and R.
Cyclopicera gracilicauda, G. S. B.
　　,,　　*nigripes*, B. and R.
　　,,　　*lata*, G. S. B.
Cyclopina gracilis, Claus.
　　,,　　*littoralis* (Brady).
Cyclops aequoreus, Fischer.
Cylindropsyllus laevis, Brady.

Dactylopus brevicornis, Claus.
　　,,　　*flavus*, Claus.
　　,,　　*minutus*, Claus.
　　,,　　*stromii* (Baird).
　　,,　　*tisboides*, Claus.
　　,,　　*tenuiremus*, B. and R.
Delavalia palustris, Brady.
　　,,　　*reflexa*, B. and R.
Doropygres normani, Brady.
　　,,　　(?) *porcicauda*, Brady.
Ectinosoma atlanticum (B. and R.).
　　,,　　*erythrops*, Brady.
　　,,　　*melaniceps*, Boeck.
　　,,　　*spinipes*, Brady.
Enhydrosoma curvatum (B. and R.).
Enterocola cruca, Norman.
Eurytemora clausii (Hoek).
Harpacticus chelifer (Müller).
　　,,　　*flexus*, B. and R.
　　,,　　*fulvus*, Fischer.
Idya furcata (Baird).
Jonesiella spinulosa.
Laophonte curticauda, Boeck.
　　,,　　*hispida* (B. and R.).
　　,,　　*horrida* (Norman).
　　,　　*lamellifera* (Claus).

Laophonte longicaudata, Boeck.
,, *serrata* (Claus).
,, *similis* (Claus).
,, *thoracica*, Boeck.
Lichomolgus arenicolus, Brady.
,, *fucicolus*, Brady.
,, *furcillatus*, Brady.
Longipedia coronata, Claus.
Mesochra lilljeborgia, Boeck.
Metridia armata, Boeck.
Misophria pallida, Boeck.
Nannopus palustris, Brady.
Normanella dubia (B. and R.).
Notodelphys allmanni, Thorell.
,, *cerulæa*, Thorell.
Oithona spinifrons, Boeck.
Parapontella brevicornis (Lubbock).
Platychelipus littoralis, G. S. Brady.
Porcellidium fimbriatum, Claus.
Pseudanthessius liber (B. and R.)
,, *thorellii* (B. and R.)
Pseudocalanus elongatus (Boeck).
Pseudocyclops crassicornis, Brady.
,, *obtusatus*, B. and R.
Pterinopsyllus insignis, Brady.
Robertsonia tenuis (B. and R.).
Scutellidium fasciatum (Boeck).
Stenhelia hispida, Brady.
,, *ima*, Brady.
Tachidius brevicornis (Müller).
Temora longicornis (Müller).
Tetragoniceps malleolata, G. S. Brady.
Thalestris clausii, Norman.
,, *harpactoides*, Claus.
,, *helgolandica*, Claus.
,, *longimana*, Claus.
,, *rufocincta*, Norman.
,, *ruforiolascens*, Claus.
,, *serrulata*, G. S. Brady.
Thorellia brunnea, Boeck.
Westwoodia nobilis, Baird.
Zaus goodsiri, G. S. Brady.
,, *spinatus*, Goodsir.
Zosime typica, Boeck.

List No. II.

This list includes those Copepods discovered in the Firth of Forth that have been added to the British fauna either as new species or new additions by investigators other than those in the service of the Fishery Boards for Scotland, since the publication of 'The Monograph of British 'Copepoda.'

Acartia bifilosus, Giesbrecht.
,, *clausi*, Giesbrecht.
,, *discaudata*, Giesbrecht.
Eurytemora affinis, Poppe.
Jonesiella hyænæ, I. C. Thompson.
Monstrilla rigida (I. C. Thompson).
Paracalanus parvus (Claus).
Stenhelia denticulata, I. C. Thompson.
,, *hirsuta*, I. C. Thompson.
Tachidius littoralis, Poppe.

List No. III.

This is a list of several described species of Copepods from the Firth of Forth that have been added to the British fauna as one of the immediate results of the investigations carried out under the direction of the Fishery Board for Scotland.

Bomolochus soleæ, Claus.
Herrmannella rostrata, Canu.
Lichomolgus agilis, Leydig.
Modiolicola insignis, Aurivillius.
Monstrilla helgolandica, Claus.
Oithona setiger, Dana.
Pseudanthessius gracilis, Claus.
,, *sauvagei*, Canu.

List No. IV.

This list contains all those species (including those in the present Report), apparently new to science, that have been discovered in the Firth of Forth during the progress of the investigations referred to.

Acontiophorus elongatus, T. and A. Scott.
**Ameira exigua*, T. Scott.
,, *exilis*, T. and A. Scott.
,, *longicaudata*, T. Scott.
* ,, *longiremis*, T. Scott.
* ,, ,, var. *intermedia*.
* ,, *reflexa*, T. Scott.
Artotrogus papillatus, T. Scott.
Canuella perplexa, T. and A. Scott.
**Cletodes curvirostris*, T. Scott.
,, *irrasa*, T. and A. Scott.
,, *lata*, T. Scott.
,, *tenuiremis*, T. Scott.
**Cyclopina elegans*, T. Scott.
Cyclopicera purpurocincta, T. Scott.
Cyclops ewarti, Brady.
Cylindropsyllus minor, T. Scott.
**Dactylopus coronatus*, T. Scott.
,, *rostratus*, T. Scott.

* New species described for the first time in this Report.

Delavalia æmula, T. Scott.
Dermatomyzon gibberum, T. and A. Scott.
Hersiliodes littoralis (T. Scott).
** *Heteropsyllus curticaudatus*, T. Scott.
* *Laophonte denticornis*, T. Scott.
*   ,,   *depressa*, T. Scott.
    ,,   *inopinata*, T. Scott.
    ,,   *littorale*, T. and A. Scott.
*   ,,   *simulans*, T. Scott.
** *Laophontodes typicus*, T. Scott.
** *Leptopsyllus typicus*, T. Scott.
Lichomolgus hirsutipes, T. Scott.
Longipedia minor, T. and A. Scott.
Neobradya pectinifer, T. Scott.
Paramesochra dubia, T. Scott.

Parartotrogus richardi, T. and A. Scott.
** *Pontopolites typicus*, T. Scott.
* *Pseudocyclopia caudata*, T. Scott.
    ,,   *crassicornis*, T. Scott.
    ,,   *minor*, T. Scott.
** *Pseudowestwoodia andrewi*, T. Scott.
Stenhelia dispar, T. and A. Scott.
Stephos minor, T. Scott.
Tetragoniceps bradyi, T. Scott.
*   ,,   *consimilis*, T. Scott.
    ,,   *incertus*, T. Scott.
    ,,   *macronyx*, T. Scott.
Thalestris forficuloides, T. and A. Scott.

In the 'Monograph of British Copepoda' 131 species are described and figured exclusive of the fresh-water forms. In list No. I., given above, 95 of these are recorded for the Firth of Forth, or nearly 72 per cent. of the total number of species, exclusive of the fresh-water forms recorded for the British Islands in 1880. It will be observed from these statements that 36 of the marine species described in the 'Monograph of British Copepoda' have not, so far, been recorded for the Forth area. On the other hand, a considerable number of species have been discovered in the Firth of Forth, in addition to those described in the Monograph referred to, and the majority of these have been recorded for the first time for the British Islands, either as species that have been described by European and other writers, or as species new to science, as shown by the following summary of the four preceding Lists :—

In List No. I. 95 species are recorded for the Forth,   .   .   95
In List No. II. 10 species are recorded,   .   .   .   10
In List No. III. 8 species (new to Britain) are recorded,   .   8
In List No. IV. 46 species (new to Science) are recorded,   .   46
                                                 ——

Total number of species of Copepoda recorded for the Forth,
     including those described in the present Report,   .   .   159

†Total number for the British Islands in 1880,   .   .   .   131

Excess of Forth species over those for the British Islands
     in 1880, ...   28

A few of the species included in this enumeration of the Copepoda of the Firth of Forth are probably somewhat doubtful, and may ultimately be set aside, as, for example, *Acartia bifilosus*, Giesbrecht, and *Oithona setiger*, Dana ; but making all due allowance for such doubtful species, it will, I think, be acknowleged that, even in the department of Marine Natural History, a fairly successful endeavour has been made to take full advantage of the great and almost unique opportunities enjoyed for the prosecution of such studies.

I will now proceed to describe the various species of Invertebrates that have been discovered in the Firth of Forth since the publication of the Eleventh Annual Report of the Fishery Board for Scotland ; and the Copepoda will be taken first in order.

* New species described for the first time in this Report.
** New genus and species described for the first time in this Report.
† This number is taken from the Index at the end of vol. iii. of the 'Monograph of British Copepoda.'

DESCRIPTION OF SPECIES.

CRUSTACEA.

1. COPEPODA.

Family CALANIDÆ.

Genus *Paracalanus*, Boeck (1864).

Paracalanus parvus (Claus).

> 1863. *Calanus parvus*, Claus (13),* p. 173, pl. xxvi. figs. 10–14 ; pl. xxvii. figs. 1–4.
> 1864. *Paracalanus parvus*, Boeck (5), p. 232.
> 1892. *Paracalanus parvus*, Canu (11), p. 169.
> 1893. *Paracalanus parvus*, I. C. Thompson (33), p. 7, pl. xv. fig. 5.
> 1894. *Paracalanus parvus*, T. Scott (30), p. 26, pl. i. figs. 1–19.

Habitat.—East of May Island, in a surface tow-net gathering. Frequent. Off St Monans. Not common.

Paracalanus parvus is easily distinguished from any other species of the British Calanidæ by the structure of the fifth pair of thoracic feet in both sexes. The only other British record for this species seems to be that of I. C. Thompson in the 'Copepoda of Liverpool Bay.' The distribution of the species appears to be world-wide ; it has been recorded from the North Sea, from the Atlantic (North and South), from the Mediterranean, and from the China Sea.

Genus *Acartia*, Dana, 1846.

Acartia clausi, Giesbrecht.

> 1889. *Acartia clausi*, Giesbrecht (17), p. 332.
> 1890. *Acartia clausi*, Canu, (10a), p. 326, pl. xxiv.
> 1893. *Acartia clausi*, I. C. Thompson (33), p. 8, pl. xv. fig. 6.
> 1894. *Acartia clausi*, T. Scott (30), p. 67, pl. vii. figs. 33–40.

Habitat.—Several parts of the Forth area. One of the chief characters that distinguish this species is the form of the fifth pair of thoracic feet. The fifth pair in the female are each armed with a stout and comparatively short apical spine in addition to the plumose seta. The following forms of Acartia are also obtained in the Firth of Forth, viz. :—*Acartia longiremis* (Lillj.), generally distributed, especially in the estuary proper. *Acartia discaudata*, Giesbrecht, off Musselburgh. *Acartia bifilosus*, Giesbrecht, West of Queensferry.

Family PSEUDOCYCLOPIIDÆ, nov. family.

Body comparatively robust. Anterior antennæ short, sixteen to seventeen-jointed. Alike on both sides in the male, and similar to those of the female. Mouth organs and swimming-feet like those of the *Calanidæ*. Fifth pair in the female simple, one-branched, and two-jointed ;

---

* The numbers in parentheses correspond with the numbers in the Bibliographical list at the end of the paper.

in the male dissimilar, and forming powerful grasping organs. The mandible-palp is comparatively small. One of the most prominent characters of this family is the structure of the anterior antennæ. The most careful observation failed to show any difference between the male right antenna and the left, and both were similar to those of the female. In my paper on the Forth Fauna, published in Part III. of the *Eleventh Annual Report of the Fishery Board for Scotland*, two species of Copepoda were described, for which it was considered necessary to institute a new genus (*Pseudocyclopia*). This new genus was placed in the family *Misophriidæ*, Brady, because of its closer affinity with that family than with the family *Calanidæ*. A further study of the characters of the genus has, however, led me to the conclusion that its position in the family *Misophriidæ* is untenable ; and as there is no other family in which it can satisfactorily be included, I propose to constitute the family *Pseudocyclopiidæ* for its reception ; and I do so with the greater confidence, as another species, now to be described, has recently been discovered, possessing all the more prominent characters that distinguish the two already described. Dr W. Giesbrecht, in his memoir ' Mitt- ' heilungen über Copepoden ' (Abdruck aus den Mittheilungen aus der Zoologischen Station zu Neapel, ii. Band 1.2. Heft), also refers to some of the peculiar characters of the two described species, and to the difference between them and those that constitute the family *Misophriidæ*.

Genus *Pseudocyclopia*, T. Scott (1892).

Pseudocyclopia caudata, sp. n.   (Pl. V. figs. 1–8.)

Female.   Like *Pseudocyclopia crassicornis*, T. Scott, in general appearance and dimension. Length, ·65 mm. ($\frac{1}{58}$th of an inch). Anterior antennæ seventeen-jointed. Basal joint very large, the others small ; the second to the fifth gradually decrease in size, the sixth is rather longer than that which preceeds or follows, while the ante-penultimate joint is more elongate than any of the others except the basal joint. The formula shows approximately the proportional lengths of all the joints :—

$$\frac{29 \cdot 8 \cdot 7 \cdot 5 \cdot 4 \cdot 7 \cdot 5 \cdot 8 \cdot 6 \cdot 9 \cdot 7 \cdot 7 \cdot 8 \cdot 10 \cdot 16 \cdot 9 \cdot 12 \cdot}{1 \cdot 2 \cdot 3 \cdot 4 \cdot 5 \cdot 6 \cdot 7 \cdot 8 \cdot 9 \cdot 10 \cdot 11 \cdot 12 \cdot 13 \cdot 14 \cdot 15 \cdot 16 \cdot 17 \cdot}$$

The posterior antennæ, mouth organs, and first and second swimming-feet are somewhat similar to those of *Pseudocyclopia crassicornis*. The basal spines of the third pair of swimming-feet, which reach to near the extremity of the outer branches, are very stout, and with the distal end boldly curved—the basal spines, very like those of *Pseudocyclopia minor*, T. Scott. The fifth pair are somewhat more robust than those of the females of the two described species, and are each armed with three stout sub-equal and setose terminal spines. The caudal stylets are nearly equal to the combined length of the last two abdominal segments, as shown in the accompanying figures.

Habitat.—Off St Monans. Scarce.

Remarks.—*Pseudocyclopia caudata* is similar in its general appearance to the two species already described ; but is readily distinguished from both, even without dissection, by the comparatively elongate caudal stylets. The difference in the proportional lengths of the joints of the anterior antennæ, and in the structure of the third and fifth pairs of thoracic feet between this and the other two species, is also of sufficient importance for diagnostic purposes.

Family MISOPHRIDÆ, Brady (1878).

Genus *Misophria*, Boeck (1864).

Misophria pallida, Boeck.
  1864. *Misophria pallida*, Boeck (5), p. 24.
  1878 *Misophria pallida*, Brady (8), vol. i. p. 79, pl. xiii. figs.
   11–16 ; pl. xviii. figs. 11–12.

Habitat.—Off St Monans, and also west of Queensferry.
Several specimens of this apparently rare species have been obtained
at both the parts of the Forth area mentioned.

Genus *Pseudocyclops*, Brady (1872).

Pseudocyclops crassiremis, Brady.
  1872. *Pseudocyclops crassiremis*, Brady (7a), vol. iv. p. 431, pl.
   xvii. figs. 1–8.
  1878. *Pseudocyclops crassiremis*, Brady (8), vol. i. p. 82, pl. vii.
   figs. 1, 2 ; pl. xii. fig. 14.

Habitat.—Off St Monans. Rather scarce.
This seems to be a perfectly distinct species, so far as could be made
out.

Family CYCLOPIDÆ.

Genus *Cyclopina*, Claus (1863).

Cyclopina gracilis, Claus.
  1863. *Cyclopina gracilis*, Claus (14), p. 104, pl. x. figs. 9–15.
  1878. *Cyclopina gracilis*, Brady (8), vol. i. p. 93, pl. xxiv.B. figs.
   1–9 ; pl. xci. figs. 10, 11.
  1892. *Cyclopina gracilis*, Canu (11), p. 181.

Habitat.—Off St Monans, and various other parts of the Forth west
to near Charleston.
This pretty and well characterised species is not uncommon within the
Forth area.

Cyclopina elegans, sp. n. (Pl. V. figs. 9–19.)
 Description.—Female. Length, ·83 mm. ($\frac{1}{30}$th of an inch). Body
elongate, slender. Forehead narrowly and evenly rounded. Anterior
antennæ rather shorter than the first cephalo-thoracic segment, and consisting of nineteen joints. The fifth to the ninth joints are very short, while
the eleventh is about equal in length to the basal joint, and considerably
longer than any of the others that precede or follow, as is shown by the
formula :—

$$\frac{12\cdot6\cdot5\cdot5\cdot3\cdot2\cdot2\cdot3\cdot11\cdot4\cdot4\cdot4\cdot4\cdot4\cdot5\cdot4\cdot6\cdot7\cdot}{1\cdot2\cdot3\cdot4\cdot5\cdot6\cdot7\cdot8\cdot9\cdot10\cdot11\cdot12\cdot13\cdot14\cdot15\cdot16\cdot17\cdot18\cdot19\cdot}$$

The fourth joint appears to be composed of two coalesced joints, as a
faint line could be observed extending from the upper margin to fully
halfway across the joint (fig. 10). Posterior antennæ four-jointed,
secondary branch obsolete or entirely absent (fig. 12). Mandibles well
developed, and furnished with a large palp. The primary branch of the
palp is elongate and three-jointed ; while the secondary branch, which
springs from the basal part of the primary branch, is short and four-
jointed (fig. 13). The basal part of the maxilla-palp, which is delated
 It

outwards, bears two distinct one-jointed branches on the exterior margin (fig. 14). Anterior foot-jaws four-jointed, with the basal joint very large comparatively, and the end joints small ; the second joint is produced interiorly so as to form a base for a stout curved spine (fig. 15). Posterior foot-jaws elongate, seven-jointed, the last five joints small, their entire length being very little greater than the length of the second joint (fig. 16). Swimming-feet robust, both branches three-jointed. The marginal spines of the outer branches of the first pair are large and dagger-shaped, and project at, or nearly at, a right angle to the outer margin. Those of the fourth pair are much smaller, but the marginal setæ on both the outer and inner branches, but especially on the inner branches, are stouter than the marginal setæ of the first pair. In the first pair, also, a stout spine springs from the inner distal margin of the second basal joint, and projects downwards beyond the end of the second joint of the inner branches (fig. 17). Fifth pair stout, each consisting of a single three-jointed branch. The breadth of the first two joints is greater than the length ; while the third joint is longer than broad, being equal to the entire length of the first two. The first joint bears a moderately long plumose seta on the outer distal angle. The second joint bears a similar seta on the inner distal angle, while the last joint bears two plumose setæ on the inner margin, and one plumose seta and a dagger-shaped spine at the apex (fig. 19). Caudal stylets elongate, being equal to the combined length of the last two segments of the abdomen ; and, in addition to the terminal setæ, each of the stylets bears one small hair near the proximal end, and another near the distal end of the outer margin. In the male the anterior antennæ are sixteen-jointed, the last six being more or less modified for grasping ; the eleventh joint is comparatively robust and hinged to the preceding one. There is also a hinged articulation between the penultimate and ante-penultimate joints (fig. 11).

Habitat.—Off St Monans. Rather rare.

Remarks.—The species now described differs in some important points from any other *Cyclopina* recorded from the British seas. It resembles *Cyclopina littoralis* in the form of the fifth pair of thoracic feet, but the proportional lengths of the joints of the anterior and posterior antennæ, and the structure of the first pair of swimming-feet, are decidedly different. It also, in this respect, differs from *Cyclopina ovalis*, Brady ; while the shorter and much fewer jointed anterior antennæ of *Cyclopina gracilis*, Claus, readily distinguish that species from the one now described.

Genus *Pterinopsyllus*, Brady (1880).

Pterinopsyllus insignis, Brady.
        1878. *Lophophorus insignis*, Brady (8), vol. i. p. 122, pl. xiii. figs. 1–10 ; pl. xv. fig. 10.
        1880. *Pterinopsyllus insignis*, Brady (8), vol. iii. p. 23. (*Lophophorus* being preoccupied, Dr Brady substituted the word *Pterinopsyllus* for this genus).
        1893. *Pterinopsyllus insignis*, T. and A. Scott (32), p. 243.

Habitat.—West of Queensferry. Frequent. This beautiful species has several times been obtained in the part of the Forth estuary referred to. In this part of the Forth there are periodic and very marked variations in the salinity of the water ; caused on the one hand by the ebb and flow of the tide, and on the other by the large volumn of fresh water from the river, which, forced back by the flowing tide, accumulates in the upper reaches, and, when the reflux takes place, rushes seawards with over-powering force, so that, during the later part of the ebb, and the early part of the flood tide, there is a large admixture of fresh water in this portion of estuary of the Forth. It may be that the peculiar alternating conditions thus produced are favourable to the welfare of this and other organisms; but whether that be so or not, it is certain that *Pterinopsyllus* is comparatively frequent here, while it is apparently exceedingly rare where the more normal marine conditions prevail. *

Family HARPACTICIDÆ, Claus (in Part).

Genus STENHELIA, Boeck (1864).

Stenhelia hispida, Brady.
>> 1880. *Stenhelia hispida*, Brady (8), vol ii. p. 32, pl. xlii. figs. 1–14.
>> 1893. *Stenhelia hispida*, I. C. Thompson (33), p. 19.

Habitat.—Of St Monans. Rather scarce.

This, which appears to be a comparatively rare species in the British seas, is apparently widely distributed. It has been obtained at Tobermory (Rev. A. M. Norman); off Hartlepool and Marsden, Durham, and at Portincross, Ayrshire (Dr G. S. Brady); in Ventry Bay, Ireland (Mr E. C. Davidson); in rock-pools at Hilbre and Puffin Islands, at Garth Ferry, and in Port Erin Bay, Isle of Man (I. C. Thompson).

Stenhelia hirsuta, I. C. Thompson.
>> 1893. *Stenhelia hirsuta*, I. C. Thompson (33), p. 20, pl. xxxi.
>> 1894. *Stenhelia hirsuta*, T. and A. Scott (31), p. 146.

Habitat.—Off St Monans and at the north end of Inchkeith Island.

Mr I. C. Thompson, of Liverpool, obtained this interesting species amongst mud dredged from 29 fathoms in the Irish Sea, and about twelve miles west from Port Erin, Isle of Man. The females of this species carry two ovisacs as shown by Thompson's figure; and several of the specimens obtained in the Firth of Forth were also provided with two ovisacs.

Stenhelia dispar, T. and A. Scott.
>> 1894. *Stenhelia dispar*, T. and A. Scott (31), p. 141, pl. viii. figs. 8–12.

Description.—Female. Length, ·55 mm. ($\frac{1}{45}$th of an inch). Rostrum of moderate length. Anterior antennæ short, moderately stout, and eight-jointed. the fifth, sixth, and seventh joints are small, but the last is about equal to the combined length of the two preceding joints, as shown by the annexed formula :—

$$\frac{27 \cdot 19 \cdot 13 \cdot 14 \cdot 6 \cdot 9 \cdot 8 \cdot 18}{1 \cdot 2 \cdot 3 \cdot 4 \cdot 5 \cdot 6 \cdot 7 \cdot 8}$$

The posterior antennæ are somewhat similar to those of *Stenhelia ima*, Brady, except that the last joint of the secondary branch bears one marginal and one apical seta instead of two apical setæ. Mouth organs similar to those of *Stenhelia ima*. First pair of thoracic feet slender, and also somewhat like those of that species. A spiniform plumose seta springs from the lower margin of the second basal joint, and close to the proximal end, interiorly, of the inner branches; there are no setæ on the inner margin of the outer branches. The fourth pair are also somewhat similar to those of *Stenhelia ima*. The basal joint of the fifth pair is large and sub-triangular, and furnished with three setæ on the inner margin and

---

* My son obtained *Pterinopsyllus* in material from the Moray Firth dredged from a depth of 40 fathoms.

two at the apex. The secondary branch is comparatively small and broadly
sub-ovate, with a somewhat bifid apex, and bears one seta on the inner
margin, one on the outer margin, and one on each lobe of the bifid apex.
Caudal stylets very short.

Habitat.—Vicinity of the Bass Rock. Rather scarce.

Remarks.—This is a small species, and resembles *Stenhelia ima* in
several of its characters ; but the short anterior antennæ, and the large
basal joint, and comparatively small and broadly ovate secondary joint
of the fifth pair of thoracic feet, readily distinguish it from that species.

Stenhelia denticulata, I. C. Thompson.

  1893. *Stenhelia denticulata,* I. C. Thompson (33), p. 20, pl. xxx.
   figs. 1–11.
  1894. *Stenhelia denticulata,* T. and A. Scott (31), p. 146.

Habitat.—Off St Monans, and in the vicinity of Inchkeith Island.

Mr I. C. Thompson obtained the two specimens, from which he described the species inside the breakwater, at Port Erin, Isle of Man.
Several specimens have been captured in the Firth of Forth. This is a
comparatively large and well-marked species. The anterior antennæ are
moderately long and slender, and eight-jointed. The second joint from
the base is armed with a strong forward-projecting tooth.

Genus *Ameira*, Boeck (1864).

Ameira reflexa, sp. n. (Pl. V. figs. 26–28.)

Description.—Female. Body moderately robust. Length, ·68 mm.
($\frac{1}{40}$th of an inch). Anterior antennæ short, stout, and eight-jointed. The
first two joints are large, the others are small, the penultimate being shorter
than any of the others. The upper distal portion of the fourth joint is
produced forwards so as to form the base of a long filament. The proportional lengths of all the joints are nearly as in the formula :—

16 ˙ 14 ˙ 6 ˙ 5 ˙ 4 ˙ 5 ˙ 3 ˙ 6 ˙
1 ˙ 2 ˙ 3 ˙ 4 ˙ 5 ˙ 6 ˙ 7 ˙ 8 ˙

Posterior antennæ short, two-jointed. A small secondary branch springs
from near the middle of the first joint, and is furnished with three
terminal setæ (fig. 22). Mandibles small. Mandible-palp one branched.
The basal joint is small, and bears a single short terminal seta. The
small one-jointed branch springs from the upper part of the external margin
of the basal joint, and bears one marginal and three apical setæ (fig. 23).
Posterior foot-jaws short, moderately stout, consisting of two sub-
equal joints, the lower rounded margin of the second joint ciliated.
Terminal claw nearly straight, and rather longer than the joint from
which it springs (fig. 24). Both branches of the first four pairs of
swimming-feet three-jointed ; the outer branches of the first pair reach
slightly beyond the end of the second joint of the inner branches. The
first joint of the inner branches is stout, and somewhat longer than the
second joint, and bears a moderately long plumose seta near the middle
of the inner margin. The second joint—which is rather smaller than the
first—is furnished with a similar seta near the distal end ; while the
last joint, which is slender, and fully one and a half times the length of
the second, is provided with one moderately short and two long setæ at
the apex, and three small spiniform setæ on the inner margin. The first
and second joints in both branches are strongly ciliated on the outer
margin. The first two joints of the outer branches are each armed with
a moderately long marginal spine, and the last joint with two shorter
marginal spines, and three plain terminal setæ (fig. 25). The outer

branches of the next three pairs are somewhat longer than the inner branches, and both branches are elongate, and moderately stout. In the fourth pair the first two joints of the outer branches are sub-equal, but the last is nearly equal to twice the length of the preceding one ; the joints of the inner branches are each somewhat longer than the one preceding, so that the last is fully twice the length of the first. Both branches have the outer margin more or less ciliated, the marginal and terminal setæ are nearly all elongate and strongly plumose, and the marginal spines of the outer branches are slender (fig. 26). The basal joint of the fifth pair is broadly foliaceous, the apical portion of which is sub-triangular and furnished with six setæ round the inner margin and end, the third one from the outside being very long. The secondary branch is comparatively small, oblong-ovate in form, and extends somewhat beyond the end of the basal joint. There are several setæ round the outer margin and apex (fig. 27). Caudal stylets short, being rather more than half the length of the last abdominal segment. The principal terminal seta of each stylet is nearly twice the length of the second (fig. 28).

Habitat.—Off Musselburgh, Firth of Forth. Rather scarce.

Remarks.—In the structure of the anterior antennæ, and of the first and fifth pairs of thoracic feet, this species differs from any others known to me. Neither the anterior antennæ nor the first swimming-feet are those of the typical *Ameira*. The end joint of the inner branches of the first pair is, in *Ameira*, usually shorter than the basal joint, but in the species now described the reverse is the case ; but otherwise there is nothing to distinguish it generically from *Ameira*.

Ameira longiremis, sp. n.   (Pl. V. figs. 29–32 ; Pl. VI. figs. 1–5.)

Description. — Female. Body elongate, robust. Length, ·74 mm. ($\frac{1}{34}$th of an inch). Anterior antennæ short, stout, eight-jointed. The first to the fifth joints gradually decrease in length, the fifth and sixth are about equal, while the sixth to the eighth gradually increase in length ; the fourth bears a long and moderately stout sensory filament. The annexed formula shows the proportional lengths of all the joints :—

$$\frac{16 \cdot 14 \cdot 11 \cdot 7 \cdot 6 \cdot 6 \cdot 8 \cdot 10}{1 \cdot 2 \cdot 3 \cdot 4 \cdot 5 \cdot 6 \cdot 7 \cdot 8}$$

Posterior antennæ comparatively large, two-jointed. Secondary branch small, one-jointed, and attached to the middle of the basal joint of the primary branch. The mandible-palp consists of a comparatively large basal joint, with a much smaller secondary one at its apex. The secondary joint bears one marginal and a few apical setæ. Posterior foot-jaws stout, the second joint somewhat gibbous below, and armed with a comparatively long terminal claw (fig. 30, pl. V.). The first pair of swimming-feet somewhat similar to those of *Ameira reflexa*, but much stouter and more elongate. In the fourth pair the inner branches are proportionally much shorter than in that species, as they scarcely reach to the end of the second joint of the outer branches. The first joint is very short, while the second and third are each rather longer than the preceding one. The armature of the fourth pair is different from that of the fourth pair in *Ameira reflexa* (fig. 5, pl. VI.). In the fifth pair the basal joint is broadly triangular, and bears two stout spiniform setæ on the distal half of the inner margin, and two apical setæ ; one of them small, the other very long, plumose, and spiniform. The secondary branch is elongate, narrow, cylindrical, being nearly five times longer than broad. Both margins are more or less ciliate. The apex is obliquely truncate, and bears several setæ (fig. 31, pl. V.). Caudal stylets very short.

Habitat.—Off St Monance. Scarce.

Remarks.—This is a robust species, with comparatively elongate swimming-feet. It is readily distinguished by the peculiar form of the fifth pair of thoracic feet.

Ameira longiremis, var. *intermedia*, nov. var. (Pl. VI. figs. 6–14.)

This form differs from the species just described, and of which it appears to be a variety, and the more important of these differences are as follows :—The anterior antennæ are shorter and less robust, and the proportional lengths of the joints are not the same, as shown by the formula :—

$$\frac{10 \cdot 14 \cdot 8 \cdot 6 \cdot 5 \cdot 6 \cdot 6 \cdot 6}{1 \cdot 2 \cdot 3 \cdot 4 \cdot 5 \cdot 6 \cdot 7 \cdot 8} \cdot$$

The last joint of the inner branches of the first thoracic feet is considerably shorter, and the spine that springs from the inner distal angle of the second basal joint is much longer than in the first pair of *Ameira longiremis*. The basal joint of the fifth pair is narrower at the proximal end, and more produced ; and the secondary branch is rather broader, and less cylindrical in form. These various differences, though scarcely marked enough to be of specific value, are yet sufficiently important to constitute a varietal difference.

Ameira exilis, T. and A. Scott. (Pl. IX. fig. 30 ; Pl. X. figs. 1–12.)

1894. *Ameira exilis*, T. and A. Scott, p. 139, pl. viii. figs. 18–20 ; pl. ix. figs. 1–3.

Description.—Female. Length, 1·4 mm. ($\frac{1}{18}$th of an inch). Body elongate, slender. Anterior antennæ slender, nine-jointed. The seventh and eighth joints very small, the others of moderate length, as shown by the formula :—

$$\frac{13 \cdot 18 \cdot 13 \cdot 10 \cdot 8 \cdot 9 \cdot 3 \cdot 3 \cdot 12}{1 \cdot 2 \cdot 3 \cdot 4 \cdot 5 \cdot 6 \cdot 7 \cdot 8 \cdot 9} \cdot$$

Posterior antennæ short, two-jointed. Secondary branch small, one-jointed, and furnished with three terminal plumose setæ. Mandibles elongate, narrow. Apex obliquely truncate, armed with a large tooth in the middle and several small ones on each side. The basal joint of the mandible-palp is considerably dilated, with the proximal end forming a narrow stalk-like attachment to the base of the mandible ; while the distal margin bears three short setæ. The secondary joint, or branch, of the palp is narrow, and about three times longer than broad, with a small marginal and four apical setæ. Posterior foot-jaws robust, and armed with a strong terminal claw (fig. 6). The first pair of swimming-feet are elongate and slender. The first joint of the inner branches, which is furnished with a plumose seta on the lower half of the inner margin, is rather longer than the second and third together. The second joint is little more than half the length of the third, and bears a small plumose seta on its inner distal angle. The outer branches, which are composed of three sub-equal joints, extend very little beyond the first joint of the inner branches. The inner branches of the next three pairs are short, and only extend to about the middle of the second joint of the elongate outer branches. The basal joint of the fifth pair is broadly triangular, and furnished with five terminal setæ ; the second one of which, counting from the outside, is very long, being more than double the length of the seta on either side of it. The secondary joint is oblong-ovate in shape, its greatest breadth being equal to about two-fifths of the length, and there are five setæ of variable lengths ranged at intervals from the middle of the outer margin to the apex, in addition to a very long intero-sub-apical seta. The apex of the basal joint reaches to about the middle of the secondary one (fig. 10). Caudal stylets shorter than the last abdominal segment, and broadly pyriform. The principal tail setæ are as long as the abdomen.

Male.—Anterior antennæ ten-jointed. The fourth joint is narrower than the one that precedes or follows, and is hinged to the fifth; the sixth, and seventh, are very short, and the joints between the eighth and ninth, and between the ninth and tenth segments, appear to be hinged so that each antenna forms a powerful grasping organ (fig. 3). The inner branches of the third pair extend to the end of the second joint of the outer branches, and is furnished, in addition to the plumose marginal setæ, with a terminal spine-like appendage (fig. 9). The form of the fifth pair is somewhat similar to those of the female, but smaller, and without the very long seta on each of the basal and secondary joints. The sixth pair of appendages are very broad and short, in form somewhat like the segment of a circle, and furnished with one long and two short setæ.

Habitat.—At Seafield near Leith. Obtained by washing some black sandy mud near low-water mark.

Remarks.—This large and fine species is readily distinguished from any other British *Ameira* by the pyriform caudal stylets; and, on dissection, by the other characters referred to in the description.

Ameira exigua,[*] sp. n.   (Pl. VI. figs. 15, 23.)

Description.—Female.   Length, ·47 mm. ($\frac{1}{33}$rd of an inch).   Rostrum prominent.   Anterior antennæ sparingly setiferous, rather slender, and eight-jointed.   The second joint is longer, and the penultimate one shorter than any of the other joints.   The proportional lengths of the joints are nearly as shown by the annexed formula :—

$$\frac{13 \cdot 16 \cdot 10 \cdot 7 \cdot 6 \cdot 6 \cdot 3 \cdot 6 \cdot}{1 \cdot 2 \cdot 3 \cdot 4 \cdot 5 \cdot 6 \cdot 7 \cdot 8 \cdot}$$

Posterior antennæ and mouth organs nearly as in *Ameira reflexa.* The first pair are moderately stout. The inner branches are of considerable length, but the outer branches are short, and do not reach the end of the first joint of the inner branches. The second joint of the inner branches is small, while the end joint is elongate and slender; but the second and third joints together are scarcely equal in length to the first joint (fig. 20). In the fourth pair the outer branches are long, and inner branches short; the inner branches scarcely extend beyond the end of the second joint of the outer branches (fig. 21). The fifth pair are foliaceous, small. The produced inner portion of the basal joint is broadly sub-conical, and furnished with five terminal and sub-terminal setæ; the second one, from the outside, is considerably longer than any of the others. The secondary joint is broadly ovate, and extends halfway beyond the end of the basal joint, and bears several setæ of unequal length round the outer margin and apex, but the inner margin is ciliated (fig. 22). Caudal stylets shorter than the last abdominal segment, and about as long as broad. The principal seta of each stylet is stout, and longer than the abdomen (fig. 23).

Habitat.—Off St Monans.   Not very common.

Remarks.—This very small but distinct species somewhat resembles *Ameira exilis* and *Ameira longipes*, Boeck. It differs from one in being not only much smaller and comparatively more robust, and in the anterior antennæ being eight instead of nine-jointed, but also in the form of the mandible-palp and the structure of the fourth pair of swimming-feet, and from the other by the structure of the mandible-palp and the fifth pair of feet.

---

[*] Exiguus = small.

Boeck, in the same year (1864) * in which he instituted the genera *Stenhelia* and *Ameira*, established a third genus, *Nitokra*, all three being closely allied to each other. Species belonging to the first two genera, but apparently none belonging to the last, have been recorded from the British seas ; at least I do not know of any record of a British *Nitokra*. In 1891 Dr W. Giesbrecht described † two new species of *Nitokra* that he had discovered amongst sea-weed in the estuary at Keil ; and in doing so he drew attention to the more important characters by which these three genera may be distinguished from each other. The characters are these :—

> For *Stenhelia*, Boeck.—' Secondary branch of the posterior antennæ
> ' three-jointed. Mandible-palp with a distinct basal part, bearing
> ' two separate branches.
> For *Ameira*, Boeck.—' Secondary branch of the posterior antennæ
> ' one-jointed. Mandible-palp one-jointed.'
> For *Nitochra*, Boeck.—' Secondary branch of the posterior antennæ
> ' one-jointed. Mandible-palp two-jointed.'

If these definitions are to be considered satisfactory, the *Ameira longipes*, Boeck, as described and figured in the ' Monograph of the British ' Copepoda,' is a *Nitokra*, as the figure of the mandible-palp exhibits two distinct joints or branches ; and probably, also, are all the species described here as *Ameira*. The subject evidently requires further study, so far as the British *Stenheliinæ* are concerned ; and meantime I prefer to adhere to the generic definition of *Ameira* in the ' Monogragh of the British Cope- ' poda. '

Genus Delavalia, Brady (1868).

Delavalia reflexa, Brady and Robertson.

1875. Delavalia reflexa, Brady and Robertson (10), p. 196.

One or two specimens of this species were obtained among material dredged off Burntisland in November last year. Several other interesting species have been discovered here, one or two of which are described in the sequel.

Genus Tetragoniceps, Brady (1880).

(?) *Tetragoniceps consimilis*,‡ sp. n. (Pl. VII. figs. 4–12.)

Description.—Female. Length, ·85 mm. ($\frac{1}{30}$th of an inch). In general appearance very like *Tetragoniceps bradyi*, T. Scott. Rostrum prominent. Anterior antennæ slender, sparingly setiferous, and eight-jointed. The length of the first joint is at least equal to the combined length of the next three, and it bears a few small but distinct blunt pointed teeth on the upper margin. The upper distal angle of the second joint is produced forward into a prominent tooth (fig. 4). The formula shows the propor- tional lengths of the joints :—

$$\frac{28 \cdot 12 \cdot 7 \cdot 8 \cdot 4 \cdot 4 \cdot 4 \cdot 6}{1 \cdot 2 \cdot 3 \cdot 4 \cdot 5 \cdot 6 \cdot 7 \cdot 8}$$

Posterior antennæ elongate, slender, three-jointed ; the last joint is nearly as long as the other two together. A rudimentary secondary branch, furnished with a single apical seta, springs from the end of the first joint. The mandibles have the truncate biting part armed with several elongate

---

* Oversigt, Norges Kyster, Copepoder, *Calanid. Cyclopid. og Harpactid.* (Christ.), 64.

† Die freilubenden Copepoden der Kieler Föhrde.

‡ Consimilis = very like.

sharp-pointed teeth, decreasing in size from the external edge. The basal part of the mandible-palp is comparatively large, becoming dilated outwardly, and bearing a few small marginal hairs. , Apical joint elongate, narrow, with two marginal and four terminal setæ ; the secondary marginal joint is very small (fig. 7). Posterior foot-jaws slender, the second joint elongate, end joint very small, terminal claw very slender (fig. 8). The first pair of thoracic feet somewhat like those of *Tetragoniceps bradyi*, but the inner branches are three-jointed, the first very long, the last two short. A small seta springs from near the distal end of the inner margin of the inner branches, and the third joint bears two moderately long and somewhat curved setæ at the apex. The outer branches are three-jointed, and do not reach the end of the first joint of the inner branches (fig. 9). Inner branches of the second, third, and fourth pairs short, and two-jointed ; outer branches three-jointed, elongate, and slender (fig. 10). Fifth pair two-jointed. Basal joint short, broad, and somewhat produced exteriorly to an angular apex armed with two short spines ; while two small setæ spring from the outer margin. Secondary joint very large and foliaceous, sub-quadrate in form, and one and a half times longer than broad, and bearing one small seta near the middle of the inner margin, and three short, stout, spiniform, and two small setæ on the obliquely truncate apex (fig. 11). Caudal stylets comparatively long and narrow, and about equal in length to the last abdominal segment, and furnished with five terminal setæ, two of them being stout and spiniform, the other three slender. One of the spiniform setæ is short, and springs from the outer edge of the stylet ; the other is elongate, and forms the third from the outside.

Habitat.—Off St Monans, Firth of Forth. Rather scarce.

Remarks.—This species closely resembles *Tetragoniceps bradyi* in general appearance as well as in a few anatomical details, but there are important differences that may require its removal from the genus in which it is provisionally placed. Two very important differences are observed in the structure of the first and fifth pairs of thoracic feet. In the first pair the inner branches are three-jointed, and the fifth pair are two-branched. These two characters indicate a considerable divergence from the typical *Tetragoniceps* ; but I prefer to leave it in that genus in the meantime, pending a further study of the species.

Genus *Laophonte*, Philippi (1840).

Laophonte depressa, sp. n.   (Pl. VI. figs. 24–31 ; Pl. VII. figs. 1–3.)

Description.—Female.   Length, ·84 mm. ($\frac{1}{30}$th of an inch). Body elongate, depressed sub-cylindrical. Breadth across the thorax equal to fully one-fourth of the length. The postero-lateral angles of the second, third, and fourth abdominal segments produced and rounded. Rostrum broad, with the apex rounded. Anterior antennæ sparingly setiferous, shorter than the first cephalo-thoracic segment, and composed of seven joints ; the third joint is longer than any of the others, the fourth and fifth are small. The proportional lengths of the joints are shown by the formula :—

$$\frac{12 \; \cdot \; 11 \; \cdot \; 15 \; \cdot \; 5 \; \cdot \; 4 \; \cdot \; 6 \; \cdot \; 7}{1 \; \cdot \; 2 \; \cdot \; 3 \; \cdot \; 4 \; \cdot \; 5 \; \cdot \; 6 \; \cdot \; 7}$$

Posterior antennæ robust ; the end joint is about equal in length to the preceding one, and bears a few stout slightly curved spines on the distal half of the upper margin and apex. A small secondary branch with four terminal plumose setæ springs from the middle of the first joint (fig. 2, Pl. VII.). Mandibles well-developed. The mandible-palp is comparatively small, and consists of a narrow elongate joint, with one marginal and one apical plumose hair, and is provided with a nearly obsolete subterminal second joint supporting three plumose setæ (fig. 25, Pl. VI.). The joints of the posterior foot-jaws are comparatively elongate and narrow. The terminal claw is strong, and slightly curved (fig. 26, Pl. VI.). The two-jointed inner branches of the first pair of thoracic feet are stout, and armed with a powerful terminal claw. The outer branches are small and three-jointed, and only extend to about the middle of the elongate first joint of the inner branches (fig. 27, Pl. VI.). The second, third, and fourth pairs are slender. The inner branches are short and two-jointed, with the first joint small. Both branches are furnished with long plumose setæ (fig. 28, Pl. VI.). The structure of the fifth pair is similar to those of *Laophonte thoracica*, Boeck; but the secondary joint is broader, the breadth being equal to about twofifths, of the length, and with the distal half sub-conical in outline (fig. 30 Pl. VI.). Caudal stylets short, about equal to half the length of the last abdominal segment.

Male.—In the male the fourth joint of the anterior antennæ is dilated and sub-rotund. The terminal joints are narrow, and form a claw-like appendage; and, being hinged to the large fourth joint, constitute an efficient grasping organ (fig. 2, Pl. VII.). The third pair of thoracic feet are moderately stout; the inner branches appear to consist of three sub-equal joints that reach to about the end of the second joint of the outer branches. An irregularly-curved spiniform appendage, extending considerably beyond the apex of the last joint, springs from the end of the second joint, as shown in the figure (fig. 29, Pl. VI.). The basal joints of the fifth pair are not produced exteriorly, but the anterior angle is furnished with two plumose setæ. The secondary joint is shorter than that of the female fifth feet, and there is proximally a lobe-like marginal process, furnished with a terminal plumose seta. On the conical end of the secondary joint there are two stout setæ on the inner and two on the outer margin, and an elongate apical seta. The sixth appendage is small, narrow, and provided with three terminal plumose setæ (figs. 31, Pl. VI.).

Habitat.—Off St Monans and also off Musselburgh. Rather scarce.

Remarks.—This species resembles *Laophonte thoracica*, Boeck, in some of its characters, as, for example, in the form of the thoracic feet; but the structure of the anterior antennæ and posterior foot-jaws, and the short caudal stylets, distinguish it at a glance from that species.

Laophonte denticornis, sp. n. (Pl. VII. figs. 13–23.)

Description.—Female. Length, ·86 mm. ($\frac{1}{29}$th of an inch). Body seen from above, slender, sub-cylindrical; the first cephalo-thoracic segment nearly equal to the combined length of the next four. The postero-lateral angles of the second, third, and fourth abdominal segments are slightly produced and rounded; while the third and fourth segments are proximally narrower than they are distally. Forehead sub-triangular. Anterior antennæ six-jointed, stout, sparingly setiferous, and about equal in length to the first cephalo-thoracic segment. The first three joints are of moderate length and sub-equal, the fourth and fifth are small. The second joint is armed on the under side with a stout conical tooth nearly at right angles to the joint (fig. 14). The proportional lengths of the joints are shown by the formula :—

$$\frac{18 \ . \ 16 \ . \ 15 \ . \ 5 \ . \ 3 \ . \ 11}{1 \ . \ 2 \ . \ 3 \ . \ 4 \ . \ 5 \ . \ 6} .$$

The fourth joint is produced on the upper side to form the base of a moderately long filament. The basal joint of the posterior antennæ is somewhat robust, and bears a short plumose seta near the middle of the upper margin. A small secondary joint, furnished with two marginal and two terminal plumose hairs, springs from the middle of the joint nearly opposite the marginal seta. The second joint is about equal in length to the first, but considerably narrower (fig. 16). The mandible-palp is small, narrow, one-jointed, and furnished with a few short plumose hairs (fig. 17). In the posterior foot-jaws the first joint is narrow; the second is considerably dilated, and furnished with a strong hooked claw, scarcely equal in length to the joint from which it springs. The inner branches of the first pair of swimming-feet are elongate and rather slender. The second joint is proportionally much longer than it is in several described species of *Laophonte*, being equal to fully one-third of the length of the first joint. Terminal claw slender, and only slightly curved. The outer branches, which consist of three sub-equal joints, are equal to about half the length of the first joint of the inner branches (fig. 19). The inner two-jointed branches of the fourth pair extend slightly beyond the end of the second joint of the outer branches (fig. 20). The inner portion of the basal joints of the fifth pair is produced into a sub-cylindrical lobe, obliquely rounded at the end, and provided with four plumose setæ arranged at nearly equal distances round the lower half and end of the inner edge. The secondary joints are small and cyclindrical in form, and furnished with five setæ on the irregular edge of the truncate apex (fig. 21). Caudal stylets narrow, and rather longer than the last abdominal segment.

Male.—The tooth on the under side of the second joint of the male anterior antennæ is scarcely so large as in the female. The terminal joints are modified somewhat like those of the male of *Laophonte depressa* (fig. 15). The inner branches of the third pair of thoracic feet are very short, and provided with a strong irregularly-curved spiniform appendage (fig. 21). The inner portion of the basal joints of the fifth pair are not produced, but are slightly bilobed, each lobe bearing a single apical seta. The secondary joints are somewhat similar to those of the female, but rather narrower (fig. 23). The sixth appendage is very small, and furnished with two short hairs.

Habitat.—Off St Monans. Rare.

Remarks.—The slender form of this species, considered in connection with the armature of the anterior antennæ, is alone sufficient to enable it to be distinguished from others of the same genus. It is unlike any other *Laophonte* known to me, except, perhaps, *Laophonte serrata* (Claus).

Laophonte littorale, T. and A. Scott.
    1893. *Laophonte littorale,* T. and A. Scott, p. 238, pl. xi. figs. 7-14.

Description.—Female. Length, ·85 mm. ($\frac{1}{30}$th of an inch). Body elongate, sub-cylindrical. Forehead very slightly produced into a bluntly rounded rostrum Anterior antennæ sparingly setiferous, and composed of seven joints ; the third joint is longer and the fifth shorter than any of the others. The proportional lengths of all the joints are shown by the formula :—

$$\frac{9 \cdot 10 \cdot 12 \cdot 5 \cdot 3 \cdot 6 \cdot 8 \cdot}{1 \cdot 2 \cdot 3 \cdot 4 \cdot 5 \cdot 6 \cdot 7 \cdot}$$

Secondary branch of posterior antennæ rudimentary, and consisting of one small joint bearing a small apical seta. Mouth organs nearly as in *Laophonte curticauda,* Boeck. Inner branches of the first thoracic feet moderately stout. The second joint is scarcely equal to one-fourth of the length of the first, and the terminal claw is stout and hooked at the apex.

Outer branches slender, three-jointed, and equal to nearly three-fifths the length of the first joint of the inner branches. Inner branches of the second, third, and fourth pairs, short, two-jointed ; those of the fourth pair scarcely reach beyond the end of the first joint of the outer branches. The outer branches of the fourth pair are short, stout, and very setose. In the fifth pair both joints are broadly foliaceous and sub-quadrate, and bear a number of strongly plumose terminal setæ. Caudal stylets equal to one and a half times the length of the last abdominal segment, and clothed with minute hairs.

Male.—Anterior antennæ very robust. The three last joints, which together form a claw-like appendage, are joined to the preceding joint by a strong hinge-like process, by which they can be folded back upon that joint, thus forming a powerful grasping organ. The inner branches of the fourth pair of thoracic feet are very rudimentary, and consist of two minute joints ; the outer branches are three-jointed and robust. The first joint is considerably longer than the next two together. The first and second joints are each armed with a stout spine at the distal end ; while the last joint bears three spines, one small and two large, as shown in the figure. Fifth pair rudimentary, consisting of a slightly produced basal portion carrying four setæ on small basal projections, and a minute spine on the inner margin. The sixth appendage consists of a small bilobed process, each lobe bearing a stout apical seta.

Habitat.—In pools of brackish water at the mouth of a small stream near Aberlady, Firth of Forth ; and in similar pools at the mouth of the River Alness, near Invergordon, Cromarty Firth.

Remarks.—This species somewhat resembles *Laophonte curticauda,* Boeck. The outer branches of the male fourth thoracic feet are not very unlike those of the male third pair of that species, but the structure of the inner branches is very different. *Laophonte littorale* appears to be confined to localities where the water is more or less brackish.

(?) *Laophonte simulans,*[*] sp. n.   (Pl. VII. figs. 24–32 ; Pl. VIII. fig. 1).

Description.—Female. Length ·43 mm. ($\frac{1}{58}$th of an inch). Body seen from above, broadly ovate, depressed. Forehead produced into a broad blunt pointed rostrum. Anterior antennæ robust, shorter than the first cephalo-thoracic segment, six-jointed. The third joint is considerably longer than any of the others, while the fourth and fifth are small ; a long slender filament springs from the fourth joint (fig. 1, Pl. VIII.). The proportional lengths of the joints are shown by the formula :—

$$\frac{14 \cdot 18 \cdot 23 \cdot 5 \cdot 3 \cdot 12}{1 \cdot 2 \cdot 3 \cdot 4 \cdot 5 \cdot 6}$$

Basal joint of posterior antennæ robust. Second joint shorter and much narrower. Secondary branch short, bearing four terminal plumose setæ (fig. 25, Pl. VII.). Mandible elongate, slender, and provided with a small one-jointed palp (fig. 26, Pl. VII.). Posterior foot-jaws stout, armed with a powerful strongly curved terminal claw, rather shorter than the joint from which it springs (fig. 27, Pl. VII.). Inner branches of the first pair of thoracic feet elongate and robust. A small blunt-pointed spine springs from near the distal end of the first joint. The second joint is short, being only equal to about one-fifth of the length of the first joint. Terminal claw stout, and strongly hooked at the end. Outer branches are slender and three-jointed, and about two thirds the length of the first joint of the inner branches. The second and third pairs are moderately stout, and have the inner branches two and the outer three-jointed,—the inner branches being shorter than the outer (fig. 28, Pl. VII.). In the fourth pair the inner

[*] Simulans = mimicking.

branches consist of one and the outer of two joints, as shown in the figure (fig. 29, Pl. VII.). The inner portion of the basal joint of the fifth pair is not developed, but looks like a flattened border to the thoracic segment to which it is attached. It is furnished with three marginal setæ ; the outer portion projects outwards in the form of an elongate narrow process carrying a slender apical seta. The secondary joint has an obovate form, and is provided with five setæ of variable lengths round the distal end (fig. 30, Pl. VII.). Caudal stylets short, with a considerable space between them. The thoracic and abdominal segments are all fringed more or less with cilia.

Habitat.—Off West Wemyss, Firth of Forth. Rare.

Remarks.—This small and very curious species was discovered within the valves of a dead *Cyprina* shell, accidentally picked out from among some trawl refuse. It seems to combine the form of one of the depressed *Dactylopus* with the anatomical characters of *Laophonte*. It differs, however, from the typical *Laophonte* in the structure of the third pair of swimming-feet, which in this species have the inner branches one and the outer branches two jointed. It also carries two ovisacs, which is very unusual among the *Laophonte*. It nevertheless agrees in so many particulars with the characters of that genus, that I propose, for the present at least, to place it among the *Laophonte*. *Laophonte simulans* is, so far, one of the smallest of the Forth species belonging to this genus.

Sub-genus *Laophontodes*, nov. sub-gen.

Like *Laophonte*, Philippi, except that the posterior antennæ have no secondary branch, and the fifth pair of thoracic feet are one-jointed. In the second, third, and fourth pairs of thoracic feet the second basal joints are elongate, and the outer branches are articulated to the apex, and the inner branches to the margin, and near the proximal end of the basal joint. The two branches are more widely apart than is usual in the genus *Laophonte*.

Laophontodes typicus, sp. n. (Pl. VIII. figs. 2–8.)

Description.—Female. Length, ·4 mm. ($\frac{1}{60}$nd of an inch). Body seen from above, narrow, elongate ; the breadth gradually decreasing towards the posterior end. All the segments more or less angular. The first cephalo-thoracic segment is nearly quadrangular in outline ; the middle part of the front margin is produced to form a broad sub-truncate rostrum. Anterior antennæ short, five-jointed. All the joints are comparatively elongate, except the fourth, which is very small. The formula shows the proportional lengths of the joints :—

$$5 \quad 6 \cdot 7 \cdot 1 \cdot 5 \cdot$$
$$1 \cdot 2 \cdot 3 \cdot 4 \cdot 5 \cdot$$

Posterior antennæ two-jointed, and of moderate size ; no secondary appendage (fig. 4). Posterior foot-jaws small, three-jointed ; the end joint very small. Terminal claw slender, slightly curved at the apex (fig. 5). First pair of thoracic feet like those of *Laophonte*. Second joint of the inner branches is scarcely equal to one-third the length of the first joint. Terminal claw slender, strongly hooked at the end. A small seta springs from the base of the claw. Outer branches three-jointed ; the first and second joints sub-equal, the third shorter, the entire outer branch is equal to fully half the length of the first joint of the inner branch (fig. 6). Inner branches of the second, third, and fourth pairs short, two-jointed, and articulated to the second basal joint near its proximal end. The first joint of the inner branches is very small, the other is elongate. Outer branches elongate, three-jointed, and articulated to end of the elongate, narrow, second basal joint; the outer and inner branches being thus widely apart (fig. 7). Fifth pair one-branched, narrow, elongate, and furnished with one seta near the middle of the inner margin, three on the outer margin, and three at the apex (fig. 8). Caudal stylets long and narrow, being equal to one and a half times the length of the last abdominal segment, and each furnished with a stout apical seta and a few minute hairs.

Habitat.—Vicinity of Inchkeith.   Rare.

Remarks.—This curious species was dredged at the north end of the Island of Inchkeith.   Very few specimens were obtained.

Genus *Normanella*, Brady (1880).

Normanella dubia (Brady and Robertson).

1875. *Laophonte dubia*, Brady and Robertson (10), p. 196.
1880. *Normanella dubia*, Brady (8), vol. ii. p. 87, pl. lxxviii. figs. 12–22.
1893. *Normanella dubia*, I. C. Thompson(33), p. 26, pl. xxi. fig. 6.

Habitat.—Off Musselburgh.   Not very scarce, but easily overlooked.
Normanella dubia seems to be widely distributed in the British seas.

Genus *Cletodes*, Brady (1872).

Cletodes irrasa, T. and A. Scott.

1894. *Cletodes irrasa*, T. and A. Scott (31), p. 141, pl. viii. figs. 8–12.

Description.—Female. Length, ·8 mm. ($\frac{1}{30}$th of an inch). Body elongate, cylindrical. All the segments, except the first, furnished with a fringe of small hairs around, but a little in front of, the posterior margin. Anterior antennæ stout, shorter than the first body segment, and composed of six joints; the second and last joints are longer than any of the others, the fifth is very small. The proportional lengths of all the joints are shown by the formula :—

$$\frac{20 \cdot 26 \cdot 20 \cdot 10 \cdot 3 \cdot 24 \cdot}{1 \cdot 2 \cdot 3 \cdot 4 \cdot 5 \cdot 6 \cdot}$$

Posterior antennæ three-jointed. A small secondary branch bearing a single apical seta springs from the end of the first joint. Inner branches of the first pair of thoracic feet, which are composed of two sub-equal joints, reach to about the middle of the second joint of the outer branches, and bear two short apical setæ. Inner branches of the fourth pair, which are also two-jointed, shorter than the first joint of the elongate outer branches. The inner portion of the basal joint of the fifth pair is in the form of a shallow lobe, carrying two long and one short seta. Secondary joint narrow and elongate, and furnished with one marginal and four terminal setæ. Caudal stylets slender, equal in length to the last abdominal segment, each with two small setæ on the inner and one on the outer margin, and a few apical setæ.

Habitat.—Vicinity of the Bass Rock.   Rather rare.

Cletodes curvirostris,* sp. n.   (Pl. VIII. figs. 18–26.)

Description.—Female. Length, ·9 mm. ($\frac{1}{28}$th of an inch). Body subcylindrical, usually arcuate when seen from the side. Rostrum short, stout, recurved. Anterior antennæ robust, short, and composed of six joints. The third is considerably larger and the fifth smaller than any of the other

---

* Referring to the recurved rostrum.

joints, as shown by the annexed formula, which indicates approximately the proportional lengths of all the joints :—

$$\frac{7 \cdot 7 \cdot 11 \cdot 4 \cdot 2 \cdot 8 \cdot}{1 \cdot 2 \cdot 3 \cdot 4 \cdot 5 \cdot 6 \cdot}$$

Posterior short and stout. Secondary appendage represented by a single small plain seta which springs from near the middle of the first joint. A small plumose seta springs from exterior margin of the same joint (fig. 20). Mandibles armed with a number of spiniform teeth, the middle one being larger than the others. Palp small, one-jointed, provided with one marginal and two terminal plumose setæ ; and a peculiar appendage, bifid at the apex, springs from a small marginal lobe (fig. 21). Posterior foot-jaws very small, with a slender terminal claw. The inner branches of the first pair of swimming-feet are composed of two nearly equal joints, the second one being only slightly longer than the other. The inner branches are about two-thirds of the length of the outer three-jointed branches. A stout setose spine springs from the inner distal angle of the second basal joint (fig. 22). Inner branches of the fourth pair small, scarcely equal in length of the first joint of the outer three-jointed branches, and composed of two joints, the first joint shorter than the other. All the swimming-feet are short and stout. Basal joints of the fifth pair small ; the produced interior portion is rather short and narrow, and provided with three stout setæ—one marginal and two terminal. A slender seta, articulated near the middle, springs from the exterior angle. Secondary joints elongate, narrow, sub-cylindrical, bearing a stout terminal seta and two smaller marginal setæ, all three being plumose (fig. 25). Caudal stylets moderately stout, rather longer than the last abdominal segment. Each stylet is furnished with a small seta near the middle of the outer margin in addition to a few terminal setæ.

Habitat.—Largo Bay. Not unfrequent.

Remarks.—This species resembles *Enhydrosoma curvata* in general appearance, but in structural details it is clearly a *Cletodes.*

* Genus *Pontopolites*, nov. gen.

Animal somewhat resembling *Dactylopus* in general form. Anterior antennæ five-jointed ; shorter than the first body segment. Posterior antennæ like those of *Cletodes*, but the secondary branch is two-jointed. Mandible-palp small ; composed of a distinct basal part and two small one-jointed branches, the posterior one being rudimentary, or nearly so. Maxillæ and anterior and posterior foot-jaws as in *Cletodes*. First pair of swimming-feet somewhat similar to those of *Attheyella pygmæa* (G. O. Sars) (= *Attheyella cryptorum*, Brady). The inner branches, which are composed of two nearly equal joints, are of about the same length as the outer three-jointed branches (fig. 14, Pl. IV.). The inner branches of the second, third, and fourth pairs consist of a single and more or less rudimentary joint. The fifth pair are one-branched. One ovisac.

Pontopolites typicus, sp. nov. (Pl. VIII. figs. 9–17.)

Description.—Female. Length, ·6 mm. ($\frac{1}{4}$rd of an inch). Body elongate, sub-cylindrical. Rostrum short, subtriangular, with the apex bluntly rounded. A minute seta springs from a small notch on each side of the apex. Anterior antennæ short, stout, five-jointed, the fourth joint very small. The proportional lengths of the joints are shown by the formula :—

$$\frac{16 \cdot 12 \cdot 14 \cdot 4 \cdot 15 \cdot}{1 \cdot 2 \cdot 3 \cdot 4 \cdot 5 \cdot}$$

The small two-jointed secondary branch of the posterior antennæ springs

* Ποντος, the sea ; Πολιτης, a citizen.

from near the base of the first joint (fig. 11). The basal part of the mandible-palp bears two spinulose setæ at the apex. The primary branch is furnished with one marginal and three terminal setæ, while the secondary branch consists of a small papilla bearing a single small hair. Posterior foot-jaws short, stout, two-jointed, and armed with a moderately long terminal claw. First pair of swimming-feet short, moderately stout ; both branches of nearly equal length ; the inner branches composed of two, the outer of three sub-equal joints (fig. 14). The first pair closely resemble those of *Attheyella pygmæa* (G. O. Sars). Inner branches of second, third, and fourth pairs small, one-jointed. Those of the fourth pair rudimentary, and bearing a short terminal spine. Outer branches elongate, the third joint being nearly equal to the combined length of the other two (fig. 15). The fifth pair consist each of a single joint (or branch), sub-quadrangular in form, but rather longer than broad, and extends obliquely outwards from its attachment to the fifth thoracic segment. A number of long stout plumose setæ fringe the oblique exterior margin, while a few plain setæ spring from the truncate apex, as shown in the figure (fig. 16). Caudal stylets very short.

Habitat.—Off Musselburgh. Not common.

Genus *Heteropsyllus,*[*] nov. gen.

Body sub-cylindrical. Anterior antennæ shorter than the first body segment, and composed of five joints. Posterior antennæ like those of *Cletodes,* but the secondary branch consists of two joints. Mandible-palp small, with a distinct basal part bearing two small branches. Mouth organs as in *Pontopolites.* Both branches of the first pair of swimming-feet three-jointed, the inner being rather longer than the outer branches. Inner branches of the second, third, and fourth pairs two-jointed, and considerably shorter than the elongate three-jointed outer branches. Fifth pair two-branched. Foliaceous. One ovisac.

Heteropsyllus curticaudatus, sp. n. (Pl. VIII. figs. 27–34 ; Pl. IX. fig. 1.)

Description.—Female. Length, ·5 mm. ($\frac{1}{50}$th of an inch). Body elongate, sub-cylindrical. Forehead produced into a short rostrum (fig. 27, Pl. VIII.). Anterior antennæ stout, shorter than the first cephalo-thoracic segment, and composed of five joints, the fourth being very short. The approximate proportional lengths of the joints are shown by the formula :—

$$\frac{10 \cdot 9 \cdot 7 \cdot 3 \cdot 14}{1 \cdot 2 \cdot 3 \cdot 4 \cdot 5} \cdot$$

The two-jointed secondary branch of the posterior antennæ is furnished with four small setæ,—one at the end of the first joint, a marginal and two terminal setæ on the second joint (fig. 29, Pl. VIII.). Mandible-palp small, but distinctly two-branched (fig. 30, Pl. VIII.). Posterior foot-jaws small, and in form somewhat similar to those of the species last described. Both branches of the first pair of swimming-feet composed of three sub-equal joints, the inner rather longer than the outer branches (fig. 32, Pl. VIII.). Outer branches of the second, third, and fourth pairs elongate and three-jointed, somewhat slender. The third joint of all the outer branches is longer than either the first or second joints. Inner branches shorter than the outer, and two-jointed ; those of the fourth pair scarcely longer than the first joint of the outer branches (fig. 33, Pl. VIII.). The basal joints of the fifth pair are large, broadly but somewhat obliquely rounded at the distal end, and provided with five terminal setæ ranged at more or less equal distances from each other round the apex and lower inner margin.

[*] Ἕτερος, different ; ψυλλος, a flea.

The seta on the outside is plain, the others are plumose. Secondary joints small, sub-cylindrical, scarcely extending beyond the end of the basal joints. Bordering upon the concave exterior margin of the basal joints, and near the proximal end of the secondary joints, there is a clearly defined lucid space of a somewhat semi-circular outline (fig. 1, Pl. V.). Caudal stylets very short, the breadth about equal to the lengths. The stylets, being comparatively narrow, are widely apart ; and as the last abdominal segment, as seen from the side, ends abruptly, the stylets, though short, are quite prominent (see figs. 27 and 34, Pl. VIII.).

Habitat.—Off Musselburgh and off Aberdour. Frequent.

Remarks.—This is a well-marked species. When mixed up with other forms it is readily distinguished by the abrupt junction of the stylets to the last abdominal segment. Though moderately frequent, both in material dredged off Musselburgh and Aberdour, no males have yet been obtained.

Genus *Nannopus*, Brady (1880).

Nannopus palustris, Brady.
        1880. *Nannopus palustris*, Brady (8), vol. ii. p. 101, pl. lxxvii. figs. 18–20.
        1892. *Nannopus palustris*, Canu (11), p. 166, pl. iv. figs. 6–21.

Habitat.—In pools near the mouth of Cocklemill Burn—a small stream that enters the Forth at the east end of Largo Bay. The mouth of this stream is comparatively narrow, but immediately beyond the outlet there is a large expanse of low-lying ground intersected by numerous furrows or ditches branching off from the main channel of the stream ; a considerable portion of this low-lying ground is covered by the sea at high-water, and especially during spring tides. This tidal lagoon, as it may be called, appears to harbour a peculiar and interesting micro-fauna, which has not hitherto been very carefully worked up. Some material was collected here by hand-net in August 1890, but it was not till a few months ago that there was sufficient leisure to attend to it. It was then ascertained that this curious species was not unfrequent in the material. *Nannopus palustris* was discovered by Dr Brady, also, in brackish-water pools, at Seaton Sluice, Northumberland, where only a few specimens were obtained. Dr Eugene Canu records this species also from Wimereaux, France, where it also occurs in somewhat similar conditions to those described above. The rudimentary structure of the inner branches of the third and fourth pairs of feet appear to be characteristic of the species.

Genus *Leptopsyllus*,* nov. gen.

Body elongate, cylindrical, somewhat similar in form to *Cylindropsyllus*, Brady. Anterior antennæ eight-jointed, short. Posterior antennæ three-jointed ; secondary branch small, one-jointed. Mandibles well-developed, the broad biting part armed with several strong teeth. Palp comparatively large, consisting of a moderately stout basal joint, and a single two-jointed branch. Other mouth appendages nearly as in *Cletodes*, except that the posterior foot-jaws are three-jointed. Both branches of the first pair of swimming-feet short and two-jointed. In the second and third pairs the inner branches are obsolete or entirely absent, but the outer branches are three-jointed. Inner branches of the fourth pair two-jointed ; outer branches three-jointed. Fifth pair foliaceous, small, two-branched.

* Λεπτος, slender ; and ψυλλος, a flea.

Leptopsyllus typicus, sp. n.   (Pl. IX. figs. 2–11.)

   Description.—Female. Length, ·74 mm. ($\frac{1}{34}$th of an inch). Anterior antennæ eight-jointed, short and robust ; the fifth, sixth, and seventh joints are shorter than any of the others. The formula shows the proportional lengths of the joints :—

$$\frac{13 \cdot 10 \cdot 6 \cdot 3 \cdot 2 \cdot 2 \cdot 2 \cdot 5 \cdot}{1 \cdot 2 \cdot 3 \cdot 4 \cdot 5 \cdot 6 \cdot 7 \cdot 8 \cdot}$$

Posterior antennæ composed of three moderately long joints ; and the small one-jointed secondary branch springs from near the end of the first joint (fig. 4). The two-jointed branch of the mandible-palp has the first joint elongate, and furnished with a marginal second joint, very short, and provided with three terminal hairs (fig. 5). Maxillæ as in *Cylindropsyllus*. Anterior foot-jaws stout, two (or three) jointed, furnished with three marginal processes, each with two small spinulose terminal setæ ; the distal end of the second joint is produced, and supports a stout curved spine. In a notch near the distal end of the second joint there is what looks like a rudimentary third joint, from which spring three small setæ (fig. 6). Posterior foot-jaws slender, three-jointed, the end joint small. The terminal claw is very long and slender, and is accompanied by a spiniform seta nearly equal to half the length of the claw (fig 7). Both branches of the first pair of swimming-feet are short and two-jointed, and somewhat widely apart ; the second joint is rather shorter than the first in both branches (fig. 8). The second and third pairs are only one-branched, the inner branches being apparently entirely absent. Each branch consists of three joints, the middle one being shorter than the one on either side. The first two joints are each armed with a moderately stout slightly curved spine on the outer distal angles, while the last joint carries two terminal spines (fig. 9). In the fourth pair the inner branches are slender and two-jointed. The end joint is small, and provided with a moderately stout terminal spine. The outer branches are three-jointed. The joints are sub-equal, and armed similar to those of the second and third pairs (fig. 10). Fifth pair small and foliaceous. Each consists of a basal and a secondary joint. The inner portion is produced into a sub-cylindrical lobe, rounded at the end, and bearing two terminal hairs. The outer portion is also produced, but not so much as the inner portion, and bears one hair. The small secondary joint is situated in the hollow between the two produced portions of the basal joint, and is furnished with three terminal hairs (fig. 11). Caudal stylets elongate, and each composed of two distinct joints. The first joint is about one and a half times the length of the last abdominal segment, and about three times longer than broad, and has the inner distal angle produced into a blunt-pointed tooth-like process. The second joint (or appendage to the stylets) is of an elongate oval form, and equal to about one-third the length of the first joint to which it is attached by a narrow hinge-like articulation.

   Habitat.—West of Queensferry. Washed from lumps of hardened mud composed of the agglutinated tubes of a species of *Sabella.* Rare. No males have yet been observed.

Genus Dactylopus, Claus (1863).

Dactylopus stromii (Baird).

         1850. *Canthocamptus stromii*, Baird (2), p. 208, pl. xxvii. fig. 3.
         1880. *Dactylopus stromii*, Brady (8), p. 3, pl. lv. figs. 1–13.
         1892. *Dactylopus stromii*, Canu (11), p. 159.
         1893. *Dactylopus stromii*, I. C. Thompson (33), p. 27, pl. xxii. fig. 4 *a, b.*

Habitat.—Off Limekilns, west of Queensferry. Washed from lumps of hardened mud. Not very common.

This species appears to be extensively distributed throughout the British Islands.

Dactylopus coronatus, sp. n. (Pl. IX. figs. 12–20.)

Description.—Female. Length, ·57 mm. ($\frac{1}{44}$th of an inch). Body moderately robust. Anterior antennæ short, stout, six-jointed, and bearing numerous elongate and stout, plain, and spinulose setæ (fig. 13). The ante-penultimate joint is rather shorter than any of the others, and bears a long stout filament. The formula shows the proportionate lengths of the joints :—

$$\frac{13 \cdot 17 \cdot 12 \cdot 8 \cdot 11 \cdot 10 \cdot}{1 \cdot 2 \cdot 3 \cdot 4 \cdot 5 \cdot 6 \cdot}$$

Posterior antennæ short and stout. Secondary branch two-jointed, strongly setiferous. Mandibles somewhat like those of *Dactylopus tisboides*, Claus. The biting part is armed with a broad trifid tooth, two spiniform teeth, and a few setiferous spines (fig. 15). Posterior foot-jaws robust. A strong setiferous spine springs from the inner distal angle of the first joint. Terminal claw stout, curved, scarcely equal in length to the joint from which it springs. Both joints are furnished with rows of cilia, as shown in the figure (fig. 16). All the swimming-feet are short and stout, but those of the first pair are more robust than the others. The first joint of the inner branches of the first pair is considerably dilated, and longer than the next two together, and furnished with a plumose seta near the middle of the inner margins. The second and third joints are sub-equal, and much narrower than the first joint. The second joint bears a seta similar to that of the first joint. The third joint bears a small marginal plumose seta, a terminal plumose seta, and two elongate curved terminal spines. The outer branches are nearly equal to the length of the first two joints of the inner branches. The first two joints are sub-equal, but the last is only about two-thirds the length of the preceding one (fig. 17). The middle joint of the outer branches is furnished with a long plumose seta near the end of the inner margin. The spines on the outer margins of the first and second joints are moderately stout and setiferous. The outer one of the five terminal spiniform setæ is about the same length as the marginal spines, but the others become gradually more elongate, so that the inner one is fully three times the length of the outer (fig. 17). The inner branches of the second, third, and fourth pairs are shorter than the outer branches, and both are furnished with several plumose setæ on the inner margins, while the outer margins are strongly ciliate. The marginal spines of the outer branches are all more or less setiferous (fig. 18). The fifth pair are each indistinctly two-branched (or two-jointed) and foliaceous. The joints are sub-equal, and are each furnished with five stout plumose setæ of various lengths round the distal end. Caudal stylets short. One ovisac.

Habitat.—Among dredged material from the Rath ground in the vicinity of the Bass Rock; also in dredged material from Largo Bay. Not common.

Remarks.—The structure and armature of the anterior antennæ are alone sufficient to distinguish this from other species of *Dactylopus.*

Genus *Thalestris*, Claus (1863).

Thalestris forficuloides, T. and A. Scott. (Pl. X. figs. 13–25.)
  1894. *Thalestris forficuloides*, T. and A. Scott (31), p. 142, pl.
    ix. figs. 4–9.

Description.—Female. Length, ·73 mm. ($\frac{1}{34}$th of an inch). Anterior antennæ nine-jointed, and bearing long slender setæ. The proportional lengths of the joints are shown in the formula :—

$$\frac{15 \cdot 18 \cdot 13 \cdot 10 \cdot 8 \cdot 11 \cdot 6 \cdot 5 \cdot 11}{1 \cdot 2 \cdot 3 \cdot 4 \cdot 5 \cdot 6 \cdot 7 \cdot 8 \cdot 9} \cdot$$

The secondary branches are composed of two moderately long joints. Mandibles with the oblique biting edge coarsely serrate. Mandible-palp with a moderately large, stout, basal part, and two small branches (fig. 16). Maxilla-palp with three small narrow branches, each bearing two to three slender apical setæ (fig. 17). Posterior foot-jaws short, moderately stout ; the last joint somewhat ovate, bearing a slender seta near the middle of the inner margin. Terminal claw slender, gently curved (fig. 19). The first pair of swimming-feet slender. Inner branches considerably longer than the outer. The inner terminal claw of both branches is comparatively very long and slender (fig. 20). In the fourth pair the inner branches reach slightly beyond the second joint of the outer branches. The last joint of the inner branches is about equal to three-fourths of the entire length of the other two joints, and the length of the last joint of the outer branches is nearly equal to that of the first and second together (fig. 21). The basal joints of the fifth pair are broadly triangular, and the apex reaches to near the middle of the secondary joints. They are furnished with three plumose setæ on the inner distal margin, and a moderately long terminal and small sub-terminal setæ. Secondary joints sub-cylindrical, and bearing eight setæ,—three on the inner and three on the outer distal margin, and two at the apex. Both margins of both joints are fringed with cilia (fig. 23). Caudal stylets very short. All the body segments are fringed with cilia round the posterior margin.

Male.—Anterior antennæ apparently eight-jointed, hinged between the second and third and between the fifth and sixth. The first three joints are sub-equal, and the fifth and last are about equal in length to that of the first three, but are narrower ; while the fourth and seventh are smaller than those that preceed or follow. The inner branches of the third pair in the male are provided with an elongate spiniform appendage, slightly hooked at the end (fig. 22). The basal joint of the fifth pair is furnished with only two small spiniform setæ. The secondary branch is somewhat like that of the female, but rather smaller.

Habitat.—At Seafield, near Leith, among mud near low-water mark. This species resembles *Thalestris forficula*, Claus, but differs in several important particulars,—it differs in the proportional lengths of the joints of nine-jointed anterior antennæ ; it also differs in the structure of the swimming-feet, as shown by the description and figures.

Genus *Pseudowestwoodia*, nov. gen.

Very like *Westwoodia* in general appearance when seen from the side. Anterior antennæ six or seven-jointed. Secondary branch of posterior antennæ small, one-jointed. Mouth appendages similar to those of *Westwoodia.* Both branches of first pair of swimming-feet two-jointed, but in general appearance the first pair resemble those of *Westwoodia.* The other thoracic appendages are similar in structure to those of that genus. The distinctive characters of *Pseudowestwoodia* are the one-jointed secondary branch of the posterior antennæ, and the first pair of swimming-feet with both branches two-jointed. All the other characters are more or less similar to those of *Westwoodia*, hence the proposed generic name for the form now under consideration.

Considering the many points of resemblance between the characters of the species now to be described and those of the genus *Westwoodia*, a strong desire was felt to include it in that genus ; but it was clearly perceived that in order to do so a considerable modification of the definition of that genus would be necessary.   In *Westwoodia* the secondary branch of the posterior antennæ is *two-jointed*, and the first pair of swimming-feet have the inner branches *three* and the outer only *one-jointed*.

Pseudowestwoodia andrewi,* sp. n.   (Pl. IX. figs. 21–29.)

Description.—Female, ·46 mm. ($\frac{1}{54}$th of an inch).   Body similar to *Westwoodia nobilis* (Baird) in general appearance, but smaller.   Anterior antennæ six-jointed, slender, and sparingly setiferous   The first three and last joints are elongate, while the fifth is smaller than any of the others.   The proportional lengths of the joints are as follows :—

$$\frac{9 \cdot 11 \cdot 11 \cdot 6 \cdot 4 \cdot 12 \cdot}{1 \cdot 2 \cdot 3 \cdot 4 \cdot 5 \cdot 6 \cdot}$$

The posterior antennæ consist of two sub-equal joints. The small one-jointed secondary branch springs from near the middle of the first joint (fig. 23). Mandible-palp with a moderately large basal joint bearing two small one-jointed apical branches, each branch furnished with several slender setæ (fig. 24). Posterior foot-jaws moderately stout. Terminal claw slender, rather longer than the joint from which it springs. The first joint bears a small seta near its inner distal angle, and a similar seta springs from about the middle of the inner margin of the second joint (fig. 25). The outer branches of the first pair of swimming-feet are about equal to half the entire length of the inner branches, and composed of two sub-equal joints. The first joint is armed with a stout spine at the outer distal angle, while the second joint bears a similar spine near the middle of the outer margin.   One small seta near the lower part of the inner margin, and two spiniform setæ at the apex. The first joint of the inner branches is moderately stout and elongate, and becomes narrower towards the distal end.   A long plumose seta springs from the inner margin at about one-third of the length of the joint from the proximal end.   The second is small and narrow, and only equal to about one-fourth of the length of the first joint, and furnished with two slender, moderately elongate and spiniform apical setæ.   A stout spine springs from both the inner and outer distal angles of the second basal joints (fig. 26).   The second, third, and fourth pairs are nearly as in *Westwoodia nobilis* (Baird).   The fifth pair are small and foliaceous.   The length of the produced inner portion of the basal joint is scarcely equal to the breadth of its proximal end.   Distal end broadly truncate, and furnished with four spiniform setæ and a short moderately stout spine—the spine being at the exterior angle.   Secondary joints small, extending little beyond the end of the produced inner portion of the basal joints ; sub-quadrangular in form, and armed with three short but strong dagger-shaped spines on the somewhat obliquely truncate end, and a small spine and a spiniform seta on the distal part of the inner margin (fig. 28).   Caudal stylets very short.   A short stout spine springs from the outer distal angle of each stylet, and the inner of the two principal tail setæ is about one third longer than the other.

Habitat.—Off Burntisland.   Frequent.   Among material dredged in three or four fathoms water.

Remarks.—This small but interesting species is readily distinguished from others, even without dissection, when examined under the microscope, by the peculiar armature of the fifth pair of thoracic feet.   It seems to have affinities with *Westwoodia* on the one hand and with *Harpacticus* on the other, and forms a connecting-link between them.

---

* The species is so named in compliment to my son, Andrew Scott, to whose painstaking and ever-ready assistance and facile pencil I owe so much of my success among the *Copepoda.*

A second species of *Pseudowestwoodia* has lately been discovered in material from Cromarty Firth. It differs from that now described in the structure of the anterior antennæ, and in the armature and form of the first and fifth feet. This species is to be described later on.

Family LICHOMOLGIDÆ.

Genus *Pseudanthessius*, Claus (1889).

[*Lichomolgus*, Thorell (Pars) 1859].

Pseudanthessius liber (Brady and Robertson).

 1875. *Lichomolgus liber*, Brady and Robertson (10), p. 197.
 1880. *Lichomolgus liber*, Brady (8), vol. iii. p. 44, pl. lxxxvi. figs. 1–13.
 1893. *Lichomolgus liber*, I. C. Thompson (33), p. 33, pl. xxv. figs. 2 *a*, *b*.

Habitat.—Vicinity of Inchkeith (north end). Scarce.

As has been pointed out by Dr Canu (*Les Copepodes du Boulonnais*, p. 241), one of the principal characters by which *Pseudanthessius*, Claus, is distinguished from *Lichomolgus* proper, is the structure of the fourth pair of thoracic feet, the inner branches of which, in *Pseudanthessius*, are only one-jointed, and furnished with two terminal setæ. In *Lichomolgus* proper, the inner branches of the same pair of feet are two-jointed, and the first joint has a seta on the inner distal angle, while the second carries two spines or spiniform setæ at the apex. According to this arrangement the following British species of the *Lichomolgidæ* will be included in the genus, *Pseudanthessius*, Claus.

 Pseudanthessius liber (Brady and Robertson), as recorded above.
 Pseudanthessius thorellii (Brady and Robertson), (= *Lichomolgus thorellii*, Brady and Robertson).
 Pseudanthessius gracilis, Claus.
 Pseudanthessius sauvagei, Canu.

The last three, as well as the first, have been obtained in the Forth, and will be referred to in the sequel.

The following British species of the *Lichomolgidæ* will be included in *Lichomolgus* proper :—

 Lichomolgus fucicolus, Brady.
 Lichomolgus furcillatus, Thorell.
 Lichomolgus forficula, Thorell.
 Lichomolgus albens, Thorell.
 Lichomolgus hirsutipes, T. Scott. (Firth of Forth.)
 Lichomolgus agilis, Leydig (= *Lichomolgus concinnus*, T. Scott. Firth of Forth.)

The following species of the *Lichomolgidæ* have also been recorded from the British seas :—

Sabelliphilus sarsi, Claparide (recorded by I. C. Thompson, Liverpool).

Species belonging to *Sabelliphilus* have the inner branches of the fourth pair of swimming-feet three jointed, and furnished with two barbed setæ on the inner margin, and two slender smooth setæ at the distal extremity. The first two joints of the anterior antennæ in *Sabelliphilus* are considerably dilated.

Herrmannella rostrata, Canu (1891) [= *Lichomolgus agilis,* T. Scott (1892). Firth of Forth.]

The structure and armature of the inner branches of the fourth pair in *Herrmannella* are somewhat similar to those of *Sabelliphilus,* but the first two joints of the anterior antennæ are cylindrical and not swollen as in that genus.

Modiolicola insignes, Aurivillius. (Firth of Forth.)

This species of the *Lichomolgidæ* somewhat resembles both *Sabelliphilus* and *Herrmannella* in the structure and armature of the fourth pair of feet. It also resembles *Herrmannella* in the form of the anterior antennæ, but it differs from *Sabelliphilus* in the first two joints of the anterior antennæ being cylindrical and not dilated ; and it differs from *Herrmannella* in the anterior foot-jaws (the external second-maxillæ) being furnished with only one internal smooth seta on the basal part of the second joint ; whereas in *Herrmannella,* there are two setæ on the inner part of the base of the second joint of the anterior foot-jaws, one seta being comparatively short and smooth, and one elongate and denticulate.

Lichomolgus arenicolus, Brady, is evidently not a true member of the genus *Lichomolgus.* In this species both branches of the fourth pair are three-jointed, and their armature is very similar to that of the second and third pairs ; but it varies also in other structural details, and to such an extent that it cannot satisfactorily be placed in any of the genera I have referred to. The same may be said of an interesting species discovered some time ago by I. C. Thompson of Liverpool, and described by him under the name of *Lichomolgus maximus ;* the species was found living as a mess-mate within the shell of the large scallop, *Pecten maximus.* That this is not a true *Lichomolgus,* according to the definition of the genus I have already referred to, will be at once apparent from the following quotation from Mr Thomson's description of the species. He says :— ' The first four pairs of swimming-feet have both branches three-jointed.' . . . . . . 'It agrees with *Lichomolgus agilis'* (= *Herrmannella rostrata,* Canu) 'in having the inner branch of the fourth pair of swimming-feet 'three-jointed.' It is evident from this description, and independent of other structural differences, that *Lichomolgus maximus* is not, any more than *Lichomolgus arenicolus,* a true member of that genus ; but the question as to which genus they should be assigned, to secure for them a satisfactory resting-place, is a matter requiring further study. It is quite possible that their structural details may not meet the conditions of any described genus. *Lichomolgus maximus,* however, appears to have a closer affinity with *Modiolicola* Aurivillius, than with any of the other genera referred to here.

Pseudanthessius thorellii (Brady and Robertson).

 1875. *Lichomolgus thorellii,* Brady and Robertson (10), p. 197.
 1880. *Lichomolgus thorellii,* Brady (8), vol. iii. p. 47, pl. lxxxviii. figs. 1–9.
 1893. *Lichomolgus thorellii,* I. C. Thompson (33), p. 33, pl. xxv. fig. 2 c.

Habitat.—Off St Monans. Scarce.

This appears to be one of the rarer species of *Lichomolgus.* I find comparatively few records of its occurrence in the British seas. Dr Brady obtained it at one or two places off the Durham and Yorkshire Coasts. Mr I. C. Thompson obtained it at Port Erin, Isle of Man ; Moray Firth amongst *Fillograna* (A. S.): these, with the present record of its occurrence in the Forth, are the only British records for this species known to me.

Pseudanthessius sauvagei, Canu.
    1891. *Pseudanthessius sauvagei*, Canu (10 *b.*), p. 481.
    1892. *Pseudanthessius sauvagei*, Canu (11), p. 243, pl. xxv.
        figs. 1–17.
    1894. *Pseudanthessius sauvagei*, T. and A. Scott (31), p. 146.

  Habitat.—Off St Monans. Rare.

Want of time has prevented the preparation of drawings of this interesting species for the present paper.

Pseudanthessius gracilis, Claus.
    1889. *Pseudanthessius gracilis*, Claus (13), vol. viii. p. 344, pl. iv.
        figs. 1–7.
    1893. *Pseudanthessius gracilis*, T. and A. Scott (32), p. 241, pl.
        xii. figs. 15–20.

  Habitat.—Off Musselburgh, among material collected in 1891. Some specimens were also taken in the Moray Firth among *Filograna implexa*. One of the Moray Firth specimens measured 1·3 mm. ($\frac{1}{19}$th of an inch). The anterior antennæ are shorter than the first body segment, and seven-jointed ; the third and the last joints are the shortest. The proportional lengths are shown by the formula:—

$$\frac{17 \;.\; 20 \;\cdot\; 8 \;\cdot\; 14 \;\cdot\; 16 \;\cdot\; 13 \;\cdot\; 9 \;\cdot}{1 \;\cdot\; 2 \;.\; 3 \;\cdot\; 4 \;\cdot\; 5 \;\cdot\; 6 \;\cdot\; 7 \;\cdot}$$

The third joint of the posterior antennæ is very small. The anterior foot-jaws are slender, and armed with a few strong teeth on the upper edge. A plumose seta springs from the inner edge near the base of the second joint. Posterior foot-jaws three-jointed. Second joint somewhat dilated, and bearing a short stout spine. The last joint very small, and armed with two terminal spines. Inner branches of the fourth pair of thoracic feet one-jointed, scarcely reaching to the end of the second joint of the outer branches, and armed with two dagger-shaped spines at the truncate apex. There is a small hook-like process near the middle of the inner margin. Fifth pair small, sub-quadrate, and furnished at the apex with an elongate dagger-like spine and a plain seta. Caudal stylets equal to about twice the length of the last abdominal segment. A small seta springs from near the middle of the outer margin of each stylet in addition to the terminal setæ.

  This species differs from *Pseudanthessius thorellii* (B. and R.), with which it is closely allied, in the form of the anterior foot-jaws in the proportional length of the inner branches of the fourth pair of thoracic feet, and in the form of the abdomen.

Family ASCOMYZONTIDÆ

Genus *Dermatomyzon*, Claus (1889).

[*Cyclopicera*, Brady (in part)].

Dermatomyzon gibberum, T. and A. Scott. (Pl. X. figs. 26–34.)
    1894. *Dermatomyzon gibberum*, T. and A. Scott, p. 144, pl. ix.
        figs. 10–14.

  Description.—Female. Length, ·5 mm. ($\frac{1}{50}$ of an inch). Cephalo-thorax broadly ovate, or pear-shaped. Abdomen very short ; its length, including that of the stylets, is scarcely equal to one-fourth of the length of cephalo-thorax. Anterior antennæ stout, seventeen-jointed. The second basal joint appears to be composed of two or three coalesced joints, as shown in the formula, which gives approximately the proportional lengths of the joints :—

$$\frac{40 \cdot (19 \cdot 10 \cdot 7) \cdot 8 \cdot 10 \cdot 12 \cdot 16 \cdot 9 \cdot 9 \cdot 16 \cdot 12 \cdot 12 \cdot 12 \cdot 12 \cdot 16 \cdot 18 \cdot 8 \cdot 26 \cdot}{1 \cdot \quad \cdot 2 \cdot \quad \cdot 3 \cdot 4 \cdot 5 \cdot 6 \cdot 7 \cdot 8 \cdot 9 \cdot 10 \cdot 11 \cdot 12 \cdot 13 \cdot 14 \cdot 15 \cdot 16 \cdot 17 \cdot}$$

Mandibles stylet-shaped, stout, and elongate. The mandible-palp consists of a single oblong joint, the length of which is greater than twice the breadth ; and three stout, moderately long and nearly equal setæ spring from its truncate apex. The maxillæ are composed of a sub-triangular primary part, furnished with three apical setæ, and a narrow cylindrical secondary part, provided with four setæ at the apex. The anterior foot-jaws have the basal part stout, but the end is slender and curved, and forms a claw-like appendage. The posterior foot-jaws and swimming-feet are somewhat like those of *Dermatomyzon nigripes* (Brady and Robertson). The fifth pair are simple and two-jointed. The first joint is short, and its breadth is about equal to twice the length, and a seta springs from its upper distal angle. The second joint is longer and narrower, the length being about twice the breadth. It is furnished with two apical setæ. The caudal stylets are nearly as long as broad, and are equal to the combined lengths of the last two abdominal segments.

Habitat.—Vicinity of the Bass Rock. Rare.

The very tumid form of the cephalo-thorax and short abdomen give to this species a very curious and striking appearance that at once distinguishes it from all other Copepods known to us.

Genus *Acontiophorus*, Brady (1880).

Acontiophorus elongatus, T. and A. Scott.
    1894. *Acontiophorus elongatus*, T. and A. Scott (34), p. 145, pl. ix. figs. 15–20.

Description.—Female. Length, 1 mm. ($\frac{1}{25}$ of an inch). This is an elongate form. The abdomen is slender, and equal to about two-thirds of the length of the cephalo-thorax. The anterior antennæ are slender, and composed of seventeen joints. The proportional lengths are as follow :—

$$\frac{24 \cdot 12 \cdot 14 \cdot 6 \cdot 6 \cdot 6 \cdot 9 \cdot 6 \cdot 8 \cdot 14 \cdot 12 \cdot 13 \cdot 13 \cdot 14 \cdot 14 \cdot 14 \cdot 24 \cdot}{1 \cdot 2 \cdot 3 \cdot 4 \cdot 5 \cdot 6 \cdot 7 \cdot 8 \cdot 9 \cdot 10 \cdot 11 \cdot 12 \cdot 13 \cdot 14 \cdot 15 \cdot 16 \cdot 17 \cdot}$$

Posterior antennæ and mouth organs are nearly as in *Acontiophorus scutatus* (Brady and Robertson). The mandibles are extremely long and slender, being about equal in length to the elongate siphon. Foot-jaws and swimming-feet also somewhat similar to those of *Acontiophorus scutatus*, but the fourth pair are armed with remarkably broad and stout dagger-shaped spines on the exterior margins of the outer branches, and broad sabre-like terminal spines on both branches. The fifth pair consists each of a single elliptical joint, furnished with three apical setæ. Caudal stylets very short.

Habitat.—Vicinity of the Rock. Frequent.

Remarks.—The slender form of this species, and especially of its elongate and slender abdomen, enables this species to be distinguished from any other described *Acontiophorus*. It differs from the typical *Acontiophorus scutatus* in the posterior antennæ having only one long and slender spine at the apex instead of two lancet-shaped apical spines, but otherwise the species now described is a true *Acontiophorus*.

II. OSTRACODA.

Family CYTHERIDÆ.

Genus *Cytheropteron*, G. O. Sars.

Cytheropteron humile, Brady and Norman.

1889. *Cytheropteron humile*, Brady and Norman (9), p. 219, pl. xx. figs. 4–7.

Habitat.—Largo Bay, and off Limekilns to the west of Queensferry. Rare.

Two specimens were obtained in a dead cyprina shell dredged in Largo Bay, and a third specimen was obtained in material collected off Limekilns. This interesting species seems to be comparatively rare. It has been taken in the Clyde, in the vicinity of Greenock ; and the Marquis de Folin has obtained it in material dredged off Vigo.

III. AMPHIPODA.

Family LYSIANASSIDÆ.

Genus *Acidostoma*, Lilljeborg (1865).

Acidostoma obesum (Spence Bate).

1862. *Anonyx obesus*, Sp. Bate (3), p. 74, pl. xii. fig. 1.
1888. *Anonyx obesus*, D. Robertson (26), p. 17.
1890. *Acidostoma obesum*, G. O. Sars (26), p. 38, pl. xiv. fig. 2.

Habitat.—Between Fidra and the Bass Rock. Not common.

Genus *Orchomene*, Boeck (1870).

Orchomene batei, G. O. Sars.

1882. *Orchomene batei*, G. O. Sars (28), i. p. 81.
1862. *Anonyx edwardsi*, Spence Bate (3), p. 73, pl. xi. fig. 5.
1892. *Orchomene batei*, Robertson (27), p. 11.

Habitat.—Vicinity of the Bass Rock. Rather rare.

Genus *Lepidepecreum*, Spence Bate (1868).

Lepidepecreum carinatum, Bate and Westwood.

1868. *Lepidepecreum carinatum*, B. and W. (4), vol. ii. p. 509.
1862. *Anonyx longicornis*, Spence Bate (3), p. 72 (♂).
1888. *Lepidepecreum carinatum* and *Anonyx longicornis*, Robertson (26), p. 91.
1891. *Lepidepecreum carinatum*, G. O. Sars (29), p. 113, pl. xxxix. fig. 1.

Habitat.—Off St Monans and other parts of the Forth. Not common. I have this species also from the Moray Firth.

Family PHOXOCEPHALIDÆ.

Genus *Harpinia*, Boeck (1876).

Harpinia crenulata, Boeck.

1870. *Harpinia crenulata*, Boeck (7), p. 56.
1891. *Harpinia crenulata*, G. O. Sars (29), p. 158, pl. lv. fig. 2.
1894. *Harpinia crenulata*, T. and A. Scott (31), p. 147.

Habitat.—Vicinity of Inchkeith Island and other parts of the Forth. Not uncommon. The second last pair of pereiopods are comparatively long. The epimeral plates of the last segment of the metasom have the lower distal angle rounded, and provided posteriorly with a single small tooth on either side, or with two or three small teeth of unequal size.

Family AMPELISCIDÆ.

Genus *Haploops*, Lilljeborg (1855).

Haploops tubicola, Lilljeborg.
    1855. *Haploops tubicola*, Lilljeborg (21), p. 134.
    1868. *Haploops tubicola*, Bate and Westwood (4), vol. ii. p. 505.
    1888. *Haploops tubicola*, D. Robertson (26), p. 22.
    1891. *Haploops tubicola*, G. O. Sars (29), p. 192, pl. lxvii.

Habitat.—Vicinity of the Bass Rock (1892), and in other parts of the Forth area. Though not recorded till now, this species has been in my possession for a considerable time. The Forth specimens are comparatively small. Another species of the Ampeliscidæ, *Byblis gaimardii*, Kröyer, has been, on the authority of Metzger,[*] recorded from the vicinity of St Abb's Head, at the mouth of the Firth of Forth (*see* Leslie and Herdman's 'Invertebrate Fauna of the Firth of Forth,' p. 105 (Appendix), and 'Revised List of Crustacea of the Firth of Forth,' by the author); but G. O. Sars, in his new work on 'The Crustacea of Norway,' does not include the British Islands in his notes on the distribution of the species. It is, perhaps, therefore right to state that I am able to corroborate Metzger's record from having been fortunate in capturing a fine specimen of *Byblis gaimardii*, Kröyer, near the May Island, in January 1890. Had there been time a description with drawings of the species would have been prepared for the present Report, but this may be done later.

Family AMPHILOCHIDÆ.

Genus *Amphilochoides*, G. O. Sars (1892).

Amphilochoides pusillus, G. O. Sars.
    1892. *Amphilochoides pusillus*, G. O. Sars (29), p. 222, pl. lxxvi.
        fig. 1.
    1894. *Amphilochoides pusillus*, T. and A. Scott (31), p. 147.

Habitat.—Vicinity of the Bass Rock. Not common. This species has no tooth at the base of the dactylus of the first pair of gnathopods, and the palm of the second pair of gnathopods is finely serrate only on the distal half; the other half is even, or nearly so, and bears a few minute setæ.

Family STENOTHOIDÆ.

Genus *Metopa*, Boeck (1870).

Metopa propinqua, G. O. Sars.
    1892. *Metopa propinqua*, G. O. Sars (29), p. 264, pl. xciii. fig. 1.

Habitat.—Off Crail, 1892. Rare.
The telson of this species is furnished with three strong denticles on each side of the upper surface and near to the margin. The second gnathopods are also moderately stout. The Forth specimen agrees with Sars' description of the species in all its principal characters. Several of the species of *Metopa* are very small and troublesome to diagnose.

    [*] Crustacea u. Mollusca v. d. Nordseefahrt d. 'Pomerania,' 1872. (Berlin, 1875.)

Family LEUCOTHOIDÆ.

Genus *Leucothoë*, Leach (1814).

Leucothoë lilljeborgii, Boeck, 1860.
    1860. *Leucothoë lilljeborgii*, Boeck, Forhandl. Skand. Naturf.
        8 de Mode.
    1888. *Leucothoë furina*, Chevreux (non furina Savigny) (12), p. 9.
    1889. *Leucothoë imparicornis*, Norman (24), p. 114, pl. x. figs. 1–4.
    1892. *Leucothoë incisa*, Robertson (27), p. 23.

Habitat.—Vicinity of the Bass Rock. Rare. This species is readily recognised by the form of the last pair of epimeral plates of the metasome.

Family PARAMPHITHOIDÆ.

Genus *Paramphithoë*, Bruzelius (1859).

Paramphithoë monocuspis, G. O. Sars.
    1893. *Paramphithoë monocuspis*, G. O. Sars (29), p. 351, pl. cxxiii.
        fig. 2.

I record this species for the Firth of Forth on the authority of Dr A. M. Norman. It was sent to him by Dr Henderson in 1884. Some time ago, when looking over a number of Forth Amphipoda, I observed one of the specimens with a single dorsal cusp, which, on further examination, was found to resemble this species fairly well: it was, however, laid aside for further study.

Paramphithoë assimilis, G. O. Sars.
    1882. *Paramphithoë assimilis*, G. O. Sars (28), p. 99, pl. v. fig. 1.
    1888. *Paramphithoë assimilis*, D. Robertson (26), p. 94.
    1893. *Paramphithoë assimilis*, G. O. Sars (29), p. 352, pl. cxxiv.
        fig. 1.

Habitat.—Found adhering to some zoophytes dredged between the Island of Inchkeith and the May. The Forth specimens belong to *Paramphithoë assimilis*, G. O. Sars, as described in 'Crustacea of Norway,' and not to *Pleustis glaber*, Boeck (*Parapleustis glaber* in 'Crustacea of 'Norway,' p. 358).

Family IPHIMEDIDÆ.

Genus *Iphimedia*, Rathke (1843).

Iphimedia minuta, G. O. Sars.
    1882. *Iphimedia minuta*, G. O. Sars (28), p. 100, pl. v. fig. 2.
    1892. *Iphimedia minuta*, D. Robertson (27), p. 23.
    1893. *Iphimedia minuta*, G. O. Sars (29), p. 379, pl. cxxxiii. fig. 1.

Habitat.—North end of Inchkeith, and in other parts of the Forth.
This species has been known to me for a considerable time as quite distinct from *Iphimedia obesa*. Though not a common species it has been obtained at several places within the estuary.

Family ATYLIDÆ.

Genus *Apherusa.*

Apherusa borealis (Boeck).
    1893. *Apherusa borealis*, G. O. Sars (29), p. 441, pl. cxv. fig. 2.

Habitat.—In various places within the Forth area, between the Island of Inchkeith and the May, as well as outside of the May Island. Sometimes frequent in the tow-net collections. The two dorsal cusps appear to be stronger in the male than in the female. The telson is of the form of narrow triangular plate with a pointed extremity, and furnished with a minute hair on each side of the apex.

ANNELIDA.

Family NEMERTIDÆ.

Genus *Cerebratulus.*

Cerebratulus angulatus (O. F. Müller).
> 1853. *Gordius fragilis*, Sir John Dalyell (15), vol. ii. p. 55, pls. vi., vii., vii*.
> 1874. *Cerebratulus angulatus*, W. C. M'Intosh (22), p. 175.
> 1894. *Cerebratulus angulatus*, T. Scott, Ann. and Mag. Scot. Nat. Hist., Part 10, p. 118.

A specimen of *Cerebratulus angulatus*, measuring about 14 inches in length, was obtained among some trawl refuse from Largo Bay (off the Wemyss). I cannot find any previous record of *Cerebratulus* having been taken in the Forth. Harry Goodsir and Sir John Dalzell publish records of this species for Scotland, but do not give the localities.

MOLLUSCA.

CONCHIFERA.

Family PANDORIDÆ.

Genus *Lyonsia*, Turton.

Lyonsia norvegica (Chemnitz).
> 1865. *Lyonsia norvegica*, J. G. Jeffreys (19), vol. iii. p. 29 (vol. v. p. 190, pl. xlviii. fig. 2.)

Habitat.—Vicinity of the Bass Rock ; depth of water 22-23 fathoms. Bottom, sandy mud. A living specimen of this mollusc was taken a short distance north-east of the Bass Rock on 30th October 1893. There does not appear to have been hitherto any very satisfactory record of the occurrence of *Lyonsia* in the Firth of Forth. There is no record of its occurrence in the Forth, either by the Rev. W. Wood (East Neuk of Fife), or by Leslie and Herdman (Invertebrate Fauna of the Firth of Forth). Dr Henderson, in his notes on the Forth Invertebrates, in the Proceedings of the Royal Physical Society of Edinburgh (1883–84), records the discovery of 'a single broken valve at Newhaven' ; but he states distinctly that 'in all cases where " Newhaven " is given as the locality, it must be ˙ understood that the specimens are from the fishermen's lines, and probably ' taken to the east of the May Island.' The present is therefore the only satisfactory record I know of for *Lyonsia norvegica* as a member of the Forth fauna.

The following is a list of some of the works more particularly referred to in this contribution towards a Natural History of the Firth of Forth :—

1. 1883.   AURIVILLIUS, C. W. S., *Bidrag till Kännedomen om Krusta-ceer, som lefva hos Mollusker och Tunikater ; Akademisk Afhandling.*

2. 1850.   BAIRD, W., 'Natural History of British Entomostraca';
             Ray Society.
3. 1862.   BATE, C. S., *Catalogue of Amphipodous Crustacea in
             the British Museum.*
4. 1863–68. BATE, C. S., and WESTWOOD, *A History of the British
             Sessile-eyed Crustacea*, vol. i. 1863, vol. ii. 1868.
5. 1864.   BOECK, A., 'Oversigt over de ved Norges Kyster jagg-
             tagne Copepoder henhorende til *Calanidernes,' Cyclopi-
             dernes Familier ;* Forhandl. i. Vidensk, Selskab i.
             Christiania.
6. 1872.   BOECK, A., *Nye Slaegter og Arter of Saltvands Cope-
             poder ;* Forhandl. i. Vidensk, Selskab i. Christiania.
7. 1870.   BOECK, A., *Crustacea Amphipoda borealis et Arctic,* Chris-
             tiania.
7a. 1872.   BRADY, *Nat. Hist. Trans.,* Northumberland and Dur-
             ham, vol. iv.
8. 1878–80. BRADY, G. S., 'A Monograph of the Free and Semi-para-
             sitic Copepoda of the British Islands,' Ray Society.
             Vol. i. 1878.
9. 1889.   BRADY AND NORMAN, 'Monograph of the Marine and
             Fresh Water Ostracoda of the North Atlantic and
             North-western Europe.'
10. 1875.   BRADY AND ROBERTSON, 'Report on Dredging off the Coast
             of Durham and North Yorkshire in 1874.' In Rep.
             45. Meet. British Ass., Adv. Ser.
10a. 1890.   CANU, EUGÈNE, 'Les Copepodes marins du Boulonnais'
             (*Bulletin Scien. de la France, et de la Belgique*, vol.
             xxii.).
10b. 1891.   CANU, EUGÈNE, 'Les Copepodes marins du Boulonnais'
             (*idem*, vol. xxiii.).
11. 1892.   CANU, EUGÈNE, 'Les Copepodes du Boulonnais, Morpho-
             logie, Embryologie, Taxonomie (*Travaux du Labor. de
             Zool. Maritime de Wimereux*, Ambleteuse, vol. vi.).
12. 1888.   CHEVREUX, E., *Bulletin de la Société d'etudis scientifiques
             de Paris* 11e année 1er semestre.
13. 1889.   CLAUS, C., 'Ueber neue oder wenig bekennte halbpara-
             sitische, Copepoden, insbesondere der Lichomolgiden,—
             und Ascomyzontiden—gruppen,' Arl. Zool. Inst. Wien.
             t. viii.
14. 1863.   CLAUS, C., *Die freilebenden Copepoden, mit besonderer
             Berücksichtig ung. der Fauna Deutschlands, der Nordsee
             und des Mittelmeeres,* Leipsic.
15. 1855.   DALYELL, Sir John, *Powers of the Creator.*
16. 1882.   GIESBRECHT, W., *Die freilebenden Copepoden, der Kieler
             Föhrde vii.tes Jahresbericht d. Commiss. f. wiss. Unters.
             d. deutschen Meere.*
17. 1892.   GIESBRECHT, W., *Fauna und Flora des Golfes von Neapel
             (Monographie Pelagische Copepoden systematik und
             faunistik).*
18. 1845.   GOODSIR, H., 'Descriptions of some Gigantic Forms of
             Invertebrate Animals from the Coast of Scotland.'
             (*Annals and Magazine of Natural History,* vol. xv.).
19. 1862–69. JEFFREYS, J. GWYN, *British Conchology.*
20. 1881.   LESLIE AND HERDMAN, *The Invertebrate Fauna of the Firth
             of Forth.*
21. 1855.   LILLJEBORG, W., *Ofvers af Vetinskap. Akad. Forhandl.*
22. 1874.   M'INTOSH, W. C., *Monograph of British Annelids.*

23. 1875. M'Intosh, W. C., *The Marine Fauna of St Andrews Bay.*
24. 1889. Norman, A. M., 'Notes on British Amphipoda—1. Megaluropes, n.g., and some Œdiceridæ.' (*Annals and Magazine of Natural History*, June.)
25. 1892. Norman, A. M., *idem*, 'Families Leucothoidæ, Pardaliscidæ, and Gammaridæ (Marine).'
26. 1888. Robertson, D., *A Contribution towards a Catalogue of the Amphipoda and Isopoda of the Clyde.*
27. 1892. Robertson, D., *A Second Contribution towards a Catalogue of the Amphipoda and Isopoda of the Clyde.*
28. 1882. Sars, G. O., *Oversigt af Norges Crustaceer*, pt. i.
29. 1890–93. Sars G. O., *An Account of the Crustacea of Norway*, pts. i.–xxi.
30. 1894. Scott, T., 'Report on Entomostraca from the Gulf of Guinea' (*Transactions of the Linnean Society, London*, 2nd ser., Zoology, vol. vi.), vol. vi. pt. i.
31. 1894. Scott, T. and A., 'On some New and Rare Crustacea from Scotland.' (*Annals and Magazine of Natural History*, ser. vi., vol. xii., February.)
32. 1893. Scott, T. and A., 'On some New and Rare Crustacea from Scotland.' (*Annals and Magazine of Natural History*, ser. vi., vol. xii., October.)
33. 1893. Thompson, I. C., *Revised Report on the Copepoda of Liverpool Bay.*
34. 1888–90. Walker, A. O., 'Higher Crustacea of Liverpool Bay.' (*Trans. Biol. Soc., Liverpool.*)
35. 1891. Walker, A. O., 'On Pherusa fucicola (Leach).' (*Annals and Magazine of Natural History*, ser. vi., vol. vii.
36. 1885. White, A., *History of the British Crustacea.*
37. 1862. Wood, Rev. W., *The East Neuk of Fife; its Histories and Antiquities: Geology, Botany, and Natural History in General.*
38. 1885. Wright, R., *On a Parasite Copepod of the Clam* (Myicola metisiensis).

DESCRIPTION OF THE PLATES.

PLATE V.

Pseudocyclopia caudata, sp. nov.

Fig. 1. Female, lateral view, × 32 diameters.
Fig. 2. Anterior antenna of the same, × 456 ,,
Fig. 3. Foot of first pair, × 456 ,,
Fig. 4. Foot of second pair, × 190 ,,
Fig. 5. Foot of third pair, × 168 ,,
Fig. 6. Foot of fourth pair, × 338 ,,
Fig. 7. Fifth pair of feet, × 570 ,,
Fig. 8. Abdomen and caudal stylets, × 95 ,,

Cyclopina elegans, sp. nov.

Fig. 9.	Female, dorsal view,	.	.	.	.	× 54 diameters
Fig. 10.	Anterior antenna—female,	.	.	.	.	× 300 ,,
Fig. 11.	Anterior antenna—male	.	.	.	.	× 300 ,,
Fig. 12.	Posterior antenna,	.	.	.	.	× 338 ,,
Fig. 13.	Mandible and palp,	.	.	.	.	× 253 ,,
Fig. 14.	Maxilla,	.	.	.	.	× 253 ,,
Fig. 15.	Anterior foot-jaw,	.	.	.	.	× 253 ,,
Fig. 16.	Posterior foot-jaw,	.	.	.	.	× 500 ,,
Fig. 17.	Foot of first pair,	.	.	.	.	× 300 ,,
Fig. 18.	Foot of fourth pair,	.	.	.	.	× 200 ,,
Fig. 19.	Foot of fifth pair,	.	.	.	.	× 380 ,,

Ameira reflexa, sp. nov.

Fig. 20.	Female, lateral view,	.	.	.	.	× 106 diameters.
Fig. 21.	Anterior antenna of the same,	.	.	.	.	× 500 ,,
Fig. 22.	Posterior antenna,	.	.	.	.	× 500 ,,
Fig. 23.	Mandible and palp,	.	.	.	.	× 500 ,,
Fig. 24.	Posterior foot-jaw,	.	.	.	.	× 760 ,,
Fig. 25.	Foot of first pair,	.	.	.	.	× 253 ,,
Fig. 26.	Foot of fourth pair,	.	.	.	.	× 135 ,,
Fig. 27.	Foot of fifth pair,	.	.	.	.	× 253 ,,
Fig. 28.	Part of abdomen and caudal stylets,	.	.	.	.	× 228 ,,

Ameira longiremis, sp. nov.

Fig. 29.	Posterior antenna,	.	.	.	.	× 380 diameters.
Fig. 30.	Posterior foot-jaw,	.	.	.	.	× 456 ,,
Fig. 31.	Foot of fifth pair,	.	.	.	.	× 380 ,,
Fig. 32.	Caudal stylets,	.	.	.	.	× 152 ,,

PLATE VI.

Ameira longiremis, sp. nov.

Fig. 1.	Female, lateral view,	.	.	.	.	× 96 diameters.
Fig. 2.	Anterior antenna of the same,	.	.	.	.	× 456 ,,
Fig. 3.	Mandible and palp,	.	.	.	.	× 380 ,,
Fig. 4.	Foot of first pair,	.	.	.	.	× 380 ,,
Fig. 5.	Foot of fourth pair,	.	.	.	.	× 380 ,,

Ameira longiremis, var. *intermedia.*

Fig. 6.	Female, lateral view,	.	.	.	.	× 53 diameters.
Fig. 7.	Anterior antenna of the same,	.	.	.	.	× 456 ,,
Fig. 8.	Posterior antenna,	.	.	.	.	× 380 ,,
Fig. 9.	Mandible and palp,	.	.	.	.	× 456 ,,
Fig. 10.	Posterior foot-jaw,	.	.	.	.	× 500 ,,
Fig. 11.	Foot of first pair,	.	.	.	.	× 380 ,,
Fig. 12.	Foot of fourth pair,	.	.	.	.	× 190 ,,
Fig. 13.	Foot of fifth pair,	.	.	.	.	× 380 ,,
Fig. 14.	Abdomen and caudal stylets,	.	.	.	.	× 190 ,,

Ameira exigua, sp. nov.

Fig. 15.	Female lateral view,	.	.	.	.	× 106 diameters.
Fig. 16.	Anterior antenna of the same,	.	.	.	.	× 500 ,,
Fig. 17.	Posterior antenna,	.	.	.	.	× 500 ,,
Fig. 18.	Mandible and palp,	.	.	.	.	× 500 ,,
Fig. 19.	Posterior foot-jaw,	.	.	.	.	× 760 ,,
Fig. 20.	Foot of first pair,	.	.	.	.	× 380 ,,
Fig. 21.	Foot of fourth pair,	.	.	.	.	× 380 ,,
Fig. 22.	Foot of fifth pair,	.	.	.	.	× 456 ,,
Fig. 23.	Abdomen and caudal stylets,	.	.	.	.	× 127 ,,

Laophonte depressa, sp. nov.

Fig. 24.	Female, dorsal view,	× 54 diameters.
Fig. 25.	Mandible and palp, .	× 380 ,,
Fig. 26.	Posterior foot-jaw, .	× 254 ,,
Fig. 27.	Foot of first pair, .	× 168 ,,
Fig. 28.	Foot of fourth pair, .	× 126 ,,
Fig. 29.	Foot of third pair—male, .	× 190 ,,
Fig. 30.	Foot of fifth pair—female, .	× 254 ,,
Fig. 31.	Foot of fifth pair—male (*a.* appendage to first abdominal segment), .	× 380 ,,

PLATE VII.

Laophonte depressa, sp. nov.

Fig. 1.	Anterior antenna of female, .	× 254 diameters.
Fig. 2.	Anterior antenna of male, .	× 254 ,,
Fig. 3.	Posterior antenna, .	× 254 ,,

Tetragoniceps consimilis, sp. nov.

Fig. 4.	Female, lateral view,	× 106 diameters.
Fig. 5.	Anterior antenna of the same,	× 254 ,,
Fig. 6.	Posterior antenna, .	× 500 ,,
Fig. 7.	Mandible and palp, .	× 380 ,,
Fig. 8.	Posterior foot-jaw, .	× 500 ,,
Fig. 9.	Foot of first pair, .	× 380 ,,
Fig. 10.	Foot of fourth pair, .	× 380 ,,
Fig. 11.	Foot of fifth pair, .	× 152 ,,
Fig. 12.	Abdomen and caudal stylets,	× 190 ,,

Laophonte denticornis, sp. nov.

Fig. 13.	Female, dorsal view,	× 80 diameters.
Fig. 14.	Anterior antenna—female, .	× 338 ,,
Fig. 15.	Anterior antenna—male, .	× 338 ,,
Fig. 16.	Posterior antenna, .	× 338 ,,
Fig. 17.	Mandible and palp, .	× 500 ,,
Fig. 18.	Posterior foot-jaw, .	× 380 ,,
Fig. 19.	Foot of first pair, .	× 338 ,,
Fig. 20.	Foot of fourth pair, .	× 380 ,,
Fig. 21.	Foot of third pair—male, .	× 250 ,,
Fig. 22.	Foot of fifth pair—female, .	× 250 ,,
Fig. 23.	Foot of fifth pair—male (*a.* appendage to first abdominal segment), .	× 380 ,,

Laophonte simulans, sp. nov.

Fig. 24.	Female, dorsal view,	× 106 diameters.
Fig. 25.	Posterior antenna, .	× 380 ,,
Fig. 26.	Mandible and palp, .	× 380 ,,
Fig. 27.	Posterior foot-jaw, .	× 380 ,,
Fig. 28.	Foot of first pair, .	× 380 ,,
Fig. 29.	Foot of third pair, .	× 380 ,,
Fig. 30.	Foot of fourth pair, .	× 500 ,,
Fig. 31.	Foot of fifth pair, .	× 380 ,,
Fig. 32.	Abdomen and caudal stylets,	× 190 ,,

PLATE VIII.

Laophonte simulans, sp. nov.

Fig. 1.	Anterior antenna—female, .	× 380 diameters.

T

Laophontodes typicus, gen. et sp. nov.

Fig. 2. Female, dorsal view,	× 190	diameters.
Fig. 3. Anterior antenna of the same,	× 500	,,
Fig. 4. Posterior antenna,	× 500	,,
Fig. 5. Posterior foot-jaw,	× 500	,,
Fig. 6. Foot of first pair,	× 500	,,
Fig. 7. Foot of fourth pair,	× 500	,,
Fig. 8. Foot of fifth pair,	× 500	,,

Pontopolites typicus, gen. et sp. nov.

Fig. 9. Female, lateral view,	× 106	diameters.
Fig. 10. Anterior antenna of the same,	× 380	,,
Fig. 11. Posterior antenna,	× 380	,,
Fig. 12. Mandible and palp,	× 760	,,
Fig. 13. Posterior foot-jaw,	× 760	,,
Fig. 14. Foot of first pair,	× 380	,,
Fig. 15. Foot of fourth pair,	× 380	,,
Fig. 16. Foot of fifth pair,	× 380	,,
Fig. 17. Abdomen and caudal stylets,	× 127	,,

Cletodes curvirostris, sp. nov.

Fig. 18. Female, lateral view,	× 70	diameters.
Fig. 19. Anterior antenna of the same,	× 380	,,
Fig. 20. Posterior antenna,	× 380	,,
Fig. 21. Mandible and palp,	× 500	,,
Fig. 22. Posterior foot-jaw,	× 760	,,
Fig. 23. Foot of first pair,	× 380	,,
Fig. 24. Foot of fourth pair,	× 386	,,
Fig. 25. Foot of fifth pair,	× 380	,,
Fig. 26. Abdomen and caudal stylets,	× 106	,,

Heteropsyllus curticaudatus, gen. et sp. nov.

Fig. 27. Female, lateral view,	× 160	diameters.
Fig. 28. Anterior antenna of the same,	× 380	,,
Fig. 29. Posterior antenna,	× 760	,,
Fig. 30. Mandible and palp,	× 760	,,
Fig. 31. Posterior foot-jaw,	× 760	,,
Fig. 32. Foot of first pair,	× 380	,,
Fig. 33. Foot of fourth pair,	× 300	,,
Fig. 34. Abdomen and caudal stylets,	× 190	,

PLATE IX.

Heteropsyllus curticaudatus, gen. et sp. nov.

Fig. 1. Foot of fifth pair,	× 380	diameters.

Leptopsyllus typicus, gen. et sp. nov.

Fig. 2. Female, dorsal view,	× 106	diameters.
Fig. 3. Anterior antenna of the same,	× 500	,,
Fig. 4. Posterior antenna,	× 500	,,
Fig. 5. Mandible and palp,	× 760	,,
Fig. 6. Anterior foot-jaw,	× 760	,,
Fig. 7. Posterior foot-jaw,	× 760	,,
Fig. 8. Foot of first pair,	× 760	,,
Fig. 9. Foot of third pair,	× 760	,,
Fig. 10. Foot of fourth pair,	× 760	,,
Fig. 11. Foot of fifth pair,	× 760	,,

Dactylopus coronatus, sp. n.

Fig. 12. Female, dorsal view, ... × 80 diameters.
Fig. 13. Anterior antenna of the same, . . × 380 ,,
Fig. 14. Posterior antenna, ... × 253 ,,
Fig. 15. Mandible and palp, ... × 253 ,,
Fig. 16. Posterior foot-jaw, ... × 380 ,,
Fig. 17. Foot of first pair, ... × 253 ,,
Fig. 18. Foot of fourth pair, ... × 190 ,,
Fig. 19. Foot of fifth pair, ... × 380 ,,
Fig. 20. Abdomen and caudal stylets, . . × 126 ,,

Pseudowestwoodia andrewi, gen. et sp. nov.

Fig. 21. Female, lateral view, ... × 106 diameters.
Fig. 22. Anterior antenna of the same, . . × 500 ,,
Fig. 23. Posterior antenna, ... × 500 ,,
Fig. 24. Mandible and palp, ... × 500 ,,
Fig. 25. Posterior foot-jaw, ... × 760 ,,
Fig. 26. Foot of first pair, ... × 500 ,,
Fig. 27. Foot of fourth pair, ... × 380 ,,
Fig. 28. Foot of fifth pair, ... × 760 ,,
Fig. 29. Abdomen and caudal stylets, . . × 380 ,,

Ameira exilis, T. and A. Scott.

Fig. 30. Male spermatophore, ... × 160 diameters.

PLATE X.

Ameira exilis, T. and A. Scott.

Fig. 1. Female, lateral view, ... × 53 diameters.
Fig. 2. Anterior antenna—female, . . . × 168 ,,
Fig. 3. Anterior antenna—male, . . . × 190 ,,
Fig. 4. Posterior antenna, ... × 254 ,,
Fig. 5. Mandible and palp, ... × 500 ,,
Fig. 6. Posterior foot-jaw, ... × 500 ,,
Fig. 7. Foot of first pair, ... × 190 ,,
Fig. 8. Foot of fourth pair, ... × 126 ,,
Fig. 9. Foot of third pair—male, . . . × 126 ,,
Fig. 10. Foot of fifth pair—female, . . × 190 ,,
Fig. 11. Foot of fifth pair—male (*a.* appendage to first abdominal
        segment), ... × 190 ,,
Fig. 12. Abdomen and caudal stylets, . . × 80 ,,

Thalestris forficuloides, T. and A. Scott.

Fig. 13. Female, lateral view, ... × 80 diameters.
Fig. 14. Anterior antenna—female, . . . × 254 ,,
Fig. 15. Anterior antenna—male, . . . × 254 ,,
Fig. 16. Posterior antenna, ... × 500 ,,
Fig. 17. Mandible and palp, ... × 500 ,,
Fig. 18. Maxilla, ... × 380 ,,
Fig. 19. Posterior foot-jaw, ... × 500 ,,
Fig. 20. Foot of first pair, ... × 254 ,,
Fig. 21. Foot of fourth pair, ... × 190 ,,
Fig. 22. Foot of third pair—male, . . . × 190 ,,
Fig. 23. Foot of fifth pair—female, . . × 127 ,,
Fig. 24. Foot of fifth pair—male, . . . × 254 ,,
Fig. 25. Abdomen and caudal stylets, . . × 80 ,,

Dermatomyzon gibberum, T. and A. Scott.

Fig. 26. Female, dorsal view, ... × 106 diameters.
Fig. 27. Anterior antenna of the same, . . × 334 ,,
Fig. 28. Mandible and palp, ... × 456 ,,
Fig. 29. Maxilla, ... × 456 ,,
Fig. 30. Anterior foot-jaw, ... × 254 ,,
Fig. 31. Posterior foot-jaw, ... × 380 ,,
Fig. 32. Foot of first pair, ... × 254 ,,
Fig. 33. Foot of fourth pair, ... × 254 ,,
Fig. 34. Foot of fifth pair, ... × 760 ,,

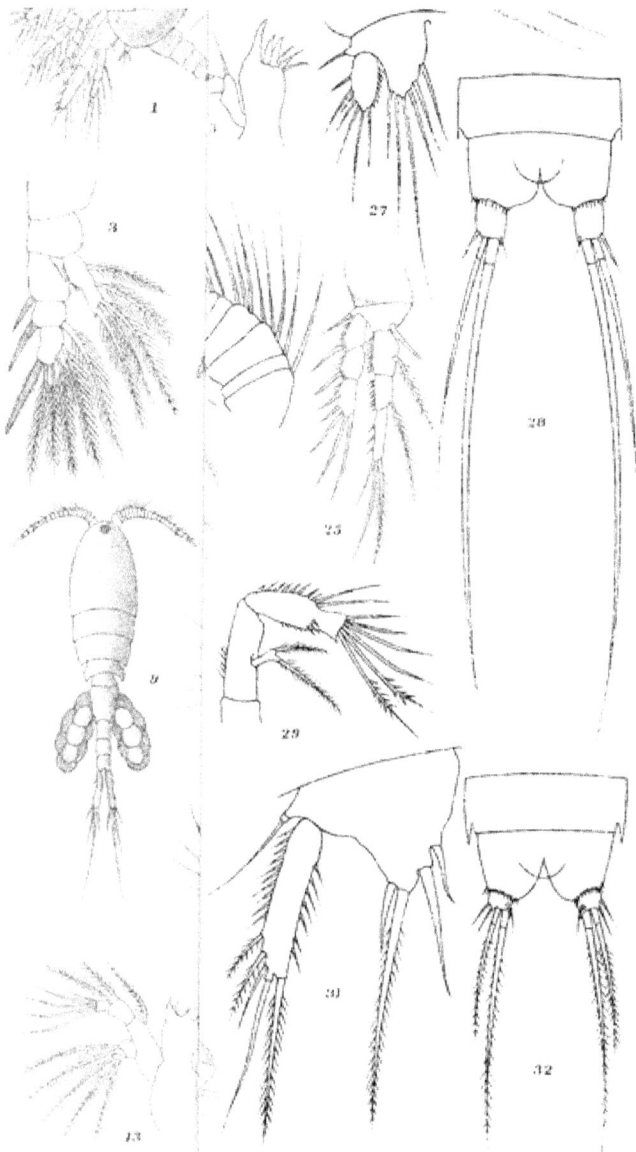

Andrew Scott, del. ad nat.   *Ridewia, sp. nov.*

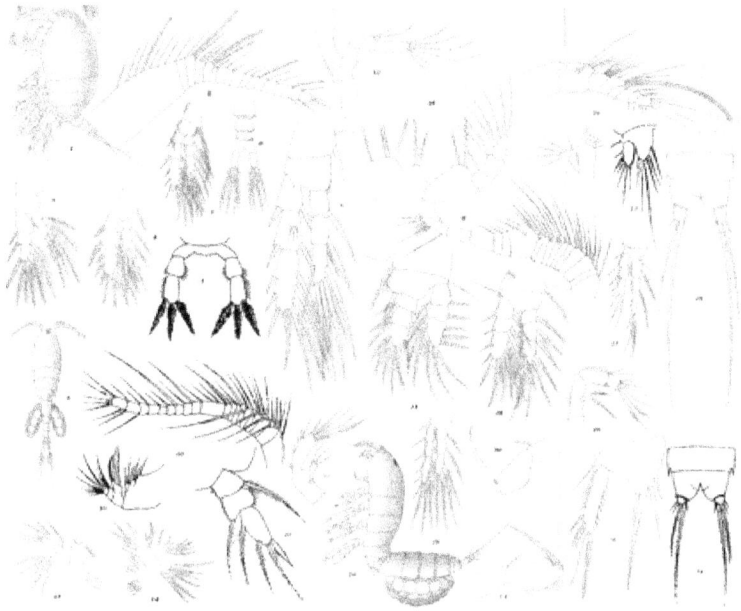

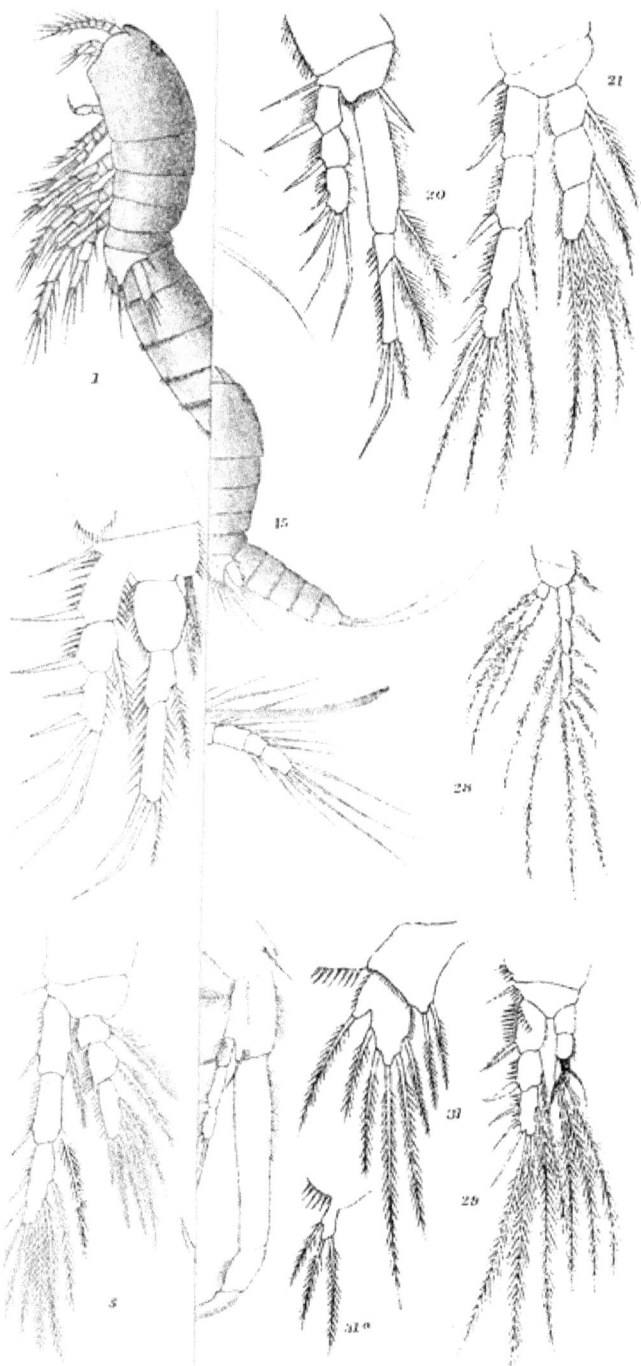

Andrew Scott, del. ad nat.   ...te depressa, sp. nov.

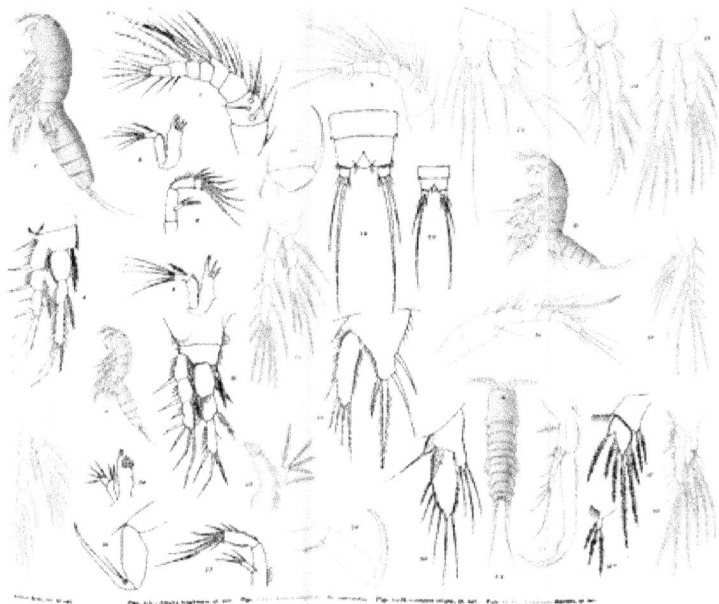

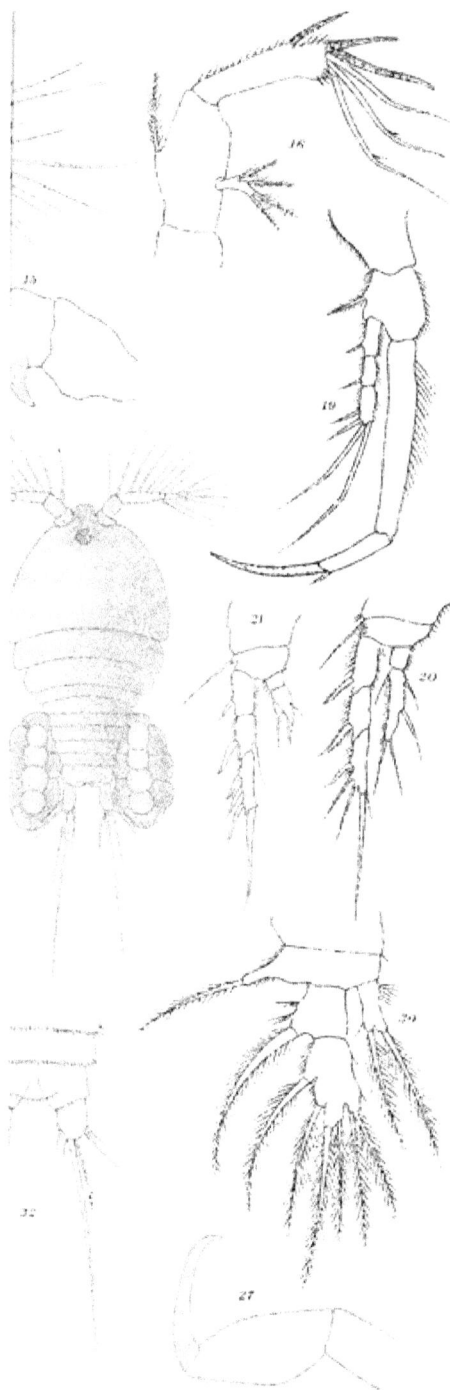

onte simulans, sp. nov.

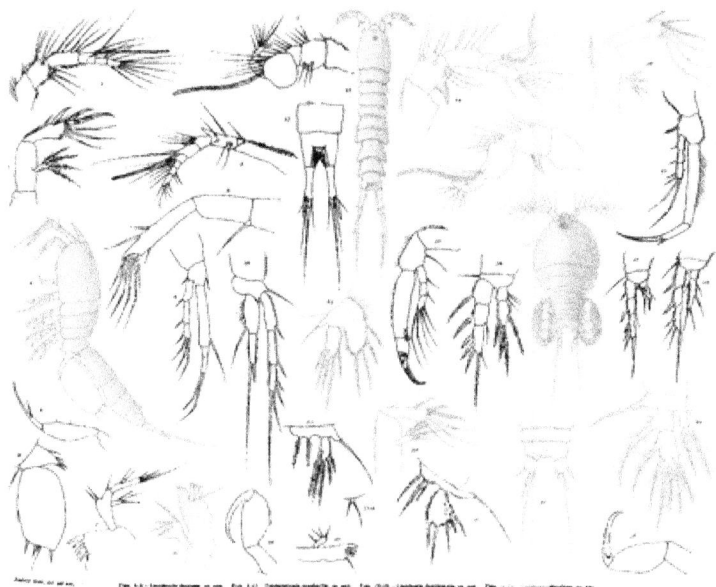

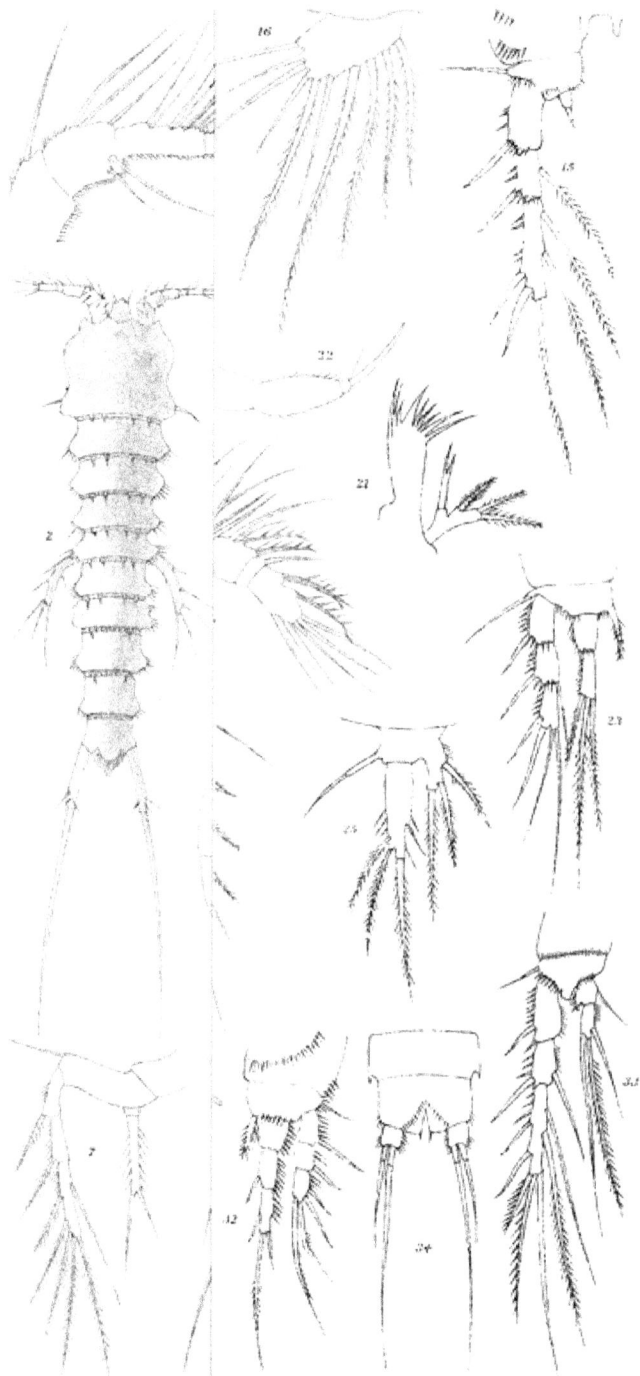

Andrew Scott, del. ad nat.   ...sius curvirostris, sp. nov.

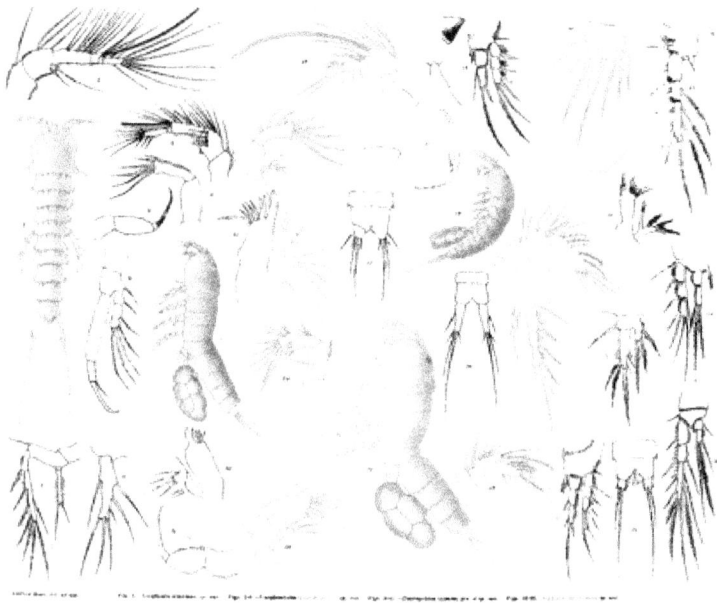

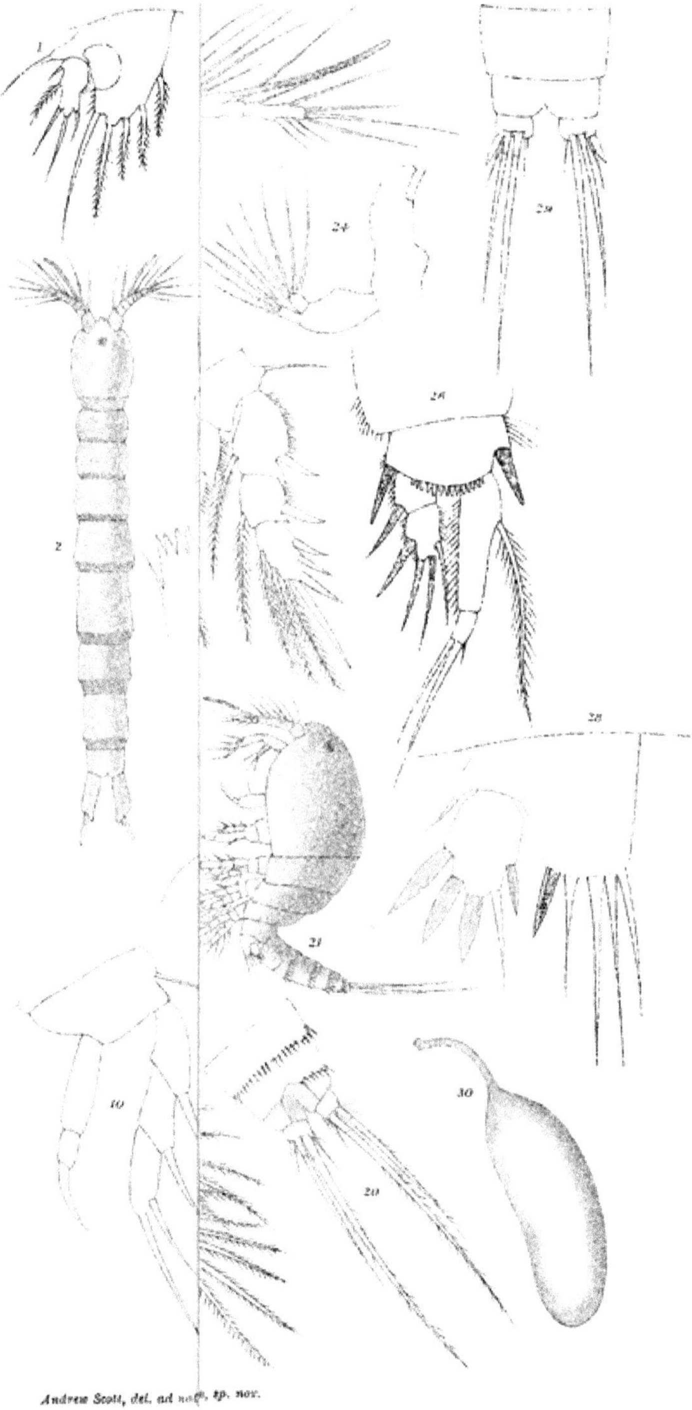

Andrew Scott, del. ad nat., sp. nov.

Fig. 1.—Diatomorphus multicaudatus, gen. n. sp. nov. Fig. 2-3.—Sophadittus lepnora, gen. n. sp. nov. Fig. 11-12.—[illegible] Fig. 13-15.—Paradiodontid nobhas, gen. n. sp. nov. Fig. 16.—[illegible] (P. et J. Sars).

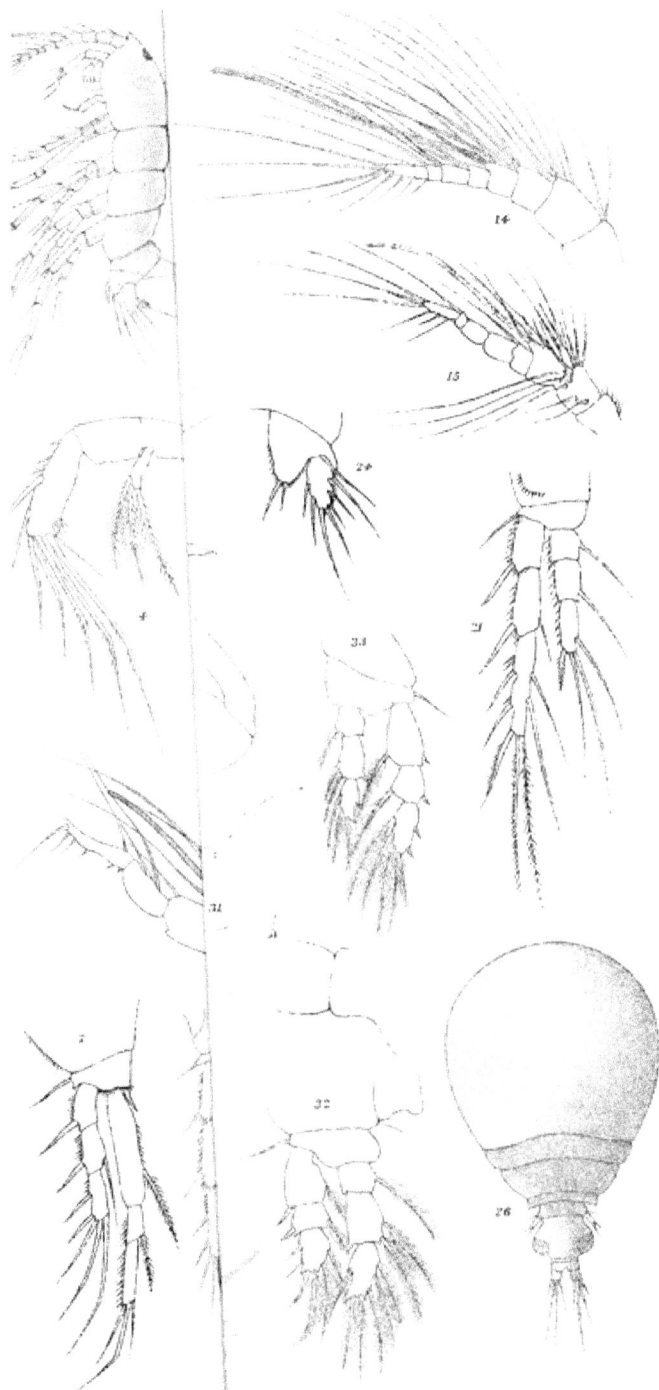

XII HR

■

IV.—THE INVERTEBRATE FAUNA OF THE INLAND WATERS OF SCOTLAND.—PART IV. By Thomas Scott, F.L.S.

In this Report on the invertebrate fauna of the inland waters of Scotland, I propose to describe, first,—the results of a partial examination of Loch Tay in Perthshire ; and second,—the results of the examination of tow-net gatherings and other material from certain Sutherlandshire lochs, collected, and forwarded to Dr Fulton, the Superintendant of Scientific Investigations, by W. S. Caine, Esq., M.P.

1.—Loch Tay, Perthshire.

Introductory.

Loch Tay belongs to the Marquis of Breadalbane and is one of the best salmon lochs in Britain.

The district about Loch Tay is known to botanists throughout the length and breadth of the land, as one of the richest in native alpine and sub-alpine plants, in the British Islands ; some of the rarest of our native ferns, mosses, lichens, as well as flowering plants, have been and may still be obtained among the gullies and rocky crevices about the summits of Ben Lawers and the neighbouring mountains, and every year people interested not only in the British flora but in other departments of natural history as well are frequent visitors in the district.

Though Loch Tay and its surroundings are thus well known not only to the mere pleasure seeker but also to the naturalist, no systematic attempt has apparently been made hitherto to investigate the invertebrate fauna of the loch, and the present contribution towards that object may, therefore, be of interest.

It was during a short visit to the beautiful and picturesque village of Kenmore, in September last year, that I had the privilege of making a partial investigation of this fine Perthshire loch. Owing to the limited time at my disposal I was only able to examine that portion of the east end of the loch extending from the East Bay where the steam-boat wharf is, westward to near Fernan. There are two islands at this end of the loch—the ' Ministers Island '—which is little more than a cairn of stones ; and Aidan's Isle or the Isle of Loch Tay.' King Donald IV. was drowned somewhere in the vicinity of Aidan's Isle ; and here Sibylla—Alexander's Queen—died, and was buried A.D. 1122.

Depth of Loch Tay.

I was unable from want of time to take soundings of the loch, but the following notes from a paper read at a meeting of the Royal Society of Edinburgh, February 20th, 1888, by Mr James S. Grant Wilson of H.M. Geological Survey, may not be out of place. Mr Wilson's paper described the results of a recent bathymetrical survey of the chief Perthshire lochs and their relation to the glaciation of the district, and was illustrated by a carefully prepared chart of the various lochs referred to.[*] In this paper Mr Wilson describes Loch Tay as being 14½ miles in length by about ⅔ of a mile in average width ; the surface level of the

[*] This paper was published in the *Scottish Geographical Magazine* for May 1888.

water, calculated from the bench mark on Kenmore Bridge, is found to be 346 feet above the sea. The general outline of the loch somewhat resembles the letter S in form but with the ends only slightly curved. The lower portion of the letter is represented by the part from Killin to Ardionaig and extends E. 25° N. about 5½ miles, the middle portion is represented by the part from Ardionaig to Fernan and extends N. 30° E. about 5¾ miles, and the upper portion is represented by the part extending from Fernan to Kenmore, and which lies in the same direction as the west end, viz.,—E. 25° N. The length of this portion is about 3½ miles. The deepest part of the loch is opposite Skiag, or a little over 5 miles in a 'bee line' from Kenmore; the depth here is 85 fathoms or 520 feet. From this deep part of the loch the bottom rises gradually but more or less irregularly towards each end. When the loch is standing at its summer level the western margin is often covered with thin patches of bright red sand which, on examination, is found to be composed almost entirely of minute fragments of garnets; numerous fragments of garnetiferous schist may also be obtained scattered about the shore at the east end of the loch.

INVERTEBRATE FAUNA OF LOCH TAY.

The Loch was examined by means of a tow-net worked from a rowing boat kindly placed at my disposal by a friend in Kenmore. No examination was made of the bottom of the Loch except where the water was shallow, that is between the steam-boat pier and the 'Minister's Island' on the south side, and from Aidan's Isle eastward on the north side. Pelagic crustacea were scarce in all the surface and under-surface gatherings, but they included one or two forms of interest, such as the curious *Bythotrephes* and the beautiful *Leptodora*. On the other hand, the material collected by dragging the tow-net through and among the bottom vegetation found growing in some of the shallower parts and especially in the vicinity of the 'Minister's Island' proved to be rich in microorganisms; over twenty species of crustacea were obtained in this way. Several species of Mollusca, Coleoptera, Arachnida, the larvæ of dragonflies and other insects, Rhizopoda, Rotifera, &c. were also observed in the same material. In the following lists I propose to give a record only of the species of Mollusca and Crustacea obtained.

THE MOLLUSCA.

Comparatively few species of *Mollusca* were observed in Loch Tay, and this paucity of molluscan species corresponds with what I have observed in the investigation of other deep fresh-water Lochs; the conditions physical or otherwise of such lochs do not seem to favour the development of the *Mollusca*.

1. LAMELLIBRANCHIATA.

Pisidium pusillum (Gmelin).	Frequent.
,, *fontinale* (Drap.).	Scarce.

2. GASTEROPODA.

Valvata piscinalis (Müller).	Frequent, } but generally of } small size.

U

CRUSTACEA.

The crustacea obtained were not only numerous, especially in the bottom material, but included several comparatively rare and interesting species.

1. COPEPODA.

Diaptomus gracilis (G. O. Sars).	Frequent.	} in tow-net gatherings.
Cyclops signatus, Koch.	Few.	in bottom material.
„ *stremuus*, Fischer.	Few.	} in tow-net gatherings.
„ *viridis* (Jurine).	Few.	}
„ *serrulatus*, Fischer.	Frequent.	
„ *magnotavus*, Cragin.	Very rare.	
„ *macrurus*, G. O. Sars.	Rare.	} in bottom material.
„ *fimbriatus*, Fischer.	Frequent.	
Attheyella crassa (G. O. Sars).	Frequent.	
„ *pygmœa* (G. O. Sars).	Few.	}

2. OSTRACODA.

Cypria serena (Koch).	Few.	}
Cyclocypris globosa (G. O. Sars).	Few.	} in bottom material.
Candona candida. (Müller)	Few.	
„ *lactea.* (Baird)	Few.	}

3. CLADOCERA.

Sida crystallina (Müller).	Scarce.	} in surface and under-surface tow-net gatherings.
Daphnia jardinii, Baird.	Frequent.	
Bosmina longirostris (Müller).	Frequent.	· net gatherings.
Ilyocryptus sordidus (Lievin).	Rare.	}
Acroperus harpæ, Baird.	Few.	
Eurycercus lamellatus (Müller).	Frequent.	
Camptocercus macrurus Müller).	Few.	} in bottom material.
Alonopsis elongata, G. O. Sars.	Frequent.	
Alona quadrangularis (Müller).	Frequent.	
Pleuroxus trigonellus (O. F. Müller).	Scarce.	}
Graptoleberis testudinaria (Fischer).	Scarce.	} in tow-net gatherings and in bottom material.
Alona guttata (G. O. Sars).	Frequent.	in bottom material.
Chydorus sphæricus (Müller).	Frequent.	in bottom material.
Polyphemus pediculus (Linné).	Frequent.	
Bythotrephes longimanus, Leydig.	Not common.	} in surface and under - surface tow-net gatherings.
Leptodora hyalina, Lilljiborg.	Moderately frequent.	

2.—Report on Tow-net and other Material from certain Suther-landshire Lochs, collected and forwarded by W. S. Caine, Esq., M.P.

Introductory Note.

The material so kindly forwarded by Mr W. S. Caine, and which on examination proved of considerable interest, comprised tow-net gatherings, samples of aquatic plants, and trout and trout's stomachs from Loch Mullach Corrie (Maol a Choire); samples of aquatic plants and trout and trout's stomachs from Loch Awe; and trout and trout's stomachs from Loch Assynt.

These lochs are all in the district of Assynt, Sutherlandshire.

Mr Caine when he forwarded the material, sent also at the same time a short and graphic description of Loch Mullach Corrie and of Loch Awe, as well as a sketch of each of these two lochs to illustrate some of the points referred to in his description, and as the description and sketches are of interest they are reproduced in the sequel.

In the following remarks the lochs are noticed in the order in which they are referred to above.

First.—Loch Mullach Corrie (Maol a Choire).

This is a small loch and is situated in a limestone district about 2½ miles from Inchnadamph Inn. Loch Mullach Corrie is reported to contain what are called 'Gilleroo' or 'Gizzard' trout, though some writers are inclined to question this, but be that as it may, the beauty of the coloration and of the form of the Loch Mullach Corrie trout are undoubted and their edible qualities are also reported to be excellent.

Description of the Loch.

Mr Caine's description of loch Mullach Corrie is as follows:—'The 'loch is commonly known as the "Gilleroo" Loch because it contains '"Gilleroo" trout. No one is able to explain how they came there; they 'are said to be the only trout of their kind found in Scotland, though 'very abundant in Irish lakes, especially in West Meath. The loch is 'about 5 furlongs in length by 2½ furlongs in width. I have sounded 'its depth in 35 places; in a direct line through its length, in the 'middle; and in two direct lines across, at equal distances apart (as 'shown in the sketch—fig. 1). Along the length the soundings are 3 'feet, 4½, 4½, 4½, 5, 5¾, 5¼, 6, 6, 6, 5¼, 5, 4¾, 4½, and 4¾ feet. Across 'the first line (*a*) the soundings are 6½ feet, 9, 8, 6, 6, 5¼, 5½, 5, and 4¾ 'feet. And across the second line (*b*) 4 feet, 6, 5¾, 5¼, 5¼, 5¼, 5½, 6, '6¼, 6, and 3½ feet; so that the average depth of the loch is about '5¼ feet. The bottom, round the edges, is gravel and stones; along one 'side the stones and gravel are clean and bright, along the other side 'they are covered with a slimy vegetable deposit. The bottom of the

Fig. 1.

'basin of the lake is soft black mud . . . from this mud spring dense 'masses of weed of three varieties. The loch is fed with spring water,

' and is very pure and transparent. There are large quantities of fresh-
' water shrimps (*Gammarus pulex*) all over the loch,—among the gravel,
' under stones, and sticking to the weeds. Leeches are also very
' abundant, but I do not find fly-life very plentiful. There are a great
' quantity of curious gelatinous substances adhering to stones and lying
' about in the gravel shoals which I never noticed before in any lake."

The crustacean fauna of the little loch, to judge from the tow-net
gatherings sent to me, is evidently very abundant and very interesting.
One of the Copepods from this loch—*Diaptomus serricornis*—is new to
the British Islands. Since the discovery of *Diaptomus serricornis*, Lilljeborg, in Loch Mullach Corrie it has been ascertained that the same species
was found by Mr David Robertson of Millport many years ago (1867) in
a pond near Lerwick, Shetland, but had remained unnoticed in print till the
discovery of the species in the Sutherlandshire loch. A variety of *Daphnia
pulex* (Linné), having the distal half of the hairs on the antennules and
also the posterior spine of a black colour, was frequent in the tow-net
gatherings from Loch Mullach Corrie; I propose to call this variety
nigrispinosa. The 'shrimp' (*Gammarus pulex*, Linné) which Mr Caine
describes as abundant in the loch is an evidence of the purity of the water.
This is a species that can only thrive in water that is more or less pure;
if transferred to impure water it very soon sickens and dies. The leeches,
referred to in the description of the loch, belong to the genus *Hæmopsis*
(the horse-leech) which is found in ponds and ditches all over the country.
Seven stomachs of trout from this loch were sent by Mr Caine and they
all contained *Gammarus* in greater or fewer numbers, in one or two of
the stomachs there were along with the *Gammarus* a few insects, or
portions of insects, chiefly of the *Phryganeidæ*. The plants that were forwarded comprised specimens of a species of *Chara*; of *Fontinalis antipyre-
ica*,—a moss common in some fresh-water streams and shallow lakes;
of *Littorella lacustris*; of *Menyanthes trifoliata* (the Bogbean); of
Millifolium spicatum (water-milfoil); and a species of *Potamogeton*, or
pond-weed. Among the animal organisms specimens of the curious
little so-called 'water-bears' (*Tardigrada*) were observed specimens of
insects chiefly of the *Phryganeidæ* and *Coleoptera*, specimens of the
Hæmopsis, already referred to, and the following crustacea were also
obtained.

AMPHIPODA.

Gammarus pulex (Linné).	Abundant.

COPEPODA.

Diaptomus serricornis, Lilljeborg.	Frequent.
Cyclops signatus, Koch.	Few.
,, *thomasi*, Forbes.	Few.
,, *virides* (Jurine).	Few.
,, *serrulatus*, Fischer.	Few.
,, *affinis*, G. O. Sars.	Few.
Attheyella crassa (G. O. Sars).	Few.

CLADOCERA.

Daphnia pulex (Linné).	Frequent
,, ,, var. nigrispinosa.	Frequent.
Bosmina longirostris (Müller).	Frequent.
Alonopsis elongata, G. O. Sars.	Few.
Alona quadrangularis (Müller).	Few.
Graptoleberis testudinaria (Fischer).	Few.
Chydorus sphæricus (Müller).	Frequent.

Second.—Loch Awe.

This is a small loch about four miles from Inchnadamph and nearly equidistant from Inchnadamph and Altnacealgach. The River Loanan flows out of this loch into Loch Assynt. The trout in Loch Awe are usually not very large—their average weight being about the one-third of a pound, their flesh is red coloured, and they are said to be of good quality. There are a number of small islands in the loch, most of them are situated near the middle, and so arranged as to divide the loch into two nearly equal portions. The loch is shallow and in some places is much overgrown with aquatic vegetation. Samples of the more common plants growing in Loch Awe were sent along with the material from Loch Mullach Corrie, they comprised specimens of *Littorella lacustris, Millefolium spicatum, Juncus bulbosus,* and of two species of *Potamogeton.* No tow-net gatherings were sent from this loch, but the following three species of crustacea were obtained by an examination of the plants mentioned above, viz.,—*Sida crystallina, Daphnia,* sp. (? *pulex*), and *Attheyella crassa.* A specimen of *Lernentoma* was obtained attached to the gills of one of the trout from this loch.

Twelve stomachs of trout from Loch Awe were examined, three of them were found to be empty; six contained the remains of insects only; two, the remains of insects and *Gammarus*; and one, the remains of the larvæ and larvæ-cases of 'Caddis-flies' (Phryganeidæ.)

Description of Loch Awe.

Mr Caine describes Loch Awe as follows:—' This is a small loch about ' three quarters of a mile long by a quarter of a mile wide, it is divided ' into two parts by a chain of wooded islands with stoney and gravelly ' beaches. The depth of the loch over all is from 5 to 7 feet at ' the distance of about 20 feet from the shelving margin. The two ' ends have stoney bottoms, quite clear of weeds for about one-third to ' one half of their area, and the shore round the islands is also free of ' weeds. There are practically two great beds of weeds—one across the ' middle of each of the two portions of the loch—which I have shaded ' on the rough sketch (from memory) given below; these weeds are the ' same growth—three varieties—as those I have sent from the "Gilleroo" ' loch, which is two and a half miles distant from Loch Awe.' Fig. 2 ' is the sketch of the loch referred, showing the beds of weeds at W.W.

Fig. 2.

Third.—Loch Assynt.

This loch, well known to the angler for the excellent sport it furnishes, is of considerable area, being about eight miles in length by one mile in breadth. Several streams, including the Loanan from Loch Awe, flow into it, and the River Inver, which, after a run of about 6 miles, falls into the sea at Loch Inver, flows out of it. Loch Assynt contains salmon, sea trout, common trout, and *Salmo ferox*—a variety of *Salmo fario*. No tow-net gatherings nor samples of acquatic plants were sent from this loch, but judging from the number and excellence of the fish contained in the loch its invertebrate fauna must be abundant, an investigation of which might be expected to yield interesting results. Twelve stomachs of trout from this loch were examined and were found to contain numerous insect and crustacean remains as shown by the following tabulated statement of the results of the examination.

Table showing the results of the examination of twelve stomachs of trout from Loch Assynt :—

No. of stomach.	Contents of stomach.
1	Remains of larvæ and larvæ-cases of 'Caddis-flies.'
2	Several specimens of *Valvata piscinalis* (a fresh-water mollusc).
3	Remains of *Limnæa peregra*, *Gammarus*, and larvæ-cases of Caddis-flies.
4	Remains of insects—species doubtful.
5	*Limnæa peregra*, *Gammarus*, remains of insects.
6	The same as 5.
7	*Valvata piscinalis*, elytra and other parts of beetles (*Coleoptera*).
8	*Valvata piscinalis*, and remains of insects.
9	Several *Limnæa peregra*, and larvæ-cases of 'Caddis-flies.'
10	One *Limnæa peregra*, remains of *Gammarus*, and 'Caddis-flies.'
11	Two *Limnæa peregra*.
12	Remains of *Gammarus*, and larvæ and larvæ-cases of 'Caddis-flies.'